# Quenched-phosphorescence Detection of Molecular Oxygen

## Applications in Life Sciences

**Detection Science Series**



*Titles in the Series:*

1: Sensor Technology in Neuroscience
2: Detection Challenges in Clinical Diagnostics
3: Advanced Synthetic Materials in Detection Science
4: Principles and Practice of Analytical Techniques in Geosciences
5: Microfluidics in Detection Science: Lab-on-a-chip Technologies
6: Electrochemical Strategies in Detection Science
7: Peroxynitrite Detection in Biological Media: Challenges and Advances
8: Biological Fluid–Surface Interactions in Detection and Medical Devices
9: Advanced Environmental Analysis: Applications of Nanomaterials, Volume 1
10: Advanced Environmental Analysis: Applications of Nanomaterials, Volume 2
11: Quenched-phosphorescence Detection of Molecular Oxygen: Applications in Life Sciences

*How to obtain future titles on publication:*
A standing order plan is available for this series. A standing order will bring delivery of each new volume immediately on publication.

*For further information please contact:*
Book Sales Department, Royal Society of Chemistry, Thomas Graham House, Science Park, Milton Road, Cambridge, CB4 0WF, UK
Telephone: +44 (0)1223 420066, Fax: +44 (0)1223 420247
Email: booksales@rsc.org
Visit our website at www.rsc.org/books

# *Quenched-phosphorescence Detection of Molecular Oxygen*

## *Applications in Life Sciences*

Edited by

**Dmitri B. Papkovsky**
*University College Cork, Ireland*
*Email: d.papkovsky@ucc.ie*

and

**Ruslan I. Dmitriev**
*University College Cork, Ireland*
*Email: r.dmitriev@ucc.ie*

ROYAL SOCIETY OF CHEMISTRY

Detection Science Series No. 11

Print ISBN: 978-1-78801-175-4
PDF ISBN: 978-1-78801-345-1
EPUB ISBN: 978-1-78801-455-7
ISSN: 2052-3068

A catalogue record for this book is available from the British Library



The Royal Society of Chemistry is a charity, registered in England and Wales, Number 207890, and a company incorporated in England by Royal Charter (Registered No. RC000524), registered office: Burlington House, Piccadilly, London W1J 0BA, UK, Telephone: +44 (0) 207 4378 6556.

For further information see our website at www.rsc.org

Printed in the United Kingdom by CPI Group (UK) Ltd, Croydon, CR0 4YY, UK

# Preface

Quenched-phosphorescence oxygen sensing has now emerged as one of the most versatile, flexible and successful sensor technologies. This is largely because of the paramount significance of $O_2$ in life and biomedical sciences and the capabilities of this detection method to quantify oxygen concentration and other related parameters - directly, reversibly, non-invasively, accurately and in various gaseous, liquid and biological samples.

Since the pioneering proofs of concept studies of Dietrich Lubbers and David Wilson and their colleagues back in the late 70s and early 80s, this sensor technology has undergone major development and diversification. This work has produced a comprehensive panel of high-performance indicator dyes, sensing materials on their basis, advanced detection schemes, analytical methodologies and measurement instrumentation. Based on this technology and materials, a diverse range of applications has been developed and demonstrated in various settings. Over the three decade-long history of optical $O_2$ sensing, life science and biomedical applications have remained as its central pillar, while many other industrial and research applications have advanced as well.

On the other hand, only in the last 5–10 years, has $O_2$ sensing technology come to a stage, when it became widely accessible and affordable for ordinary users (researchers, clinicians, industry), and transferrable from high technology research and development labs into real-life environment, practical research and industrial applications. A number of commercial instruments and systems designed for routine laboratory use and also suitable for large-scale and large-volume applications such as biological screening, packaging, environmental and process control, quality assurance, have been introduced.

In many research labs, solid-state $O_2$ sensors have already become routine analytical tools used on a daily basis, like pH meters or UV-Vis spectroscopy.

Detection Science Series No. 11
Quenched-phosphorescence Detection of Molecular Oxygen: Applications in Life Sciences
Edited by Dmitri B. Papkovsky and Ruslan I. Dmitriev

Published by the Royal Society of Chemistry, www.rsc.org

Further development of bioanalytical systems, which rely on quenched-phosphorescence oxygen sensing continues with great pace. At the same time, the level of awareness of both the research community and industry about such systems, their analytical and research capabilities and potential benefits of their adoption and use still remain low. These factors limit wider uptake and practical use of these technologies. More demonstrational and educational work needs to be conducted to break these barriers.

In this book we aimed to assemble a comprehensive collection of papers covering all core aspects of quenched-phosphorescence oxygen sensing technology and its applications. The first group of chapters describe the fundamentals and core aspects of $O_2$ sensing technique, the range of most common sensor dyes (Pt(II)-porphyrins, Ru(II)- and Ir(III)-cyclometallated complexes), materials (solid-state sensors, soluble probes and nanoparticle formulations), fabrication technologies and dedicated instrumentation.

The second group of chapters describes specific life science applications and customized experimental setups and sensor systems (*e.g.* marine research and aquatic organisms, live cell analysis, oxygen imaging and oxygen transport in tissue, *in vivo* oxygen sensing and imaging) and application areas (cancer and stem cell research, photodynamic therapy, vascular biology and clinical applications, bio-imaging, microfluidic biochips, food packaging and safety). Some of these applications have already been commercialized or are close to this stage and therefore available for ordinary users.

The leading experts in respective areas, who have extensive hands-on experience with corresponding sensor systems and biological applications and know their merits and limitations, provided the chapters for the book. Altogether, this gives a comprehensive picture of the current status in this area, performance and capabilities of the different sensor systems and applications, and future avenues of research and development in this field.

The book is targeted at potential new users and young researchers who are not very familiar with these technologies and applications, but who can benefit from them. It is also of significant interest for established experts and researchers in $O_2$ sensing and current users of adjacent life science applications. Presenting them with the recent achievements in the broader area and related applications can further stimulate their work. We are hoping to attract many new researchers, young scientists and end-users to this very exciting technology.

Dmitri B. Papkovsky
Ruslan I. Dmitriev

# Contents



Detection Science Series No. 11
Quenched-phosphorescence Detection of Molecular Oxygen: Applications in Life Sciences
Edited by Dmitri B. Papkovsky and Ruslan I. Dmitriev

Published by the Royal Society of Chemistry, www.rsc.org

CHAPTER 1

# *Fundamentals of Quenched Phosphorescence $O_2$ Sensing and Rational Design of Sensor Materials*

SERGEY M. BORISOV

Graz University of Technology, Stremayrgasse 9, 8010, Graz, Austria
*E-mail: sergey.borisov@tugraz.at

## 1.1 Introduction

In the last decades, optical oxygen sensors (oxygen optodes) became indispensable analytical tools, which are nowadays widely applied in academia and industry. Their popularity is explained by the numerous advantages offered by the optical detection method such as the absence of electromagnetic interferences, minimally invasive character (measurement though a transparent wall of a reactor), versatility of sensor formats varying from planar foils and fiber-optic sensors to nanoparticles, suitability for imaging *etc.* Optical oxygen sensors do not consume the analyte, which favourably distinguishes them from electrochemical sensors such as Clark electrode or galvanic cells. Optodes allow for oxygen measurement in gases and solutions

Detection Science Series No. 11
Quenched-phosphorescence Detection of Molecular Oxygen: Applications in Life Sciences
Edited by Dmitri B. Papkovsky and Ruslan I. Dmitriev

Published by the Royal Society of Chemistry, www.rsc.org

with dynamic ranges, which can be adjusted over many orders of magnitude. Finally, optical oxygen sensors are very useful for measurement of air pressure on surfaces (pressure sensitive paints) or as transducers for enzymatic sensors making use of oxygen consumption such as glucose or lactate sensors.

## 1.2 Mechanism of Oxygen Quenching

Oxygen is one of the most powerful luminescence quenchers. Quenching of fluorescent dyes (excited singlet state, $S_1$) and phosphorescent dyes (excited triplet state, $T_1$) is spin-allowed. Moreover, the energies of excited states of oxygen ($^1\Sigma_g^+$ and $^1\Delta_g$) are lower than the energies of the excited states of most organic dyes and metal complexes (Figure 1.1), which makes quenching *via* energy transfer favourable.

The mechanism of oxygen quenching is rather complex and the exact pathways and formed products depend on many factors.[1] Electron-exchange Dexter-type energy transfer is the predominant mechanism of oxygen quenching. Quenching of fluorescent dyes (D) can result in the formation of the dye in the triplet excited state or in the ground state:[1]

$$D(S_1) \rightarrow D(S_0) + h\nu$$

$$D(S_1) + O_2(^3\Sigma_g^-) \rightarrow D(T_1) + O_2(^3\Sigma_g^-)$$

$$D(S_1) + O_2(^3\Sigma_g^-) \rightarrow D(T_1) + O_2(^1\Delta_g)$$

$$D(S_1) + O_2(^3\Sigma_g^-) \rightarrow D(S_0) + O_2(^3\Sigma_g^-)$$

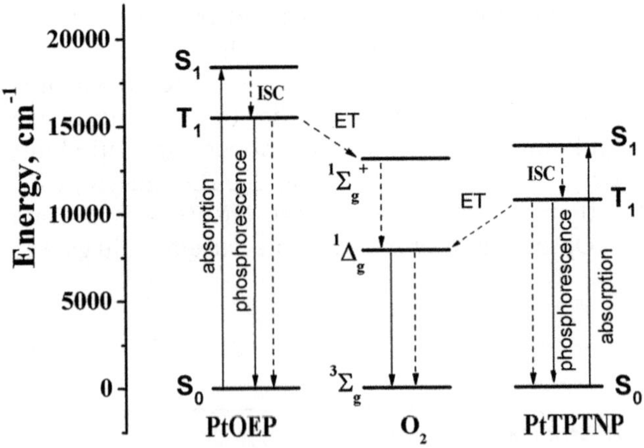


**Figure 1.1** Energy diagram for two phosphorescent oxygen indicators: platinum(II) octaethylporphyrin (PtOEP) and platinum(II) tetraphenyltetranaphthoporphyrin (PtTPTNP).

The triplet state of the dye is deactivated to the ground state:

$$D(T_1) + O_2(^3\Sigma_g^-) \rightarrow D(S_0) + O_2(^3\Sigma_g^-)$$

$$D(T_1) + O_2(^3\Sigma_g^-) \rightarrow D(S_0) + O_2(^1\Delta_g)$$

For the quenching of phosphorescence, the dye is deactivated into the ground state and singlet oxygen is formed:

$$D(T_1) + O_2(^3\Sigma_g^-) \rightarrow D(S_0) + O_2(^1\Sigma_g^+)$$

$$D(T_1) + O_2(^3\Sigma_g^-) \rightarrow D(S_0) + O_2(^1\Delta_g)$$

Depending on the triplet energy of the dye, formation of singlet oxygen either only in the $^1\Delta_g$ state (*e.g.* for PtTPTNP, Figure 1.1) or in both $^1\Delta_g$ and $^1\Sigma_g^+$ states (*e.g.* for PtOEP) is possible. Notably, $O_2(^1\Sigma_g^+)$ deactivates very fast into $O_2(^1\Delta_g)$ state.

Apart from the energy transfer, electron transfer leading to superoxide is also possible:

$$D(T_1) + O_2(^3\Sigma_g^-) \rightarrow D^{\bullet+} + O_2^{\bullet-}$$

This process can play a significant role for metal complexes with strong reducing properties (particularly in the excited state), for instance Ir(III) cyclometalated complexes.[2] Rapid back electron transfer can result in the formation of singlet oxygen and the sensitizer in the ground state.

Importantly for all these processes, singlet oxygen represents one of the main products. Since its deactivation to the triplet state regenerates the analyte, optical oxygen sensors do not consume the analyte in theory. However, the lifetime of singlet oxygen in polymers can be much longer compared to that in the aqueous phase (~3 µs), which can be sufficient for it to react with the sensor components (dye or polymer), see Chapter 1.7.

Independent of the quenching mechanism, the quenching behavior for dissolved dyes is described by the Stern–Volmer equation:

$$\frac{I_0}{I} = \frac{\tau_0}{\tau} = 1 + k_q\tau_0[O_2] = 1 + K_{SV}[O_2] \tag{1.1}$$

where $I_0(\tau_0)$ and $I(\tau)$ are the luminescence intensity (decay time) in the absence and in the presence of oxygen, respectively, $k_q$ is the bimolecular quenching constant and $K_{SV}$ is the Stern–Volmer constant.

From eqn (1.1) it is evident that the efficiency of quenching depends on both the bimolecular quenching constant and the decay time of the luminophore $\tau_0$. The $k_q$ constant is determined mostly by oxygen diffusion since the diffusion of the much larger dye is significantly slower already in solution and is virtually non-existant for immobilized dyes. The $k_q$ constant often approaches the diffusion-controlled limit $k_{diff}$ for quenching of fluorescence[1] but is lower for quenching of phosphorescence. For common phosphorescent indicators such as Pt(II) porphyrins or Ru(II) polypyridyl complexes, it is usually close to 1/9 of $k_{diff}$,[2] where 1/9 is the spin statistical factor accounting for the formation of both products in the singlet state. However, $k_q$ is sometimes higher than this value even in the case of purely energy transfer-based quenching[1] and may be even higher if electron transfer is involved.[2]

Clearly, the $\tau_0$ has a much stronger influence on the Stern–Volmer constant than $k_q$. In fact, assuming a $k_q = k_{diff} = 2.1 \times 10^{10}$ M$^{-1}$ s$^{-1}$ for an air-saturated toluene solution ($C(O_2) \approx 1.8$ mM) of a typical fluorescent dye with $\tau_0$ of 4 ns, the $I_0/I$ value calculated with eqn (1.1) is only 1.15. On the other hand, for a phosphorescent indicator such as PtOEP ($\tau_0$ = 85 μs) even with much lower $k_q$ 2.4 × 10$^9$ M$^{-1}$ s$^{-1}$ (1/9$k_{diff}$) the luminescence intensity and decay time decrease 367-fold in the same conditions. Since the diffusion of oxygen is significantly slower in the polymers compared to the solution, it is evident that only phosphorescent indicators will provide the required resolution when embedded in common polymeric matrices. Additionally, whereas the tunability of fluorescence decay times is usually limited by 1–2 orders of magnitude, the phosphorescence decay time can vary from several microseconds to hundreds of milliseconds. This provides virtually unlimited flexibility in designing oxygen-sensing materials for very different applications.

## 1.3 Requirements for Phosphorescent Indicators

In order to navigate among hundreds of reported oxygen indicators it is useful to define the important parameters, which should be considered when the indicators are selected. These include:

(i) *Spectral properties (absorption and emission maxima).* In contrast to fluorescent dyes, phosphorescent indicators possess large Stokes shifts, which simplifies signal separation and reduces interferences caused by scattering and autofluorescence. Nevertheless, indicators excitable and emitting at longer wavelength are preferable for the same reasons. In photosynthetic systems, however, such excitation can result in much higher levels of autofluorescence and dyes with other spectral properties can be a better choice. Clearly, autofluorescence can be completely eliminated in the time domain measurement, but it does interfere with the measurement in the frequency domain unless multi-frequency measurement is performed.

Compatibility of the indicator with the light sources, photodetectors and other optical components should also be considered. Whereas a wide range of light sources is available for the whole spectral range, the detectors are mainly limited to avalanche photodiodes, CCD-arrays and photomultipliers. Although the sensitivity of PMTs is generally very high, it deteriorates fast in the NIR part of the spectrum. In the case of fiber-optic sensors, the quartz glass fibers are compatible with all oxygen indicators. In contrast, much cheaper plastic fibers show strong absorption in the NIR part of the spectrum, which limits the practically useful length to 1–2 meters.

(ii) *Brightness of luminescence.* Brightness can be defined as the product of molar absorption coefficient $\varepsilon$ and luminescence quantum yield $\phi$. Clearly, in case of phosphorescent indicators high efficiency of inter-system crossing (ISC, $S_1 \rightarrow T_1$ transition, Figure 1.1) is one of the prerequisites for bright phosphorescence. Bright indicators allow for thinner sensing layers and therefore for faster response times. In the case of very bright but less photostable indicators the operational lifetime of the sensor can be extended by using low intensities of the excitation light nevertheless allowing for acceptable signal to noise ratio.

(iii) *Luminescence decay times.* The luminescence decay times along with permeability of the polymer govern the sensitivity of the sensor. In order to obtain oxygen sensors for physiologically relevant conditions on the basis of polymeric matrices such as polystyrene (moderate oxygen permeability) the optimal decay times are about 20–100 μs, the requirement met by most Pt(II) porphyrins. Indicators with shorter decay times would deliver lower resolution, whereas the sensors based on the indicators with much longer lifetimes (*e.g.* Pd(II) porphyrins) will only be suitable for trace sensing. On the other hand, if trace sensing applications are intended, the long luminescence lifetimes (≫1 ms) are highly desirable.

(iv) *Photostability.* Photodegradation of the indicator results in sensor drift. This can be less critical for the lifetime measurements if no other luminescent species are formed since the decay time remains constant independently on the amount of the remaining dye and only the *S*/*N* ratio is affected. In contrast, much stronger drift is expected for intensity-based measurement. This is true for ratiometric measurements where photobleaching of the reference dye will be as critical as the bleaching of the indicator itself unless both bleach with approximately the same speed. It goes without saying that the higher the photostability, the better. However, for many applications indicators with moderate photostability can be fully adequate even for long-term measurements. On the other hand, photobleaching can be an issue if the light densities are very high. For instance, the Eddy correlation technique which gained extreme popularity in the oceanographic community in the last years[3] relies on the use of fiber-optic microsensors with very fast response. Here not only is the density of light exceptionally high, but also high frequency of measurements is required (several times per second). As a consequence, sensors based on even highly photostable dyes will have limited lifetime.

In respect to photobleaching, two aspects should be mentioned. First, photobleaching rates can be significantly different in the absence of oxygen and under air saturation. In the second case, oxidation of the dye by photosynthesized singlet oxygen may be the predominant mechanism of photodegradation. Second, drift of the sensor properties can be observed not only due to oxidation of the dye but also due to degradation of the matrix polymer resulting in accumulation of the moieties acting as quenchers of phosphorescence of the indicator. In contrast to the first mechanism (photobleaching of the dye itself) characterized by degradation of the luminescence intensity but relative stability of the decay time, in the second mechanism both parameters will be affected to approximately the same degree. Although this effect can be overcome in theory by using an oxidation-resistant polymer, the choice of such polymers is very narrow and the compatibility of the indicator and the polymer can represent a serious problem.

(v) *Solubility in polymers.* Low compatibility of indicator and the polymer is likely to result in dye aggregation. For instance, even rather hydrophobic indicator dyes based on conjugated structures readily aggregate in silicone rubber. Low solubility of the dye in the matrix can also result in its migration into a polymeric support (*e.g.* commonly used poly(ethylene terephthalate)). This effect is accelerated at elevated temperatures, *e.g.* during autoclaving. Covalent grafting of the indicator into the polymer overcomes aggregation, migration and leaching issues but is often synthetically demanding since it requires the introduction of functional groups into the indicator molecule or in both the dye and the polymer.

Only a few indicator classes show the properties which fulfil most of the requirements. Dyes with the best combination of properties will be good indicators for general use covering the majority of applications. However, for some applications one or the other requirement can be a cut-off criterion whereas other properties are of less importance. Here a specialized tailor-made indicator dye may be a much better choice.

## 1.4 Brief Overview of the Most Common Indicators

An impressive number of phosphorescent dyes were reported in the last decades and most of them qualify as oxygen indicators.[4,5] However, there is a significant gap between chemists who design the sensing materials and end-users such as biologists, oceanographers, geologists *etc.* Therefore, the most widely used indicators are either commercially available or can be easily prepared in the lab. Fortunately, some comparably new indicators such as benzoporphyrin complexes have been also commercialized recently.

The structures of the most common oxygen indicators are shown in Figure 1.2 and their photophysical properties are summarized in Table 1.1 Ruthenium tris(4,7-diphenyl-1,10-phenanthroline) (Ru-dpp) is the brightest among the polypyridyl complexes and was applied in many sensing materials. This dye can be synthesized in only one step. The limitations

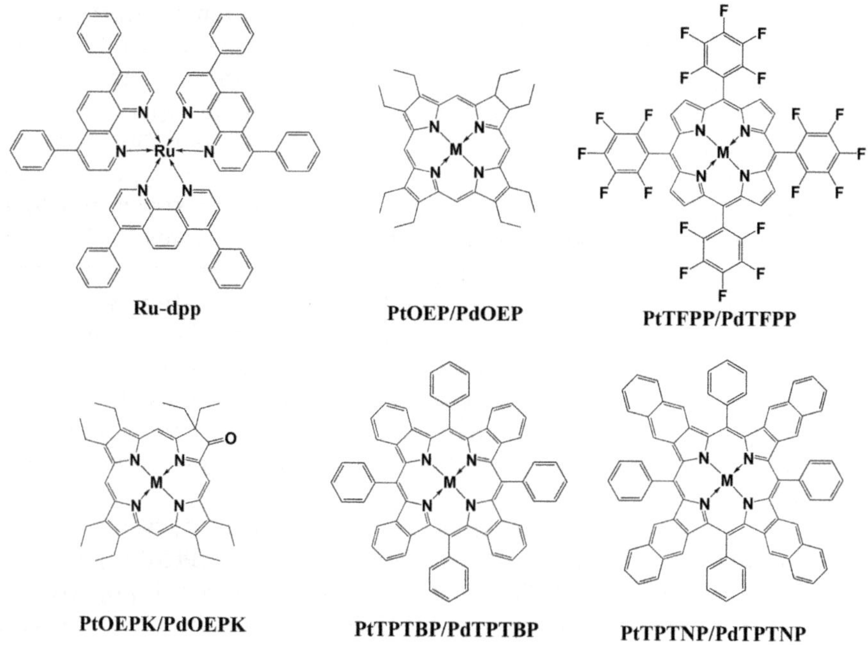


**Figure 1.2** Chemical structures of popular phosphorescent oxygen indicators. M = Pt(II) or Pd(II)

**Table 1.1** Photophysical properties of selected oxygen indicators (room temperature, anoxic solutions in organic solvents).

| Complex | $\lambda_{max}$abs, nm | $\varepsilon$, $M^{-1}$ $cm^{-1}$ | $\lambda_{max}$em, nm | $\tau_0$, µs | $\phi$, % | Reference |
|---|---|---|---|---|---|---|
| Ru-dpp | 463 | 28 600 | 618 | 6.4 | 37 | 6 |
| PtOEP | 382 | 214 000 | 649 | 75 | 41 | 7 |
| | 536 | 42 500 | | | | |
| PdOEP | 395 | 127 000 | 669 | 1220 | 19 | 8 |
| | 547 | 32 000 | | | | |
| PtTFPP | 394 | 220 000 | 648 | 60 | 9[a] | 9 |
| | 540 | 18 100 | | | | |
| PdTFPP | 407 | – | 670 | ~900 | 3[a] | 10 |
| | 553 | – | | | | |
| PtOEPK | 398 | 86 200 | 758 | 60 | 12 | 11 |
| | 592 | 55 100 | | | | |
| PdOEPK | 410 | 82 600 | 789 | 455 | 1 | 11 |
| | 603 | 53 500 | | | | |
| PtTPTBP | 430 | 205 000 | 770 | 47 | 51[b] | 12 |
| | 614 | 136 000 | | | | |
| PdTPTBP | 443 | 416 000 | 800 | 286 | 21[b] | 12 |
| | 628 | 173 000 | | | | |
| PtTPTNP | 436 | 95 000 | 883 | 8.5 | 22[b] | 13 |
| | 689 | 150 000 | | | | |
| PdTPTNP | 463 | 144 000 | 937 | 65 | 6.5[b] | 14 |
| | 705 | 168 000 | | | | |

[a]The values might be underestimated.
[b]The values are likely to be overestimated.

include moderate luminescence brightness and high temperature dependency of the luminescence decay time, which is also comparably short so that immobilization in highly oxygen-permeable matrices is essential to achieve optimal sensitivity. Platinum(II) and palladium(II) complexes with porphyrins and their derivatives are extremely popular. The main reasons are simple synthesis and often commercial availability, acceptable photophysical properties and versatility of the porphyrin structure which allows for numerous modifications. Pt(II) and Pd(II) complexes with octaethylporphyrin (OEP) possess good brightness upon excitation in the Soret band (UV range). Molar absorption coefficients for excitation with green light are much lower (Table 1.1). The photostability of OEP complexes is, however, moderate. Therefore, for most applications they have been substituted by the complexes of the same metals with meso-(pentafluorophenyl)porphyrin (TFPP) which are known to be very photostable. As a trade-off, the brightness of these dyes is about 2-fold lower than for the OEP complexes.

As was mentioned above, NIR dyes are advantageous for several reasons. Platinum(II) complex with ocaethylporphyrin-ketone[11] (PtOEPK) enjoyed great popularity as a NIR emitting oxygen indicator. Spectral properties of the Pt(II) complex with tetra(pentafluorophenyl)porpholactone[15] are very similar to that of the PtOEPK. Although these dyes feature absorption and emission spectra, which are bathochromically shifted compared to the parent porphyrins, the position of the Q-band is not fully adequate for *in vivo* measurements. Moreover, the brightness of the dyes is moderate. Interestingly, the Pd(II) complexes with porphyrin-ketones and porphyrin lactones show very weak phosphorescence, which is different from the general trend shown by the porphyrins (~2–3 fold lower brightness of the Pd(II) complexes, Table 1.1).

Although NIR-emitting metal complexes of benzoporphyrins were reported more than two decades ago,[16] only in the last decade have they become popular indicator dyes. This is explained by the development of convenient synthetic methods leading to a great variety of structures (pyrrole condensation under Lindsey conditions with oxidative aromatization elaborated by Vinogradov and co-workers[17] and more recently modified template method[18]) and also of increasing interest to researchers in *in vivo* applications. Indeed, the absorption of the Q-band for the Pt(II) and Pd(II) complexes of tetraphenyltetrabenzoporphyrin (TPTBP) peaks at 614 and 628 nm, respectively, which is very close to so called "NIR optical window". The molar absorption coefficients are exceptionally high, which in combination with good quantum yields, results in excellent brightness of these compounds. It should be noted that some recent investigations[19] indicate that the quantum yields are not as high as previously reported by various groups, but the dyes nevertheless remain the brightest in this spectral range. The group of Vinogradov designed various benzoporphyrin-based dendrimeric probes for extracellular measurements and *in vivo* applications[20–22] some of them also suitable for two-photon excitation. In an alternative approach, lipophilic complexes of TPTBP can be embedded into water-dispersible nanoparticles with or without cell penetration capabilities.[23–25] The absorption and emission spectra of tetranaphthoporphyrins such as TPTNP (Figure 1.2) are even more bathochromically shifted and fully match the NIR optical window. Unfortunately, the luminescence quantum yields and photostability are significantly lower than those of the benzoporphyrins. Additionally to these main classes several molecular hybrids were reported. Hybrids of benzo- and naphthoporphyrins[26] possess spectral properties tuneable over a broad range resulting in tailor-made dyes, matching different available LED sources. The hybrids of benzoporphyrins and phthalocyanines (aza-benzoporphyrins)[27] share very high photostability and comparably narrow absorption bands of phthalocyanines and possess very large Stokes shift. A large palette of NIR oxygen indicators is very useful when designing sensor arrays for *in vivo* applications, such as dual sensors for simultaneous subcutaneous monitoring of glucose (*via* an oxygen transducer) and oxygen.[28]

The indicators described above can be viewed as general-purpose indicators. However, for some specific oxygen sensing applications more exotic dyes can be a better choice. These include, for instance, cyclometalated iridium(III) coumarin complexes featuring ultrabright emission,[29] visible light-excitable europium(III) complexes which possess typical narrow band emission of Eu(III),[30] or luminescent chelates of Gd(III),[30] Al(III) and $BF_2$[31] which possess long phosphorescence decay times for the design of trace oxygen sensors.

## 1.5 Rational Design of Optical Sensing Materials

Two main groups of the oxygen-sensing materials can be distinguished: (i) molecular probes including water-soluble dyes, dendrimers, dye-protein conjugates *etc.* and (ii) materials based on oxygen indicators immobilized in some matrix (mostly polymers or sol–gels) or adsorbed on/covalently coupled to the surface of a (non)porous material.

The materials of the first group are not reusable so they are predominantly used for microscopic applications and for assays in samples of small volumes such as microwells in plate readers. In order to be useful, such probes must possess excellent solubility in water. Although this criterion is met by water-soluble phosphorescent dyes, they are poorly suited to practical applications for several reasons: (i) their phosphorescence is typically weak due to efficient radiationless deactivation in water; (ii) the response to oxygen is highly influenced by the environment and is likely to vary significantly upon interaction of the probe with biological components such as proteins; (iii) such probes are more prone to cross-talk to ionic species which can act as quenches; (iv) they can show high cytotoxicity due to close proximity to cells. Therefore, in order to overcome these limitations, a phosphorescent indicator is usually protected by creating a polymer shell (dendrimers),[22] immobilization into nanoparticles or covalent coupling to a protein[32] or a peptide.[33] A polymeric shell or a protein not only protects the dye from undesired interference but also tunes the oxygen permeability of the environment thus helping to design a probe with an optimal dynamic range.

The second group of materials is used for the preparation of reusable sensors. Here, the indicator is immobilized into a polymeric or sol–gel matrix, which acts a solvent and a support for the dye and as a permeation-selective barrier. Due to variety of the available polymers, polymer-immobilized dyes are more common than those based on silica or Ormosils (organically modified silica).[34] Among the vast group of polymers, hydrophobic ones are preferable since potentially interfering species (ions, proteins, water) are prevented from interaction with the indicator.

The following aspects should be considered in respect to the polymer choice:

(i) Oxygen permeability of the polymer. This parameter along with the phosphorescence decay time of the dye is essential for the design of the sensors with the desired dynamic range (see below);
(ii) Compatibility of the dye and the polymer. Compatibility can be low if both components differ significantly in polarity. For instance, most dyes will aggregate readily in silicone rubber or in amorphous perfluorinated polymers (such as Teflon AF) so that modification of the dyes which improves their solubility in such matrices cannot be avoided.
(iii) Physical properties and chemical stability of the polymer. Whereas many polymers will form a mechanically stable layer, the sensors prepared from some crystalline polymers will tend to build cracks and detach from the support. Cracking of the sensing layers is also common for sol–gel materials. The glass-transition point ($T_g$) is another important parameter. One can expect a significant change in the sensing properties of the material if the $T_g$ point is located within the measurement temperature range or the temperature corresponding to the $T_g$ point is reached during sterilization by autoclaving. Finally, stability of the polymers towards oxidation *via* photosensitized singlet oxygen can be very important in some cases and particularly for trace sensors. Unfortunately, the number of truly oxidation-resistant polymers is very limited and most of them do not fulfil the requirements (i) and (ii).

One should mention that the materials of the second group often contain additional components. For instance, scattering substances such as titanium dioxide can be added into the sensing layer (or in the next layer above the sensing layer) to increase the brightness. A layer containing carbon black can be added to provide optical isolation *i.e.* to avoid saturation of the detector and photobleaching of the dye due to ambient light and to eliminate the disturbance of the excitation light in case of photosynthetic systems.

The design of oxygen sensitive nanoparticles is very similar to the probes of the second group *i.e.* they are based on phosphorescent indicators immobilized into polymer or a sol–gel. Often functional groups on the surface of the nanoparticle are introduced to improve dispersibility in water and provide additional properties, *e.g.* cell penetration ability. On the other side, the application area of such nanosensors is identical to that of the materials of the first group: intra- and extracellular imaging, assays in respiration vials and microplates.

## 1.6 Sensitivity and Dynamic Range of Oxygen Sensors

Sensitivity and dynamic range of optical oxygen sensors is determined by two main factors: the phosphorescence decay time of the indicator $\tau_0$ (Figure 1.3a) and the oxygen permeability of the polymer (Figure 1.3b). This, in turn, can be viewed as a product of oxygen solubility in the polymer and the diffusion coefficient. In other words, the properties of the indicator and the polymer are equally important for design of the sensing material.

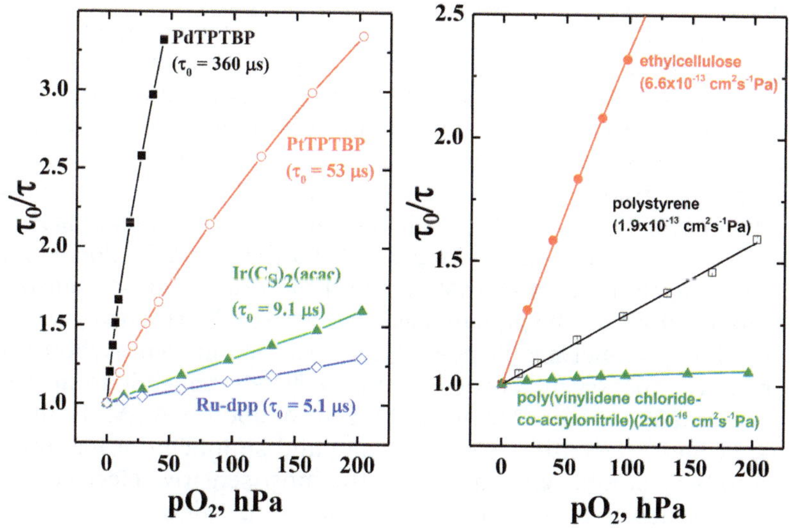


**Figure 1.3** Stern–Volmer plots for the optical oxygen sensors based on different luminescent dyes immobilized in polystyrene (left) and on $Ir(Cs)_2acac$ embedded in different polymers (right). PdTPTBP/PtTPTBP = palladium(II)/platinum(II) tetraphenyltetrabenzoporphyrin; $Ir(Cs)_2acac$ = iridium(III) bis-(benzothiazol-2-yl)-7-(diethylamino)-coumarin-(acetylacetonate); Ru-dpp = ruthenium(II) tris-4,7-diphenyl-1,10-phenanthroline.

For instance, Figure 1.3 shows that the sensitivity of the sensor based on an Ir(III) coumarin complex ($Ir(Cs)_2acac$) immobilized in highly oxygen-permeable ethylcellulose is similar to that of the material based on Pt(II) tetraphenyltetrabenzoporphyrin ($\tau_0$ much longer than for the Ir(III) complex) embedded into less oxygen-permeable polystyrene.

Among the polymeric matrices poly(trimethylsilyl propyne) shows the highest oxygen permeability[5] but tends to significantly change the properties over time. Silicon rubber and perfluorinated amorphous Teflon AF belong to stable polymers with very high oxygen permeability. Oxygen permeability of ethylcellulose, some Ormosils and Hyflon AD is moderately high. Polymers with moderate oxygen permeability such as polystyrene, polysulfone and poly(methyl methacrylate) are the most commonly used matrices due to their good chemical stability, low cost, good solubility in various organic solvents and optimal sensitivity at ambient conditions in combination with Pt(II) porphyrins. Polymers with low oxygen permeability such as poly(vinylidene chloride) and poly(acrylonitrile) are not useful matrices but some of them are excellent support materials (poly(ethylene terephthalate)).

As a rule of thumb, one can say that the dynamic range of optical oxygen sensors based on bright indicators is about three orders of magnitude in $pO_2$/concentration of dissolved oxygen. The lower level is determined by the slope of the curve (Stern–Volmer constant) and noise of the read-out instrument. At the upper limit, decrease of the phosphorescence intensity due to quenching is too high to deliver an acceptable S/N ratio. Sensors based on less bright indicators will have a shorter dynamic range in the same conditions. As a consequence, it is impossible to design an oxygen sensor matching all possible applications. As was mentioned above, the combination of Pt(II) porphyrins with polymers of moderate oxygen permeability results in sensors resolving from about 0.1 to 100 kPa $pO_2$. The sensitivity can be improved by roughly 1 order of magnitude *via* substitution of the Pt(II) porphyrin *via* a Pd(II) analogue. However, even this sensor is not suitable for some applications, for example for the quantification of oxygen in oxygen minimum zones[35] (nM concentration range). Trace sensors for these applications can be prepared *via* combining Pd(II) (benzo)porphyrins and highly oxygen-permeable polymers such as Teflon AF or Hyflon AD.[36,37] For instance, an oxygen sensor based on PdTFPP embedded into Hyflon AD 60 ("LUMOS sensor")[36] showed the LOD of about 0.5 nM. In terms of detection limit, resolution, sampling rate and simplicity of manufacture this optode favourably compares with the most sensitive electrochemical Clark-type sensor reported to date (STOX sensor),[38] Figure 1.4. This direct comparison undoubtedly illustrates the high potential of the optode technology for oxygen sensing.

As can be observed from Figure 1.3, quenching behaviour in polymers is more complex compared to the solutions where strictly linear Stern–Volmer plots are always observed. In contrast, in most polymers Stern–Volmer

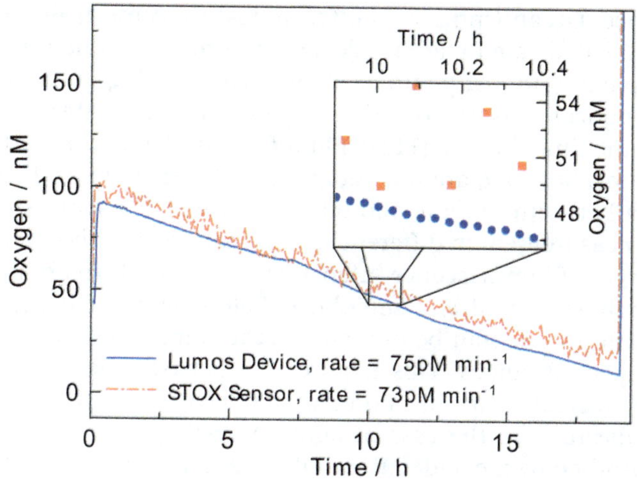


**Figure 1.4** Respiration in a deep sea sample simultaneously monitored with an oxygen optode LUMOS and a STOX electrochemical sensor. Reproduced from P. Lehner *et al.*, *PLoS One*, 2015, **10**(6), e0128125,[36] https://doi.org/10.1371/journal.pone.0128125, published under a CC-BY 4.0 Licence, https://creativecommons.org/licenses/by/4.0/.

plots are not linear which can be explained by localization of the indicator in microenvironments of different oxygen permeability. The so called "two site model"[39] assumes the existence of only two microenvironments:

$$\frac{I}{I_0} = \frac{f_1}{1 + K_{SV}^1[O_2]} + \frac{f_2}{1 + K_{SV}^2[O_2]} \tag{1.2}$$

$f_1$ and $f_2$ - fractions of the total emission for each component ($f_1 + f_2 = 1$); $K_{SV}^1$ and $K_{SV}^2$ - the Stern–Volmer constants for each component. Eqn (1.2) is known to fit both the intensity and the lifetime Stern–Volmer plots very well, despite that, in the second case, the fit is not physically meaningful.

## 1.7 Referenced Oxygen Sensing and Imaging

Luminescence intensity is notoriously known to be a rather ambiguous parameter, which depends on many factors including intensity of the excitation light, sensitivity of photodetector, scattering in the sample *etc.* Therefore, referencing in sensing and imaging is essential for precise quantification of $pO_2$. Fortunately, phosphorescence decay time represents a self-referenced parameter, which is not affected by the above-mentioned factors. Phosphorescence decay time can be conveniently measured with low cost compact phase fluorometers. Several major suppliers (PreSens,

Pyro Science, Ocean Optics *etc.*) offer optical oxygen sensors in combination with the dedicated read-out devices and these products are nowadays widely applied in industry and academia. For imaging of oxygen distribution with optical sensors both the frequency domain (FLIM camera from PCO) and the time domain (FLIM/PLIM from Becker and Hickl) are useful. Many set-ups, however, are not equipped with detectors for lifetime imaging but rather with several photodetectors, which enable simultaneous intensity measurement in different spectral ranges. In this case, referenced measurement of luminescence intensity is very useful. To obtain a suitable material a fluorescent dye is immobilized along with the oxygen indicator. Clearly, both dyes should be excitable at the same wavelength but emit in different spectral ranges. This requirement is easily met since the Stokes shift is much smaller for the fluorescent dyes compared to phosphorescent oxygen indicators. In the case of light harvesting systems, the excitation light is absorbed by the added fluorophore and the energy is transferred to the oxygen indicator *via* Förster Resonance Energy Transfer (FRET) mechanism. This approach not only yields ratiometric intensity measurement (utilizing residual fluorescence of the fluorophore) but also allows for significant increase of fluorescence brightness due to high molar absorption coefficient of the added dye.[40] It should be kept in mind that the ratiometric referencing scheme does not compensate for photobleaching or leaching of the oxygen indicator or the reference. The reference dye might show much higher photobleaching rates in the presence of oxygen indicator due to the formation of highly reactive singlet oxygen. Leaching can be critical in the case of nanoparticles due to their high surface to volume ratio and therefore much more efficient interaction with the environment. Here, covalent immobilization of indicators is strongly preferable. Polymeric materials based on conjugated polymers (combining dual function of the matrix and fluorophore), where an oxygen indicator is covalently grafted into the polymer and acts as a FRET acceptor, were also shown to be highly promising for imaging applications.[24,41] Importantly, ratiometric character is also preserved under two-photon excitation since the conjugated polymer acts as an excellent two-photon antenna.[24]

## 1.8 Artefacts in Oxygen Sensing

Although optical oxygen sensors belong to very robust analytical tools, precise oxygen determination can be compromised by a number of errors, cross-sensitivities and artefacts. As for all chemical sensors, oxygen optodes are cross-sensitive to temperature which affects the phosphorescence decay time of the indicator ($\tau_0$ decreases due to thermal quenching) and the Stern–Volmer constant (increases in most cases due to faster diffusion of oxygen). These effects must be compensated for in an additional temperature measurement. Calibration might be another critical issue. When the exact form of the Stern–Volmer plot is known, two-point recalibration is almost always sufficient. Whereas anoxic conditions can be easily obtained using high purity gases or a solution of an oxygen scavenger (sodium sulfite, sodium dithionite, glucose solution with added glucose oxidase), the calibration in air is more critical. If the sensor is calibrated in the gas phase, temperature, atmospheric pressure and relative humidity should be measured. Although the last parameter is constant in water, it must be ensured that 100% air saturation at the given temperature is obtained. Colder or warmer solutions brought to the calibration temperature might be over- or undersaturated, respectively.

Evidently, the above errors can be avoided by careful calibration; however, the measurements can still be compromised by several artefacts that originate from the nature of the sensing material itself. As was discussed above, optical oxygen sensors, in theory, do no consume oxygen, which is highly advantageous over other analytical methods. However, this assumption might not be true in reality. Photosynthesized singlet oxygen belongs to highly reactive species and may react with the polymer before it deactivates back to the triplet state. As a consequence, not only is the matrix chemically modified (which may be accompanied *e.g.* by the change in oxygen permeability or accumulation of luminescence quenchers) but the analyte is also consumed. In fact, as was demonstrated for materials bearing an additional gas-blocking layer, most of the commonly employed polymers do react with singlet oxygen.[42] However, due to abundance of the analyte at air saturated conditions oxygen consumption might not be detectable at all for comparably fast responding sensors. On the other hand, the situation is radically different in the case of trace sensors. Here, oxygen consumption can dramatically affect the results.[43] For instance, the sensor based on a phosphorescent $BF_2$-chelate immobilized in polystyrene and equilibrated with 0.05% vol. $O_2$ showed a more than 10-fold decrease in quenching upon increasing the intensity of the excitation light by about 20-fold. This effect was even more pronounced at lower $pO_2$ so that the Stern–Volmer plot showed a characteristic upward curvature. Notably, the same matrix is fully adequate for measurements at physiological conditions even at comparably high light intensities typical *e.g.* for fiber-optic microsensors.

Evidently, oxygen consumption can be eliminated by making use of oxidatively robust matrices. Amorphous perfluorinated polymers such as Teflon AF, Hyflon AD and Cytop represent excellent candidates for this purpose. Unfortunately, conventional phosphorescent indicators are poorly soluble in these matrices and tend to aggregate even at low concentrations. Modification of indicators with perfluoroalkyl chains[31,37] renders them compatible with the perfluorinated polymers but the synthesis and purification of the dyes are not straightforward.

Trace, and particularly ultratrace, oxygen sensors also show several effects, which alter the form of the response curve.[43] First, the very long emission lifetime of the oxygen indicator can result in accumulation of the dye in the triplet state and therefore depopulation of the ground state. This effect affects the luminescence intensity plots but not the decay time plots. It can be minimized by using lower light intensities. Second, since dye concentration in polymers is typically high (0.01 mol $l^{-1}$) accumulation of the indicator in the triplet state can result in triplet–triplet annihilation, *i.e.* a process generating one dye molecule in the $S_1$ state and one molecule in the $S_0$ state upon interaction of two molecules in the $T_1$ state. Triplet–triplet annihilation results in shorter lifetimes and lower luminescence intensities. Since the efficiency of this process is proportional to the concentration of the dye in the triplet state, the negative effect of triplet–triplet annihilation can be minimized by using lower intensities of excitation light and by decreasing dye concentration in the polymer. Finally, the lifetime of singlet oxygen in some polymers can be rather long (>1 ms in perfluorinated polymers)[43] which can convert part of the analyte into the form not available for luminescence quenching. This effect can become significant at high light intensity (=high concentration of triplets) in combination with low oxygen concentration since one oxygen molecule has to quench several dye molecules within their lifetime. This effect can be minimized by using lower dye concentration resulting in higher oxygen to dye ratio. It can be concluded that such undesired effects can be minimized by using lower dye concentrations in the matrix and lower excitation intensities (the latter also helping to minimize oxygen consumption), but the trade-off is the lower signal-to-noise ratio. Here dyes with the highest phosphorescence quantum yields and optimized set-up, which minimizes the light losses, are beneficial. Fortunately, the above effects are marginal under "normal conditions" *i.e.* high $pO_2$ values and comparably low light intensities.

## 1.9 Conclusions and Outlook

It can be concluded that rational design of optical oxygen sensing materials is necessary for them to show the required performance in envisaged applications. Although it is not feasible to create an oxygen optode suitable for all potential applications, a great variety of oxygen indicators and polymeric matrices allow for virtually unlimited possibilities providing great flexibility in respect to optical properties and dynamic range of the sensors. Whereas many existing sensors already enable reliable measurements in many fields of science and technology, new materials for some demanding and emerging applications have yet to be developed. These include, for instance, a sensor operating in organic solvents with performance comparable to that of the state-of-the-art sensors for aqueous phase or a fast responding microoptode showing no drift upon prolonged usage at very high measurement rates (5–10 Hz).

## References

1. C. Schweitzer and R. Schmidt, *Chem. Rev.*, 2003, **103**, 1685–1758.
2. P. I. Djurovich, D. Murphy, M. E. Thompson, B. Hernandez, R. Gao, P. L. Hunt and M. Selke, *Dalton Trans.*, 2007, 3763.
3. M. Holtappels, C. Noss, K. Hancke, C. Cathalot, D. F. McGinnis, A. Lorke and R. N. Glud, *PLoS One*, 2015, **10**, e0116564.

4. M. Quaranta, S. M. Borisov and I. Klimant, *Bioanal. Rev.*, 2012, **4**, 115–157.
5. X. Wang and O. S. Wolfbeis, *Chem. Soc. Rev.*, 2014, **43**, 3666–3761.
6. P. C. Alford, M. J. Cook, A. P. Lewis, G. S. G. McAuliffe, V. Skarda, A. J. Thomson, J. L. Glasper and D. J. Robbins, *J. Chem. Soc., Perkin Trans. 2*, 1985, 705–709.
7. A. K. Bansal, W. Holzer, A. Penzkofer and T. Tsuboi, *Chem. Phys.*, 2006, **330**, 118–129.
8. K. Koren, S. M. Borisov, R. Saf and I. Klimant, *Eur. J. Inorg. Chem.*, 2011, **2011**, 1531–1534.
9. S.-W. Lai, Y.-J. Hou, C.-M. Che, H.-L. Pang, K.-Y. Wong, C. K. Chang and N. Zhu, *Inorg. Chem.*, 2004, **43**, 3724–3732.
10. W.-P. To, Y. Liu, T.-C. Lau and C.-M. Che, *Chem.–Eur. J.*, 2013, **19**, 5654–5664.
11. D. B. Papkovsky, G. V. Ponomarev, W. Trettnak and P. O'Leary, *Anal. Chem.*, 1995, **67**, 4112–4117.
12. S. M. Borisov, G. Nuss, W. Haas, R. Saf, M. Schmuck and I. Klimant, *J. Photochem. Photobiol., A*, 2009, **201**, 128–135.
13. J. R. Sommer, R. T. Farley, K. R. Graham, Y. Yang, J. R. Reynolds, J. Xue and K. S. Schanze, *ACS Appl. Mater. Interfaces*, 2009, **1**, 274–278.
14. J. E. Rogers, K. A. Nguyen, D. C. Hufnagle, D. G. McLean, W. Su, K. M. Gossett, A. R. Burke, S. A. Vinogradov, R. Pachter and P. A. Fleitz, *J. Phys. Chem. A*, 2003, **107**, 11331–11339.
15. G. Khalil, M. Gouterman, S. Ching, C. Costin, L. Coyle, S. Gouin, E. Green, M. Sadilek, R. Wan, J. Yearyean and B. Zelelow, *J. Porphyrins Phthalocyanines*, 2002, **6**, 135–145.
16. S. A. Vinogradov and D. F. Wilson, *J. Chem. Soc., Perkin Trans. 2*, 1995, 103.
17. O. S. Finikova, A. V. Cheprakov, I. P. Beletskaya, P. J. Carroll and S. A. Vinogradov, *J. Org. Chem.*, 2004, **69**, 522–535.
18. L. H. Hutter, B. J. Müller, K. Koren, S. M. Borisov and I. Klimant, *J. Mater. Chem. C*, 2014, **2**, 7589–7598.
19. T. V. Esipova, H. J. Rivera-Jacquez, B. Weber, A. E. Masunov and S. A. Vinogradov, *J. Phys. Chem. A*, 2017, **121**, 6243–6255.
20. I. B. Rietveld, E. Kim and S. A. Vinogradov, *Tetrahedron*, 2003, **59**, 3821–3831.
21. T. V. Esipova, A. Karagodov, J. Miller, D. F. Wilson, T. M. Busch and S. A. Vinogradov, *Anal. Chem.*, 2011, **83**, 8756–8765.
22. A. Y. Lebedev, A. V. Cheprakov, S. Sakadžić, D. A. Boas, D. F. Wilson and S. A. Vinogradov, *ACS Appl. Mater. Interfaces*, 2009, **1**, 1292–1304.
23. V. Tsytsarev, H. Arakawa, S. Borisov, E. Pumbo, R. S. Erzurumlu and D. B. Papkovsky, *J. Neurosci. Methods*, 2013, **216**, 146–151.
24. R. I. Dmitriev, S. M. Borisov, H. Düssmann, S. Sun, B. J. Müller, J. Prehn, V. P. Baklaushev, I. Klimant and D. B. Papkovsky, *ACS Nano*, 2015, **9**, 5275–5288.
25. J. Napp, T. Behnke, L. Fischer, C. Würth, M. Wottawa, D. M. Katschinski, F. Alves, U. Resch-Genger and M. Schäferling, *Anal. Chem.*, 2011, **83**, 9039–9046.

26. F. Niedermair, S. M. Borisov, G. Zenkl, O. T. Hofmann, H. Weber, R. Saf and I. Klimant, *Inorg. Chem.*, 2010, **49**, 9333–9342.
27. S. M. Borisov, G. Zenkl and I. Klimant, *ACS Appl. Mater. Interfaces*, 2010, **2**, 366–374.
28. B. Nacht, C. Larndorfer, S. Sax, S. M. Borisov, M. Hajnsek, F. Sinner, E. J. W. List-Kratochvil and I. Klimant, *Biosens. Bioelectron.*, 2015, **64**, 102–110.
29. S. M. Borisov and I. Klimant, *Anal. Chem.*, 2007, **79**, 7501–7509.
30. S. M. Borisov, R. Fischer, R. Saf and I. Klimant, *Adv. Funct. Mater.*, 2014, **24**, 6548–6560.
31. P. Lehner, C. Staudinger, S. M. Borisov and I. Klimant, *Nat. Commun.*, 2014, **5**, 4460.
32. C. O'Donovan, J. Hynes, D. Yashunski and D. B. Papkovsky, *J. Mater. Chem.*, 2005, **15**, 2946–2951.
33. K. Koren, R. I. Dmitriev, S. M. Borisov, D. B. Papkovsky and I. Klimant, *ChemBioChem*, 2012, **13**, 1184–1190.
34. I. Klimant, F. Ruckruh, G. Liebsch, A. Stangelmayer and O. S. Wolfbeis, *Microchim. Acta*, 1999, **131**, 35–46.
35. A. Paulmier and D. Ruiz-Pino, *Prog. Oceanogr.*, 2009, **80**, 113–128.
36. P. Lehner, C. Larndorfer, E. Garcia-Robledo, M. Larsen, S. M. Borisov, N.-P. Revsbech, R. N. Glud, D. E. Canfield and I. Klimant, *PLoS One*, 2015, **10**, e0128125.
37. M. Larsen, P. Lehner, S. M. Borisov, I. Klimant, J. P. Fischer, F. J. Stewart, D. E. Canfield and R. N. Glud, *Limnol. Oceanogr.: Methods*, 2016, **14**, 784–800.
38. N. Revsbech, L. H. Larsen, J. Gundersen, T. Dalsgaard and O. Ulloa, *Limnol. Oceanogr.: Methods*, 2009, **7**, 371–381.
39. E. R. Carraway, J. N. Demas, B. A. DeGraff and J. R. Bacon, *Anal. Chem.*, 1991, **63**, 337–342.
40. T. Mayr, S. M. Borisov, T. Abel, B. Enko, K. Waich, G. Mistlberger and I. Klimant, *Anal. Chem.*, 2009, **81**, 6541–6545.
41. H. Shi, X. Ma, Q. Zhao, B. Liu, Q. Qu, Z. An, Y. Zhao and W. Huang, *Adv. Funct. Mater.*, 2014, **24**, 4823–4830.
42. B. Enko, S. M. Borisov, J. Regensburger, W. Bäumler, G. Gescheidt and I. Klimant, *J. Phys. Chem. A*, 2013, **117**, 8873–8882.
43. P. Lehner, C. Staudinger, S. M. Borisov, J. Regensburger and I. Klimant, *Chem.-Eur. J.*, 2015, **21**, 3978–3986.

CHAPTER 2

# New Polymer-based Sensor Materials and Fabrication Technologies for Large-scale Applications

EFE ARMAGAN[a], DMITRI B. PAPKOVSKY[b] AND CLAUDIO TONCELLI*[a]

[a]Empa, Swiss Federal Laboratories for Materials Science and Technology, Laboratory for Biomimetic Membranes & Textiles, Lerchenfeldstrasse 5, CH-9014 St. Gallen, Switzerland; [b]School of Biochemistry and Cell Biology, University College Cork, College Road, Cork, Ireland
*E-mail: claudio.toncelli@empa.ch

## 2.1 Introduction

The continuous progress in the synthesis of new phosphorescent reporter dyes and nanostructures with sensitivity to molecular oxygen ($O_2$) has produced a growing toolbox of optical transducers with a variegated response to $O_2$, brightness, lifetime ranges, photostability, accessible functional groups, sensor material types and formats.[1–4] In liquid media, such long-decay emitting transducers can undergo quenching by triplet–triplet annihilation (once they are at close inter-molecular distances) as well as their photoluminescence being affected by other quenchers or environmental parameters, thus preventing the generation of strong optical signals and predictable operation

Detection Science Series No. 11
Quenched-phosphorescence Detection of Molecular Oxygen: Applications in Life Sciences
Edited by Dmitri B. Papkovsky and Ruslan I. Dmitriev

Published by the Royal Society of Chemistry, www.rsc.org

as $O_2$ sensors. On the other hand, encapsulation of $O_2$ phosphors in solid matrices helps to overcome these problems and facilitate sensor handling and reuse, thus providing a convenient solution for the translation of the basic $O_2$ quenching concept to commercial sensors.[1,3–6] Phosphor type and encapsulation process can be tailored to the required application, with variable sensor shape, fine structure, and sizes spanning the nano-, micro and macroscale.

Whilst nanosensors are usually engineered with a sophisticated surface chemistry to interact with a specific interface,[3] the development of planar film sensors, sensor dots, coated optical fibres or assay vessels is a compromise between the generation of a homogenous sensor material and a cost-efficient fabrication and integration processes.[1,5,6] However, trade-offs are usually required between sensor complexity and analytical performance, easiness, controllability and scaling up capabilities of sensor fabrication procedure, integration of oxygen sensors in specific application and measurement setup, especially for large scale packaging applications.[1,5,6]

Such need is particularly prominent to control the exposure of $O_2$-sensitive items and evaluate any packaging failures. Pharmaceutical, electronic and food industries routinely produce consumer goods that need to be protected and maintained in an inert atmosphere.[6,7]

Modified atmosphere packaging (MAP) in food is often applied to reduce the $O_2$ content in packaging, typically below 0.5%.[5] However, fresh red meat is packaged under high $O_2$ levels (40–60%) to improve its color and appearance. Improper gas flushing, trapped air, poor sealing or permeation of $O_2$ *via* the packaging material or accidental damage can alter $O_2$ levels and lead to fast product degradation.[8] Integrated $O_2$ sensors could help identifying packaging failure at any stage, in a non-destructive manner and in each single packaged product. This application requires low cost, calibration-free sensors working accurately and robustly on disposable and continuous basis, and which allow easy integration in various types of packages and products, throughout the whole supply chain (from farm to fork).

The encapsulation medium regulates $O_2$ accessibility, enhances and preserves phosphor stability under different physical, mechanical, thermal or light conditions, does not interfere with phosphor operation while shielding it from external interferences and stabilizing with covalent or non-covalent interactions from leaching, migration or aggregation.[1,4–6,9]

A customized solution, which includes sensor design, fabrication and integration, is often needed to optimally match the key requirements and operational specifications of each particular application. For example, sensors for trace $O_2$ detection or near-anoxic conditions[10] use different phosphors and encapsulation media than the ones utilized in standard industrial process control and packaging[4–6] where a broader $O_2$ dynamic range is required. In addition, different temperature and humidity operational windows, as well as spatial homogeneity and spatial resolution are needed. While mapping is not generally required for $O_2$ sensors in packaging,[5–7] it is fundamental for the development of $O_2$ sensitive paints.[1]

Sensors for large-scale applications require continuous, high volume production processes, minimal number of components and fabrication steps, use of environmentally friendly solvents, biocompatible and safe materials and ingredients, low energy consumption during the entire sensor fabrication cycle, high reproducibility and uniformity of individual sensors to enable calibration-free (or factory-calibrated) operation, high robustness and flexibility.[1,5,6] Casting of a physical mixture of phosphor and polymer solution (liquid precursor 'cocktail') onto a substrate to produce thin film solid-state sensor coatings can satisfy many of these requirements.[1,4–6] However, performance of many such sensors, including those produced commercially, is often jeopardized by micro-heterogeneities, cross-sensitivity, drift or instability of calibration, difficulties in process control and production of standardised and identical batches of sensors.[6] This obviously affects sensor robustness, reproducibility, accuracy, signal intensity, spatial homogeneity and production costs.

More sophisticated material processing, namely solvent crazing,[11,12] encapsulation in mesoporous membranes,[13,14] electrospinning[15–18] and electrophoretic deposition,[19] despite requiring more complex equipment facilities, have the potential to improve homogeneity of sensor micro-environment, enlarge the operational range and sensitivity of conventional $O_2$ sensors. Parallel to this, research in enhancing the matrix/phosphor compatibility by covalent or coordinative chemical bonding between the two components has produced sensibly lower phosphor leaching, formation of mixed polymer phases with support material, higher spatial and in-depth homogeneity of the resulting $O_2$ sensors.[20]

## 2.2 Physical Entrapment of Phosphors Within Inorganic and Organic Matrices

The concept of physical entrapment of optical transducers within a matrix is generally simpler and more cost-efficient than systems involving a covalent or coordinative bridging between the phosphor and the matrix.[1,2,4,5,9] Here, a number of approaches and processing methods have been exploited (Table 2.1).

Traditionally, sensor coatings were fabricated by casting a liquid 'cocktail' containing a hydrophobic phosphor, hydrophobic polymer or its organosilica precursor in a non-polar organic solvent onto a suitable substrate, such as the surface of glass slide, the tip of an optical fibre or a polyester film.[2] After solvent evaporation, the phosphor becomes efficiently entrapped within the polymer, which forms a thin $O_2$-permeable film on the corresponding substrate. Recently, addition of water-soluble components within the sensor formulation has enabled the simultaneous formation of mesoporous $O_2$ sensing coatings.[21,22]

**Table 2.1** Common types of solid-state $O_2$ sensors and processes used for their fabrication.

| Sensor type | Description | Sensor components | Fabrication steps | Equipment required | Reference |
|---|---|---|---|---|---|
| *Conventional sensor materials and methods* | | | | | |
| 1 | Phosphor dissolved in polymer cocktail, applied as a thin film on planar substrate and dried to form a thin coating | 1. Phosphor<br>2. Polymer<br>3. Organic solvent<br>4. Support material<br>5. Additives, *e.g.* scattering particles (optional) | 1. Making cocktail<br>2. Casting on substrate<br>3. Drying<br>4. Annealing (optional) | Dip-coater, pipette (for drop casting process), knife coater, spin coater, inkjet and tampon printer | 31–85 |
| 2 | Phosphor dissolved in ormosil precursor(s), applied on planar substrate and polymerised to form xerogel or aerogel coating | 1. Phosphor<br>2. Ormosil monomers<br>3. Catalyst<br>4. Solvent<br>5. Support | 1. Making cocktail<br>2. Casting<br>3. Drying/polymerising<br>4. Curing | Spin coater, knife-coater, spray coater | 86–133 |
| 3 | Phosphorescent micro-beads dispersed in silicone components, applied on planar substrate and cured to form coating | 1. Phosphor<br>2. Silicone precursor(s)<br>3. Curing agent<br>4. Solvent<br>5. Support material | 1. Making cocktail<br>2. Dispensing dispersion onto substrate<br>3. Drying/polymerising<br>4. Curing | Dip-coater, spin coater, knife coater pipette (for drop casting process) | 136–156 |
| *New sensor materials and methods* | | | | | |
| 4 | Microporous polymer membrane impregnated with a phosphor by diffusion or soaking method | 1. Phosphor<br>2. Solvent<br>3. Microporous membrane | 1. Dissolve dye in solvent<br>2. Soak polymer membrane in solution 1, incubate, heat<br>3. Wash the membrane, dry | Pipette (for drop casting processes), dip coater, spin coater | 14, 24, 60, 157–170 |

| | | | | | |
|---|---|---|---|---|---|
| 5 | PP, PE, PPS film or fiber substrate impregnated with phosphor by solvent crazing method (fully or locally – spot-crazing) | 1. Phosphor<br>2. Solvent<br>3. Semi-crystalline polymeric film or fiber (PP, PE or PPS) | 1. Dissolve dye in solvent<br>2. Submerge the film/fiber in solution 1<br>3. Apply tensile drawing to the substrate<br>4. Wash, dry<br>5. Anneal (optional) | Customized tensile drawing/polymer orientation machine (for bulk crazing), or puncture test equipment (for spot crazing) | 11, 12, 171–174 |
| 6 | Phosphor-polymer cocktail electrospun to yield nanofibers | 1. Phosphor<br>2. Polymer<br>3. Solvent<br>4. Support substrate | 1. Cocktail preparation<br>2. Nanofibre formation<br>3. Membrane washing, drying | Electrospinning equipment | 15–18, 175–185 |
| 7 | Polymeric nanoparticles dyed with phosphor | 1. Phosphor<br>2. Solvent<br>3. Polymer base<br>4. Support material | 1. Inclusion of the phosphor in cross-linked PS microparticles<br>2. Electrophoretic deposition | Electrophoresis equipment | 19 |
| 8 | Layer-by-layer deposition of polycation and polyanion from solution, with phosphor dissolved in one of them. | 1. Phosphor<br>2. Polycation solution<br>3. Polyanion solution<br>4. Washing solution<br>5. Support material | 1. Dispense solution of polycation with phosphor<br>2. Remove excess, rinse with water<br>3. Dispense polyanion solution<br>4. Remove excess, rinse with water<br>5. Repeat steps 2-4 (to produce coatings with high signals) | Dip-coater, spin-coater | 25, 186–189 |

A modification of this approach is when a phosphor dissolved in organic solvent (*i.e.* without matrix polymer) is applied directly on a micro-structured solid polymeric matrix (*e.g.* mesoporous microparticles[23] or membranes[13]) and let the phosphor molecules diffuse into its depth. The process, which can be facilitated by solvent optimisation to enhance polymer swelling and/or by high temperature accelerating phosphor diffusion, should be designed such that it does not alter the final mechanical integrity and properties of the matrix (solvent-compatibility) and provides uniform dyeing of the matrix. Such microporous,[13] microfibrous[18] and microparticle[23] materials, acting as both sensor substrate and encapsulation matrix, on the one hand help to match the mechanical properties of the coating with the substrate, therefore providing higher adhesion at the interface, whereas on the other hand provide homogeneous micro-environment and/or dispersability (powder form) of the phosphor emitter.

Indeed, often the mechanical and adhesion properties of the preferred type of sensor coating do not match the ones required by the application, as well as $O_2$ permeability, sensitivity, *etc*. Then it is necessary to add an intermediate step where the sensor material is produced in a micro-encapsulated form (*e.g.* phosphorescent microparticle powder) and then dispersed in a carrier media such as silicone rubber.[24] The latter provides the sensor coating good $O_2$ permeability, mechanical and chemical stability, adhesion to the support material.

High density networks of nanocavities can also be formed *in situ* by stretching a semi-crystalline polymer film or fibre in a physically active liquid environment (PALE), the so-called solvent crazing.[11] By dispersing the phosphor in the PALE, efficient encapsulation of the phosphor and staining of the host polymer can be achieved by simple means, driven by diffusive processes.

The phosphor can be also be entrapped in a matrix *via* layer-by-layer assembly of oppositely charged polymeric layers doped by phosphor molecules (LbL method).[25] Furthermore, electrically-induced adhesion to charged substrates *via* electrophoretic deposition[19] and induced fibre formation by electrospinning[15] under the applied external electrical field produced sensors with high $O_2$ accessibility, higher sensitivity and fast response.

A number of $O_2$-sensitive nanophosphors were recently described.[26] Their integration in sensor material usually proceeds by one of the above methods; however, since often these sensors do not exhibit phosphorescence prior to matrix integration, reactivity of the functional moieties during anchoring them to the matrix and phosphor's surface chemistry should be taken into account.

Spatial resolution in sensor fabrication and patterning of individual sensors (*e.g. via* photolithography processes) is a crucial requirement for their integration in microfluidic devices.[27–29] This, however, is not such an issue for sensors aimed at large scale production, and therefore we do not include this in the review.

### 2.2.1 Casting of Polymer 'Cocktails'

Casting of polymer/phosphor cocktails represents a useful fabrication method and it is still the most widely used route to producing large quantities of sensor films and coatings on planar supports or partly on optical fibres.[1,4–6,9] The latter are utilized when there is a need to detect $O_2$ in cumbersome regions or where the use of conventional planar sensors is precluded.[30] Coating of curved surfaces is obviously more complex than the planar counterparts; therefore, it requires a tighter control of the mechanical properties of the matrix (*i.e.* adhesion, flexibility to name a few).

Different casting techniques are employed in the fabrication of conventional $O_2$ sensors, including drop coating, knife casting, dip coating and spin coating. Drop coating and spin coating involve the deposition of cocktail aliquots on clean substrate surface with subsequent spreading and passive drying by solvent evaporation. The sensor quality and thickness in this case will be mainly determined by the cocktail viscosity, polymer concentration, solvent polarity, the support surface chemistry, the volume of cocktail applied and environmental parameters (temperature, humidity, convection, *etc.*). Knife casting and spin coating usually produce better results than drop coating as they permit a higher homogeneity, better control of the coating thickness avoiding the so-called 'coffee stain' effects, which is the presence of higher phosphor concentration at the coating edges. However, as spin coating relies on the spreading of the droplet *via* rotation of the substrate, this has to show sufficient surface tension to avoid diffusion on the substrate prior to the rotational spinning.

Dip-coating consists of submerging the substrate and retrieving it at a pre-set speed whereas, with knife-coating (or slit-bar coating), the polymeric cocktail is positioned on a substrate and its surface is levelled by moving a blade along the substrate length. As with previous techniques, their viability is strictly related to the polymeric cocktail viscosity. With the dip-coating process, a relatively higher inherent viscosity would lead to thicker coatings with strong thickness variation along the z-axis of the substrate. Knife-coating of polymer cocktails generate a smooth coating only when the solution is not spread before that the blade has homogenised the coating thickness.

Dip-coating and knife-coating are viable for large-area sensor productions, which is crucial for PSP applications. More recent modifications of fabrication of $O_2$ sensor coatings include tampon and jet printing.

$O_2$ sensors in packaging applications have some add-on requirements, such as the production of discrete, small-size sensors in specific locations of the substrate (to reduce sensor price and facilitate its visualisation and access for the measurement), correct integration within packaging multi-layer laminates, without any negative impact on sensor performance.[6] This imposes stricter guidelines on sensor materials and manufacturing processes, and major restrictions on price (cent per sensor range).

Many polymeric materials have been utilized as matrices for $O_2$-sensitive phosphors, the most common being polystyrene, ethyl cellulose, polysulfone and fluorinated polymers (Figure 2.1).

Polystyrene (PS) is still one of the most common polymers used for sensor fabrication,[31–33] because of its moderate $O_2$ permeability (suitable for 0–21 kPa or 0–100 kPa $O_2$ range with many common phosphor reporters), relatively high mechanical and thermal stability, hydrophobicity (contains aromatic rings and dissolves well many synthetic organometallic phosphors with aromatic ligands) and no significant cross-sensitivity to humidity.

Researchers often use this matrix to test new phosphors,[34–41] as polystyrene is soluble in a wide range of organic solvents (*i.e.* tetrahydrofuran,[38,42,43] toluene,[35,44] chloroform,[34,36,37,39–41,45,46] butyl acetate[47]). Polystyrene can be utilized with mixtures of different emitters or coupled with organic photodetectors,[44] systems operating in ratiometric mode,[45] additives (singlet oxygen scavengers) which reduce signal drift,[42] or integrated in a multi-sensing device.[47] Sensor sensitivity can be tuned by changing the substituent in the *para* position of the phenyl ring or introducing fluorine atoms in both *o*-positions.[48] Thus, polystyrene modified with *t*-butyl groups showed approximately 4-fold higher sensitivity than unsubstituted polystyrene. The presence of fluorine atoms significantly increased sensor photostability, which is already high for fluorinated phosphors.[4] Due to the ease of functionalization of phenyl halogens on the polymer skeleton, referenced dyes can be coupled to the backbone to obtain Förster resonance energy transfer pairs between the embedded temperature-sensitive and oxygen-sensitive phosphors, respectively.[49]

Ethyl cellulose (EC) is another promising material with high optical transparency, solubility in many organic solvents and $O_2$ permeability similar to polystyrene.[50–56] However, this hydrophilic polymer shows strong cross-sensitivity to humidity, with $O_2$ sensitivity reduced by 50% at 85% RH if compared to 0% RH. Indeed, EC varies its $O_2$ permeability upon the entrapment of water molecules within the matrix.[57] Also cross-sensitivity to temperature is more pronounced than Ormosils-based coatings (see Section 2.2.3).[58]

Polysulfones (PSU) are also promising alternative to PS and EC based optical oxygen sensors as this polymer matrix show high optical clarity (though only when the membranes are not produced in air),[59] chemical stability, moderate $O_2$ permeability and low cross-sensitivity to humidity.[60] It has been shown[59,61–63] that polysulfone membranes can be utilized to determine the thermodynamic parameters of the oxygen quenching process and model sensor calibration. Polysulfone based coatings with high mechanical and thermal stability can withstand sterilization by autoclaving,[64] without changing their $O_2$ sensing properties.

Plasticised poly(vinyl chloride) membranes were employed to develop a dual analyte dissolved $O_2$/pH sensor based on one phosphorescent porphyrin phosphor with multi-receptor functionalities. In this case, proton permeability of the membrane was enhanced by adding a borate salt within the formulation.[65] However, the use of plasticizers and other additives makes

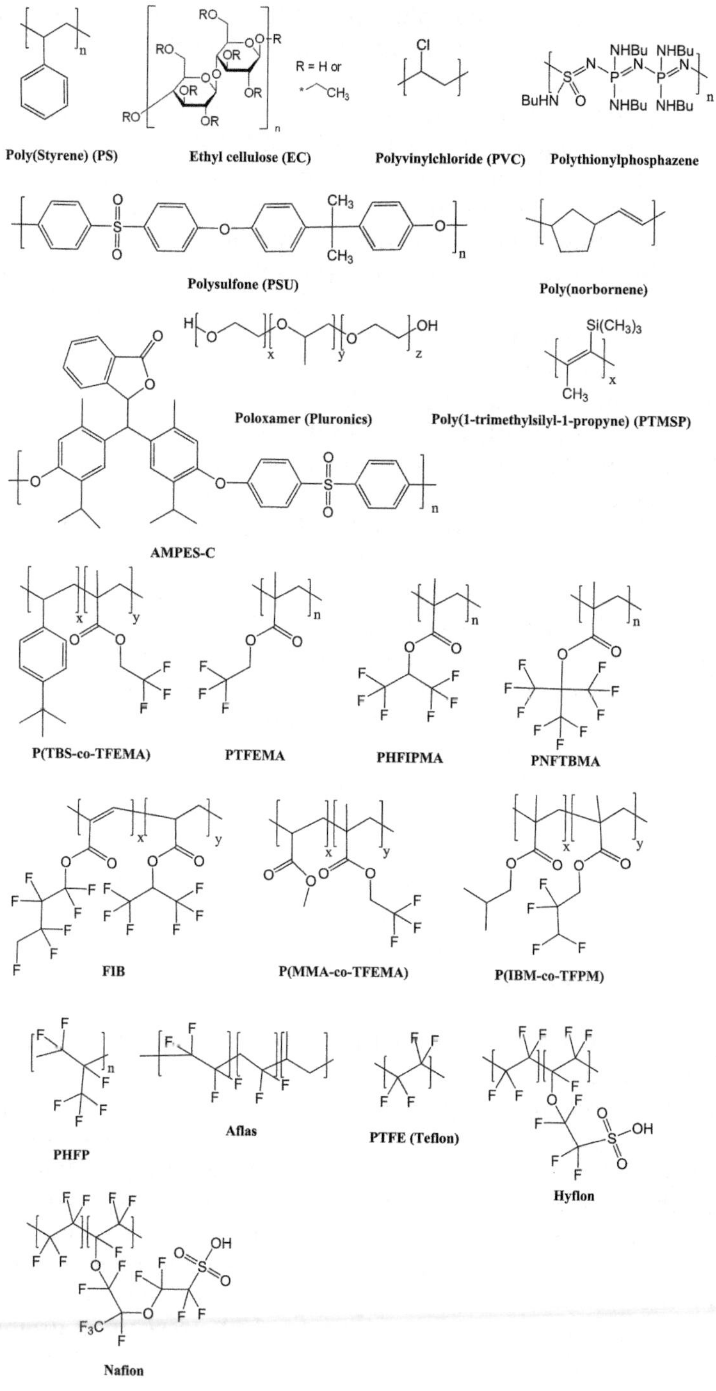


**Figure 2.1** Chemical structures of common polymer matrices utilized for phosphor encapsulation.

the sensor structure more complex and difficult to reproduce and stabilize (batch to batch variability and signal drifting by leaching of low-molecular weight compounds).

Other polymers, namely poly(norbornene),[66] poly(thionylphosphazene),[67] cellulose acetate butyrate,[68] AMPES-C[69] and poly(1-trimethylsily-1-propyne),[70,71] offer a possibility to change $O_2$ permeability and allow $O_2$ sensing in different dynamic ranges. In particular, poly(1-trimethylsilyl-1-propyne) (PTMSP) have shown ten-fold higher $O_2$ permeability than conventional gas permeable poly(dimethylsiloxane) (PDMS) films.[70] Non-functionalised poly(norbornene) has similar $O_2$ permeability to PS, which can be tuned by varying the steric hindrance of the side groups.[66]

Fluorinated methacrylates[72–77] and other fluoropolymers such as poly(hexafluoropropylene),[78] poly(tetrafluoroethylene) (PTFE),[10] Aflas,[79,80] Hyflon[10,81] and Nafion[80] have remarkably enhanced the sensitivity to $O_2$ for a given phosphor and led to solid sensors with better photostability. Indeed, they display extremely high $O_2$ permeability, thanks to the limited inter-molecular interactions due to shorter C–F bond lengths compared to C–H ones. Increasing the content of fluorinated methylene side groups of polymethacrylates further enhances their sensitivity to $O_2$ and reduces cross-sensitivity to temperature.[75] Since fluoropolymers show high chemical inertness, they are particularly useful for the detection of $O_2$ in chemically aggressive environments.[79] The combination of an aluminium chelate with a PTFE matrix has recently achieved the highest sensitivity to $O_2$ with a limit of detection of 5 ppb in the gas phase or 7 pM in solution.[10]

A few studies utilized polymer/phosphor coatings on optical fibres, mainly based on PS deposited on the tip of stainless steel needles,[82] onto the fibre surface[83] with the light being conveyed through the capillary wall and coupled with an organic photodiode, or dip-coated onto a glass capillary[84] to produce optically responsive fibres. Polymethacrylates with different methylene side chains and Pt(II) porphyrin phosphor were coated onto optical fibres for the measurement of $O_2$ saturation in arteries.[85] Response time and $O_2$ sensitivity increase at longer aliphatic ester substitutions. Bulkier side chains do not allow a tight polymer packing and consequently gas permeability is increased.

As a general remark, despite casting of a physical mixture of polymer and phosphor permits an easy upscaling, simple and cost-efficient processability and a variegated response to $O_2$, accuracy, robustness and reproducibility of the resulting sensors are quite average and it is rather difficult to generate factory-calibrated sensors by this technique. Indeed, variations in sensor morphology and micro-environment, interactions between the sensor and support materials (mixed polymer phases), large volume contraction and associated mechanical stress upon drying of the polymeric cocktails, non-uniformity of coatings and other factors lead to pronounced heterogeneity, complex sensor behaviour (non-linear Stern–Volmer plots and $O_2$ calibration), variable read-out of $O_2$ concentration, possible delamination.

### 2.2.2 Physical Mixing of Phosphors with Ormosils

Compared to one or two component silicon rubbers, organically modified silicon precursors (Ormosils) represent a more convenient route for encapsulation of fluorophores as they allow a versatile modulation of the network stiffness by varying the functionality of the precursors.[86–89]

The pioneering investigations on xerogel from Ormosils precursors were with phosphor-doped sol–gel solutions deposited on tapered optical fibres,[86,87,89] or on planar sensors.[88] Tetra-functional cross-linkers were originally utilized, although they generally produced 'stiff' films cracking very easily. In order to improve the film quality, several strategies were applied: the addition of a non-ionic surfactant (Triton X-100) to the sol–gel,[90,91] use of tri-functional precursors,[92–94] ionic liquids (strong cross-talk to humidity is expected for hydrophilic ionic liquids)[95] or a mixture of ethylene glycol and citric acid to obtain a silicon/polyester matrix.[96] Synthetic xerogel can also be mechanically ground to produce $O_2$-sensitive microparticles.[97–100]

$O_2$ sensitivity of such solid-state sensors is affected by the many sol–gel and polymerisation process parameters. Indeed, it increases at lower water/organosilane ratio (*e.g.* higher hydrolysis rate) as higher porosity of the network is generated.[101,102] The pH of the solution also affects the kinetics of sol–gel formation and optical properties of the phosphor (*e.g.* lifetime, emission wavelength): below the isoelectric point of silica (pH 2), the gel is characterised by fast hydrolysis and weakly branched microporous structures, whereas the opposite happens above the isoelectric point.[102–104] Sensor sensitivity can be improved by a post-curing process at high temperatures since it releases water and other small molecules from the matrix (although it might induce microphase segregation of the phosphor)[105] or by adding silver-coated silica nanoparticles to generate a plasmonic enhancement of the sensor signal.[106]

It has also been shown[107] that a trifunctional organo-silane remarkably increases the $O_2$ quenching due to increased hydrophobicity of the matrix, and this effect was enhanced by extending the aliphatic chain of the trifunctional Ormosil.[108–110] Subsequent study with trifunctional Ormosils with different aliphatic side chains has demonstrated a linear correlation between matrix polarity and sensor sensitivity to $O_2$.[111] The trifunctional Ormosils with phenyl groups were also tested as $O_2$-sensitive matrix[112] and a significant increase of the sensor photostability was attained although the mixture had to be mixed with a monofunctional precursor to avoid any film cracking.

The fabrication of sol–gel $O_2$ sensors was accomplished by adding a trifunctional organo-silane with a long aliphatic chain (*e.g.* octyl group) to tetraethyl orthosilicate (TEOS), which resulted in a noticeable improvement of the film quality, sensor long-term stability, sensitivity to $O_2$ and homogeneity of phosphor micro-environment (linear Stern–Volmer calibration).[113,114]

Using a combination of fluorinated and non-fluorinated trifunctional precursors, higher sensitivity,[115–118] shorter sensor response time,[115–117] smaller humidity cross-talk[119] and enhanced photostability[120,121] were attained compared to conventional xerogels. Aromatic fluorinated precursors further increased the $O_2$ quenching.[122]

Although temperature cross-sensitivity has been relatively low between 21 °C and 35 °C for conventional xerogels[123] and between 25 °C and 45 °C for fluorinated ones,[119] significant changes in oxygen diffusion and solubility are expected at higher temperature ranges. Temperature compensation was obtained by developing a modified Stern–Volmer model,[124] or by encapsulating within the xerogel a dual sensing mechanism, where semi-conductor quantum dots, 7-amino-4-trifluoromethyl coumarin or carboxyfluorescein were exclusively sensitive to temperature and a ruthenium(III) dye or a Pt(II)-porphyrin was sensitive to both $O_2$ and temperature.[125–128]

Referencing was also achieved by encapsulating a second $O_2$-insensitive dye into the matrix and developing a ratiometric intensity based $O_2$ sensor.[129,130] The chemical inertness of the xerogel matrix can be improved by its blending with a fluoropolymer.[131]

Aerogel-based $O_2$ sensors were also prepared by a one-step supercritical extraction technique, although such materials showed poor stability in the presence of water, mechanical properties and non-linear Stern–Volmer calibration.[132,133]

### 2.2.3 Phosphor Integration in Silicone Rubbers

Silicone networks, thanks to their limited inter-macromolecular interactions and the low rotation energy of the Si–O bond, display remarkably high gas permeability (which decreases with substitution with bulkier functional groups in the side and in the backbone chains of the silicone polymers[134,135]) and high optical clarity. Deposition techniques do not differ from the polymer 'cocktail' casting described in the previous section, although dip-coating and spin-coating were generally preferred due to the high viscosity of the sols.

Commercial mono- and bi-component silicone rubbers were among the first matrices utilized to encapsulate phosphors.[136–138] Although the results were encouraging, the hydrophilic and charged metal–organic Ru(II) complexes were seen to undergo microphase separation, which resulted in complex non-linear Stern–Volmer (SV) calibrations. Improved sensor characteristics were observed when dispersing a more hydrophobic phosphor, namely a pyrenebutyric acid, within the silicon rubber, which led to linear SV calibration plots.[139] Subsequently, a structure–performance study by varying the nature of the substituent in pyrene derivatives and the type of silicon rubber demonstrated that silanol-terminated silicon rubber and 1-pyrene butanol show the highest compatibility among the tested resins and phosphors.[140] However, this study was not very comprehensive to evaluate the extent of $O_2$ quenching by measuring only under nitrogen and in air, without calibration points at intermediate $O_2$ pressures. Despite a few promising results, this approach cannot be extended to the whole library of phosphors described in the literature, thus different systems were later evaluated to enhance the sensor compatibility within the silicon matrix.

Indeed, a method to control phosphor dispersion is to use an intermediate encapsulation step within mesoporous micro- and nanoparticles, such as alumina, silica or titania[141] to stabilize the dye, $O_2$ response and enhance the matrix compatibility.

Such particles retain their mechanical integrity upon encapsulation and the phosphor can be physically immobilized in their mesoporous structure. Thus, in ref. 142, Zeolite Y (Faujasite) having pore diameter of 1.2 nm was used to encapsulate a ruthenium dye. The pore size of the zeolite was chosen to adsorb either the whole metalloorganic complex or simply the metal ion and then chelate the latter with light antenna ligands. Both types of phosphor-loaded zeolites were dispersed in a silicone rubber, which cures in the presence of humidity. An approximately 50% difference in $O_2$ quenching constant was observed for these two samples, which retained their emission properties throughout one year of storage.[142]

In a similar investigation, the zeolite was replaced by fumed silica. Higher luminescence intensity, larger sensitivity towards $O_2$ and less curved Stern–Volmer calibrations were observed for such phosphor-adsorbed silica dispersed in silicone rubber, as compared to phosphor loaded directly in the silicone rubber.[143,144] Mesoporous silica nanoparticles MCM 41 and SBA 15 were used for the inclusion of a trinuclear starburst Ru(III) complex, showing $I_0/I_{100}$ between 3 and 5.5.[145] The same approach was repeated for carbon nanodots,[146] two Eu(III),[23,147] two Cu(I),[148,149] two Re(I)[150,151] complexes, and a Pt(II) porphyrin.[152] Pyrene encapsulation in porous metalloorganic frameworks (MOFs) was investigated to develop pressure-sensitive paints. Dispersion of these MOFs in silicone rubber, which could be sprayed or painted on metallic surfaces, showed remarkable photostability of the sensor even at 160 °C, linear Stern–Volmer plots and sensitivity tunable by the phosphor loading within the MOFs.[153]

An $O_2$ sensor based on triplet–triplet annihilation-based upconversion and the DLR calibration approach was recently described, which was produced by encapsulating in porous glass microbeads a fluorinated Pd-benzoporphyrin and a substituted perylene dye, and subsequently dispersing these beads in silicone rubber.[154]

Core–shell silica nanoparticles doped with a Pt(II) porphyrin and subsequently embedded in xerogels showed high sensitivity to $O_2$ – up to $I_0/I_{100}$ of 166.[155] Dual temperature/$O_2$ sensors were also produced by loading the phosphor in the shell of quantum dots/silica core–shell nanoparticles and dispersing the latter in fluorinated xerogels. Fast signal response in both adsorption and desorption cycles were observed with a dynamic $O_2$ range of 0–100 kPa $O_2$.[156]

### 2.2.4 Impregnation of Microporous Membranes and Microparticles with Phosphor Molecules

An alternative procedure to fabricate $O_2$ sensors is the direct deposition of phosphor molecules in an appropriate microporous polymeric matrix. Different materials have been utilized, namely nanoporous alumina,[157–161] poly(dimethylsiloxane) in combination with pyrogenic fumed silica,[162,163] filter paper[164] and non-woven polyolefin membranes.[14,60,165] All these materials display a certain degree of porosity or nano/micro structure to accommodate the phosphor within the matrix.

Nanoporous alumina with high optical clarity was prepared by mixing an aluminium oxide/hydroxide solution with poly(vinyl alcohol) (*e.g.* the porogen component) under acid catalysis conditions followed by deposition onto a Mylar polyester film *via* curtain coating. Afterwards, a solution of the phosphor in chloroform was deposited *via* spin coating. Due to the nanoporosity of the matrix, the Stern–Volmer constant was increased by a factor of 100 compared to the homologous polystyrene-based $O_2$ sensors.[157,159] In addition, the intensity signal was stable upon different sterilization procedures and after long-term storing at room temperature up to twelve months. This process is compatible with fluorinated Pt(II) porphyrins,[159] Ir(III)[158,160] and Cu(I) complexes.[161]

Commercial silicone resins generated by hydrosilylation reaction of a pre-polymer and a curing agent in the presence of a palladium catalyst, thanks to their high oxygen permeability, optical transparency, represent an alternative to sol–gel chemistry. In this case, the microporous network is formed in the absence of the phosphor followed by its diffusion in the swelled matrix. As the phosphor could not be sufficiently dispersed in this matrix, pyrogenic silica particles were added as fillers.[162,163]

It was also demonstrated,[164] that fluorescence of the matrix can be utilized in ratiometric $O_2$ sensors, as long as this referencing signal does not overlap with phosphor emission. To develop this concept, common filter paper and a gadolinium dye were used.

$O_2$ sensors were also produced from commercial grafted (*e.g.* with a hydrophilic surface chemistry) and ungrafted microstructured polypropylene membranes (common non-woven fabric materials). Two different methods of encapsulation were employed: swelling in organic solvent containing a phosphor followed by solvent evaporation[60,165] and applying drops of a ternary mixture water/ethyl acetate/Kolliphor P188 surfactant.[14] The use of polyolefin membranes as scaffold for oxygen sensing showed high phosphorescence intensity signals, optimal lifetime signals (22–30 μs at 21 kPa and 50–60 μs at 0 kPa $O_2$), linear Stern–Volmer plots, low temperature dependence between 10 °C and 30 °C and low cross-sensitivity to humidity.[165]

The use of an intermediate step of encapsulation prior to the sensor fabrication is particularly beneficial when using hydrogel as a sensor matrix. Indeed, hydrogels as sensor matrices offer several advantages: high flexibility and mechanical stability, adhesion to various surfaces including optical fibers, the possibility to vary their viscosity by temperature (thermo-gelating) or shear stress (thixotropic behavior). However, their network structure, when prepared with conventional polymer cocktail formulations, cannot stably immobilize many phosphors. Therefore, similar to silicone rubbers and xerogels, the phosphor needs first to be encapsulated in microbeads.

Ormosil microparticles provide an effective encapsulation media for phosphors bearing lipophilic anions, they can then be embedded in a polyurethane hydrogel.[166] The incorporation of pH and $O_2$ sensitive microbeads with tailored perm-selectivities allows sensing of $O_2$ and pH by phase measurements at two different modulation frequencies, according to a modified dual lifetime referencing (DLR) method. Some other multi-sensing films and multi-layer structures were produced using microbeads dispersed in poly(urethane) hydrogel matrix, particularly using microbeads prepared by a co-precipitation method with styrene and acrylonitrile[167,168] or styrene and vinylpyrrolidone as reactive monomers.[169,170]

A water-sprayable thermogelating $O_2$ sensor was also fabricated by immobilizing phosphor-loaded cross-linked polystyrene microparticles in a block copolymer of poly(ethylene glycol) and poly(propylene glycol) units (poloxamer).[24]

### 2.2.5 Solvent Crazing

The solvent crazing method involves tensile drawing of a polymeric substrate (film[11,12,171,172] or fiber[173]) in a physically active liquid environment (PALE), which leads to the development of a specific nanoporous structure with overall volume porosity of up to 55–60%, pore sizes ranging from 5–15 nm, and narrow pore size distribution. The process is controlled by the nature of the polymer (chemical structure, degree of crystallinity and topological distribution of the crystalline phase) and by the deformation conditions (tensile strain, applied stress, type of solvent, strain rate). By dissolving an $O_2$-sensitive dye in the PALE, its highly efficient one-step encapsulation in the sensing material can be achieved, driven by the diffusion process. Phosphors that are not readily soluble in the bulk polymer can be incorporated in such a manner.

Following the initial studies with submerged polypropylene and polyethylene substrates in heptane,[171] spatial control and localized impregnation of the polymer with high in-depth and spatial homogeneity of the resulting sensor material were attained.[12] The latter process was called spot-crazing. In order to use more environmentally friendly solvents, spot-crazing of poly(phenylene sulfide) matrix and impregnation with Pt(II)-benzoporphyrin dye were also performed in 2-butanone.[11] The resulting sensors showed a dynamic linear range of 0.1–100 kPa $O_2$, low temperature and humidity cross-sensitivity and good reproducibility (RSD 1.5% at 21 kPa and 0.5% at zero $O_2$). Later on, the spot-crazing

$O_2$ sensor technology was automated by the use of standard puncture test equipment, which can also be used for sensor integration in conventional packaging materials.[172]

Encapsulation of Pt-benzoporphyrin dye in the pore-structure of a hard elastic polypropylene film produced $O_2$ sensors which show a reversible response to tensile drawing *via* changes in phosphorescence intensity and lifetime and $O_2$ quenching constant.[174] Such materials can be used as mechanically tunable $O_2$ sensors or combined $O_2$ and strain sensors.

### 2.2.6 Electrospinning

Electrospinning uses an electric field between sample dispenser and collector to draw charged threads of polymer solution. Nano- and micro-fibres produced by this technique display high $O_2$ sensitivity and short response time due to their large surface area-to-volume ratio and nanoporous scaffold architecture.[175] Electrospinning is also compatible with large-scale production of sensor membranes.

$O_2$ sensors were first produced by using a cocktail of polystyrene and a Eu(III) complex as the feeding electrospinning solution.[176] High sensitivity ($I_0/I_{100} = 15.56$), good linear Stern–Volmer characteristics and short response/recovery time (t↓ of 7 s and t↑ (s) of 14 s) were obtained. Sensible improvement of the response time and sensitivity to $O_2$ were later confirmed for Cu(I),[177] Ir(III)[16,17] or Eu(III)[176,178,179] complexes. Comparison with conventionally cast films of polystyrene/fluorinated Pd(II) porphyrins revealed the response time was decreased by two orders of magnitude (from several seconds to less than 40 ms) while sensor sensitivity and signal intensity were unaltered.[180]

Two complexes of Re(I) were integrated in poly(vinylpyrrolidone) electrospun fibers, albeit, due to the water solubility of the polymer, the sensors were strongly affected by humidity.[181,182] Additives, such as plasticizers, water miscible and immiscible ionic liquids and perfluorinated compounds (*e.g.* which enhance $O_2$ solubility in the matrix) allowed the electrospinning of ethyl cellulose and poly(methylmethacrylate) formulations.[183] The resulting sensors showed high stability over three years, probably due to the stabilization of the ruthenium oxidation state by the perfluorinated compound. Encapsulation of a referencing dye for ratiometric intensity sensing mode is also compatible with the electrospinning technique.[184]

Co-axial electrospinning was performed to obtain a fibre with a poly(dimethylsiloxane) (PDMS) core in which the phosphor was encapsulated and a poly(caprolactone) (PCL) shell which preserved the mechanical integrity of PDMS during the curing process and also enhanced biocompatibility of the fibre.[15] Core–shell fibres can also be used to encapsulate in different layers up-converting nanoparticles and $O_2$-sensitive dye in an electro-optical handshake fashion. The use of up-converting nanoparticles minimize light absorption and scattering effects of biological tissues on the sensor.[185] It is important to mention that even such high nanostructure complexity can be achieved in a one-step process by utilizing different electrospinning setup and feeding devices.

Finally, nanoporous poly(caprolactone) fibers were impregnated by soaking them in a solution of an $O_2$-sensitive dye, to generate a phosphorescent scaffold suitable for cell growth, contactless monitoring and live imaging of $O_2$.[18]

### 2.2.7 Electrophoretic Deposition

Electrophoretic deposition (EPD) is the electrically-driven migration of molecules on a charged substrate and it is commonly utilized as a coating method in the automotive industry. Although only one study has been reported so far, EPD of hybrid polymeric nanoparticles prepared by the co-precipitation method represents a promising method for spatially controlled deposition, and mass-production of $O_2$ sensors.[19] The resulting sensor showed excellent response time (<0.4 s), Stern–Volmer constants ranging between 45 $bar^{-1}$ and 52 $bar^{-1}$ for the gas phase and 20.72 $bar^{-1}$ for aqueous phase, and the absence of particle leaching from the surface.

### 2.2.8 Layer-by-layer Deposition (LbL)

The electrostatic layer-by-layer self-assembly technique is based upon the electrostatic attraction of charged ionic sites on a polymeric molecule to oppositely charged species. Entrapment of phosphors within poly-ion multilayers can be accomplished by pre-mixing the dye with an oppositely charged poly-ion to form a stable ionic complex. In optical $O_2$ sensing this approach was demonstrated by embedding a Ru(II) complex in a multilayer of poly(sodium styrene sulfonate), poly(diallyl dimethlyammomium) chloride and poly(ethylenimine).[25] A similar method was developed for coating of optical fibres,[186] with low water solubility of PtTFPP dye circumvented by its solubilization in micelles also used in the LbL process. Comparing the different polycations (*i.e.* poly(allylamine hydrochloride) (PAH), poly(diallyldimethylammonium chloride) and polyethyleneimine), the highest sensitivity of the sensor was achieved with PAH, which also showed high coating roughness.[187] The optimal number of spacing layers, which reduced phosphor self-quenching, depended on the type of cationic polyelectrolyte used.[188]

A variation of the co-precipitation method is the encapsulation in calcium carbonate nanoparticles stabilized with polyvinylsulfonate and embedded in a LbL multi-layer nanocapsules of poly(styrene sulfonate) and poly(diallyldimethylammonium chloride), in which a limit of detection of 7.62 μM and dynamic calibration range up to 139.31 μM of dissolved $O_2$ was attained.[189]

## 2.3 Covalent and Coordinative Binding of Phosphors to Substrates

As already discussed above, physical entrapment of a phosphor in a matrix is often insufficient to prevent phase segregation, self-quenching and leaching of the phosphor. Hence, coupling of the phosphor to the matrix material can be used to overcome all this. The phosphor can be adsorbed on planar or nanoparticle surfaces (*e.g.* liquid–solid interface), but this leaves it accessible for other quenchers and interfering species. A more attractive strategy is phosphor's covalent or coordinative binding to an organic or inorganic matrix and subsequent preparation of sensor coating. Such coupling can be done by post-functionalization or grafting to reactive polymers or inorganic materials, or by co-polymerization of matrix monomer and the phosphor.

One approach to covalently bind phosphors to the substrate network is through functionalization of ormosil precursors, which can then react to produce xerogels.[190,191] Another approach is to use organo-silanes containing thiol groups, coupled *e.g.* with PtTFPP dye by click modification.[20]

A Ru(II) complex with amino-functionalized 1,10-phenanothroline ligand was also covalently attached to the backbone of polythionylphosphazene.[192] Aldehyde-amine coupling reaction was also suggested for phosphor immobilization, *e.g.* by reductive amination of a water-soluble amine-functionalized poly(siloxane) with 4-(2-pyridyl)benzaldehyde coordinatively bound to an Ir(III) complex.[193,194] An aqueous dispersion of these materials was used as a pressure-sensitive paint. By blending the phosphor-immobilized polymer with microcrystalline cellulose, a nearly 30% increase in $O_2$ sensitivity was observed.

Linear poly(dimethylsiloxane) chains were anchored to a cyclometallated Ir(III) complex by hydrosilylation reaction with the terminal alkene present on a pyridine ligand.[195–197] The same approach was used with a Tb(III) complex, although this sensor showed non-linear Stern–Vomer plots.[198] NIR-emitting styrene-functionalized Pt(II)- and Pd(II) benzoporphyrins were similarly embedded in PDMS, with their $K_{SV}$ values reaching 0.082–0.141 $hPa^{-1}$ for Pt(II) benzoporphyrins and 0.921 $hPa^{-1}$ for Pd(II) benzoporphyrins.[199]

Since many phosphors consist of a metal ion surrounded by aromatic ligands, complexation can be obtained *in situ*, between the metal ion and ligands present as side groups on the matrix backbone. This has been demonstrated by coordinative linkage to bovine albumin serum functionalized with an imidazole ligand and subsequent chelation with the metal ion.[39]

Insertion of functional groups as side moieties anchored onto the polymer backbone may be attained not only by post-modification reactions, but also by utilizing functional monomers during the polymerization. Pt(II) porphyrins bearing acrylate side groups were co-polymerized with isobutyl methacrylate and 2,2,2-trifluoroethyl methacrylate, 2-hydroxyethyl methacrylate and acrylamide,[77,200–202] The coupling was carried out either *via* pentafluorophenyl substituents reacting with nucleophiles, or by ester formation of a carboxy-functionalized porphyrin molecule with hydroxyethylmethacrylate. The latter can also be covalently attached to a thermo-responsive polymer, to display temperature and $O_2$ sensitivity.[203] By co-polymerising styrene and pentafluorostyrene in the presence of di-thiols and PtTFPP, click-coupling between the porphyrin and matrix can be achieved.[20] Br-substituted benzoporphyrins were also used as functional monomers for the functionalization of reactive polymers *via* Suzuki coupling.[204] Formation of a hydrogel where phosphor aggregation was avoided by the spatial constraining 3D mesh was demonstrated using a carboxy-functionalized Pt(II) porphyrin and poly(ethylene glycol) with telechelic amine groups.[205] A Pt(II) porphyrin with polyimide side groups was first synthesized by means of a triazine linker which was then incorporated in a linear polymer chain.[206]

Generally, sensors based on conjugated polymers are susceptible to photooxidation and irreversible degradation, however, the ones based on poly(9,9-dioctylfluorene) showed reversible quenching and good potential as a solid-state $O_2$ sensor.[207]

Phosphors can be anchored not only on the backbone of macromolecules or Ormosil precursors, but also on planar and micro/nano surfaces. Covalent binding to metal surfaces can be attained by utilizing phosphors with pendant thiol groups whereas carboxylic-functionalized phosphors are prone to be immobilized on a metal oxide substrate. As an example, an anodized alumina layer was used for the chemisorption of carboxy-functionalized pyrenes with different spacer arms. A higher sensitivity to $O_2$ was observed for shorter spacers although the response time was relatively slow (50 s).[208] This can be explained by π-stacking of pyrene molecules to form aggregates and regions with different accessibility for $O_2$ molecules. Intermediate layers of polyanion (*e.g.* poly(styrene sulfonate) and poly(acrylic acid),[209] addition of long chain carboxylic acids[210] or perfluorinated carboxylic acids,[211] were applied to overcome these issues, however response time and signal heterogeneity were not sensibly improved.

Binding of Fe(III) porphyrins to a semi-conductor surface was obtained by integrating a six-membered cyclic disulfide and two extended imidazolyl residues binding the phosphor as axial ligands (the variation in the semiconductor photoluminescence was obtained upon binding of oxygen on the porphyrin ligand).[212] Covalent anchoring of Ormosil functionalized metallo-porphyrins to glass or silicon substrates was also demonstrated.[213]

Another promising strategy is covalent attachment to the surface of mesoporous and non-porous microbeads and nanobeads. Similar to the above-mentioned synthetic approaches, phenanthroline ligands conjugated with Ormosils permit a catalyst-free conjugation of linked Ru(II)-based oxygen sensors to silica surfaces, namely MCM-41,[214,215] SBA-15,[215] MSU-3[216,217] and silica nanoparticles.[218] A sol–gel reaction was also developed for upconverting nanoparticles.[219]

Pentafluorophenyl-amine coupling has been utilized for the covalent assembly of fluorinated metalloporphyrins on an amino-silica nanoparticle surface.[220] Amino-silica microbeads were utilized to develop a ratiometric probe by covalent attachment of a carboxy-terminated metalloporphyrin as oxygen sensing unit and a carboxy-functionalized quantum dot as reference dye.[221]

## 2.4 Conclusions and Outlook

More than thirty years of research in optical $O_2$ sensors has produced a plethora of phosphors, encapsulation matrices, designs and processes for the fabrication of active elements for $O_2$ sensing. Physical encapsulation of the phosphor in organic or inorganic matrices through 'cocktail' formulations and casting of sensor coatings, although it looks simple, has limitations in terms of sensor reproducibility, heterogeneity, phosphor leaching, process control and overall manufacturing costs. Covalent linkage of the phosphor to matrix polymer or its embedding in nano- and microparticles subsequently dispersed in a carrier media such as silicone, can address some of these issues, however these methods are not simple and their viability and benefits for large-scale production of high performance $O_2$ sensors is still questionable.

Several emerging techniques, such as dyeing of micro-structured nonwovens (PE and PP fabric), solvent-crazing of conventional polymer films (PE, PP, PPS), electrospinning, electrophoretic deposition and the LbL process, help to simplify sensor fabrication, enhance their homogeneity by providing more uniform sensor materials with high surface area and physical entrapment of the phosphor in the nanostructures (pores, fibres), and ensure stronger retention of phosphors by the matrix. These production methods also allow up-scaling, continuous production of high quality $O_2$ sensors with stable and pre-determined working characteristics. Some of them also enable production of discrete sensors, patterns of sensors, incorporation of individual/discrete $O_2$ sensors in specific locations and integrating them with targeted samples, sample vessels or sensing applications. However, more work is required to further optimised and fine-tune the chemical structure of the phosphors, matrix and substrate materials, in order to enable fabrication of high-performance $O_2$ sensors on a large scale and at affordable costs. Printable and extrudable $O_2$ sensors are under active development and we expect to see more of such materials, related publications and industrial applications in the near future.

Yet another promising avenue is the development of highly porous metalloorganic frameworks (MOFs), which can act as both matrix and sensor. Thus, a limit of detection of 0.07 ppm and $I_0/I_{100}$ = 410 was achieved with a Cu(I)-triazolate MOF.[222–226] However, modified MOFs as $O_2$ sensors often suffer from poor signal reversibility, likely due to structural changes that hinder the oxygen desorption process, and, at the present stage, full desorption can be attained only by utilizing vacuum to strip off $O_2$ from the matrix.

Stringent requirements on the price of sensor units in large scale applications have recently been tackled by utilizing rare metal-free sensor nanostructures. Indeed, carbon nanodots, consisting of an internal para-crystalline carbon with angstrom-sized polyaromatic carbon domains, displayed strong room temperature phosphorescence after embedding them in rigid matrices, such as potash alum,[227] polyurethane,[26] composites based on urea–biuret.[228] Unfortunately, their very long emission lifetimes (in the ms–s range) translate in a low $O_2$ range, only suitable for near-anoxic environments and yet not viable for large-scale applications. The same is true for all-organic phosphors, such di-fluoro boron β-diketonates covalently linked to poly(lactic acid) (PLA), which are tunable for $O_2$ sensitivity by PLA molecular weight and halide substituent, but still do not provide useful brightness and sensing range (0–21 kPa $O_2$).[229,230] Finally, encapsulation in polymer gratings and photonic crystals has application potential, showing up to 320-fold enhancement of the photoluminescent signal and used of matrix nano-structuring for tuning $O_2$ sensitivity.[231,232]

## Abbreviations

| | |
|---|---|
| DLR | Dual lifetime referencing |
| EC | Ethyl cellulose |
| EPD | Electrophoretic deposition |
| LbL | Layer-by-layer |
| MAP | Modified atmosphere packaging |
| MOF | Metal-organic framework |
| Ormosils | Organically modified silicon precursors |
| PAH | Poly(allylamine hydrochloride) |
| PALE | Physically active liquid environment |
| PCL | Poly(caprolactone) |
| PDMS | Poly(dimethylsiloxane) |
| PE | Polyethylene |
| PLA | Poly(lactic acid) |
| PP | Polypropylene |
| PPS | Poly(phenylene sulfide) |
| PS | Polystyrene |
| PSU | Polysulfone |
| PTFE | Poly(tetrafluoroethylene) |
| PTMSP | Poly(1-trimethylsilyl-1-propyne) |
| PtTFPP | Platinum(II)-5,10,15,20-tetrakis-(2,3,4,5,6-pentafluorphenyl)-porphyrin |
| RH | Relative humidity |

## Acknowledgement

Financial support from the Science Foundation Ireland, grant 12/RC/2276 (DBP), is gratefully acknowledged.

## References

1. O. S. Wolfbeis, *Adv. Mater.*, 2008, **20**, 3759–3763.
2. M. Quaranta, S. M. Borisov and I. Klimant, *Bioanal. Rev.*, 2012, **4**, 115–157.
3. D. B. Papkovsky and R. I. Dmitriev, *Chem. Soc. Rev.*, 2013, **42**, 8700–8732.
4. X. Wang and O. S. Wolfbeis, *Chem. Soc. Rev.*, 2014, **43**, 3666–3761.
5. A. Mills, *Chem. Soc. Rev.*, 2005, **34**, 1003.
6. S. Banerjee, C. Kelly, J. P. Kerry and D. B. Papkovsky, *Trends Food Sci. Technol.*, 2016, **50**, 85–102.
7. S. J. Lee and A. T. M. M. Rahman, *Intelligent Packaging for Food Products*, Elsevier Ltd, 2013.
8. S. J. Lee and A. T. M. Rahman, *Innovations in Food Packaging*, Elsevier, 2014.
9. Y. Amao, *Microchim. Acta*, 2003, **143**, 1–12.
10. P. Lehner, C. Staudinger, S. M. Borisov and I. Klimant, *Nat. Commun.*, 2014, **5**, 4460.
11. C. Toncelli, O. V. Arzhakova, A. Dolgova, A. L. Volynskii, J. P. Kerry and D. B. Papkovsky, *J. Mater. Chem. C*, 2014, **2**, 8035–8041.
12. C. Toncelli, O. V. Arzhakova, A. Dolgova, A. L. Volynskii, N. F. Bakeev, J. P. Kerry and D. B. Papkovsky, *Anal. Chem.*, 2014, **86**, 1917–1923.
13. C. A. Kelly, C. Toncelli, J. P. Kerry and D. B. Papkovsky, *J. Mater. Chem. C*, 2014, **2**, 2169–2174.
14. C. A. Kelly, C. Toncelli, J. P. Kerry and D. B. Papkovsky, *Sens. Actuators, B*, 2014, **203**, 935–940.
15. R. Xue, P. Behera, J. Xu, M. S. Viapiano and J. J. Lannutti, *Sens. Actuators, B*, 2014, **192**, 697–707.
16. X. Wang, Y. Yang and B. He, *Sens. Actuators, B*, 2017, **241**, 957–966.
17. R. Kai, W. Jun and J. Huali, *Sens. Actuators, B*, 2017, **240**, 697–708.
18. G. Yazgan, R. I. Dmitriev, V. Tyagi, J. Jenkins, G.-M. Rotaru, M. Rottmar, R. M. Rossi, C. Toncelli, D. B. Papkovsky, K. Maniura-Weber and G. Fortunato, *Sci. Rep.*, 2017, **7**, 158.
19. M. Marín-Suárez, S. Medina-Rodríguez, O. Ergeneman, S. Pané, J. F. Fernández-Sánchez, B. J. Nelson and A. Fernández-Gutiérrez, *Nanoscale*, 2013, 263–271.
20. K. Koren, S. M. Borisov and I. Klimant, *Sens. Actuators, B*, 2012, **169**, 173–181.
21. W. Cui, R. Liu, E. Manna, J. M. Park, F. Fungura, J. Shinar and R. Shinar, *Anal. Chim. Acta*, 2015, **853**, 563–571.
22. S. Lee and J.-W. Park, *Sens. Actuators, B*, 2017, **249**, 364–377.
23. Q. Zuo, B. Li, L. Zhang, Y. Wang, Y. Liu, J. Zhang, Y. Chen and L. Guo, *J. Solid State Chem.*, 2010, **183**, 1715–1720.
24. X. D. Wang, R. J. Meier, C. Schmittlein, S. Schreml, M. Schäferling and O. S. Wolfbeis, *Sens. Actuators, B*, 2015, **221**, 37–44.
25. D. A. Chang-Yen and B. K. Gale, *Lab Chip*, 2003, **3**, 297.

26. J. Tan, R. Zou, J. Zhang, W. Li, L. Zhang and D. Yue, *Nanoscale*, 2016, **8**, 4742–4747.
27. C. S. Burke, O. McGaughey, J.-M. Sabattié, H. Barry, A. K. McEvoy, C. McDonagh and B. D. MacCraith, *Analyst*, 2005, **130**, 41–45.
28. V. Nock, M. Alkaisi and R. J. Blaikie, *Microelectron. Eng.*, 2010, **87**, 814–816.
29. J. Ehgartner, P. Sulzer, T. Burger, A. Kasjanow, D. Bouwes, U. Krühne, I. Klimant and T. Mayr, *Sens. Actuators, B*, 2016, **228**, 748–757.
30. O. S. Wolfbeis, *Anal. Chem.*, 2006, **78**, 3859–3874.
31. P. M. Gewehr and D. T. Delpy, *Med. Biol. Eng. Comput.*, 1993, **31**, 11–21.
32. P. Hartmann, M. J. P. Leiner and M. E. Lippitsch, *Anal. Chem.*, 1995, **67**, 88–93.
33. D. B. Papkovsky, G. V Ponomarev, W. Trettnak and P. O'Leary, *Anal. Chem.*, 1995, **67**, 4112–4117.
34. S. M. Borisov, G. Zenkl and I. Klimant, *ACS Appl. Mater. Interfaces*, 2010, **2**, 366–374.
35. S. M. Borisov, R. Fischer, R. Saf and I. Klimant, *Adv. Funct. Mater.*, 2014, **24**, 6548–6560.
36. S. M. Borisov, S. M. Borisov, I. Klimant and I. Klimant, *Anal. Chem.*, 2007, **2**, 7501–7509.
37. S. M. Borisov, G. Nuss and I. Klimant, *Anal. Chem.*, 2008, **80**, 9435–9442.
38. C. Karakus, L. H. Fischer, S. Schmeding, J. Hummel, N. Risch, M. Schäferling and E. Holder, *Dalton Trans.*, 2012, **41**, 9623–9632.
39. K. Koren, S. M. Borisov, R. Saf and I. Klimant, *Eur. J. Inorg. Chem.*, 2011, 1531–1534.
40. S. M. Borisov, R. Saf, R. Fischer and I. Klimant, *Inorg. Chem.*, 2013, **52**, 1206–1216.
41. S. M. Borisov, A. Alemayehu and A. Ghosh, *J. Mater. Chem. C*, 2016, **5822**, 5822–5828.
42. A. J. Palma, J. Lopez-Gonzalez, L. J. Asensio, M. D. Fernandez-Ramos and L. F. Capitan-Vallvey, *Sens. Actuators, B*, 2007, **121**, 629–638.
43. N. López-Ruiz, A. Martínez-Olmos, I. M. Pérez de Vargas-Sansalvador, M. D. Fernández-Ramos, M. A. Carvajal, L. F. Capitan-Vallvey and A. J. Palma, *Sens. Actuators, B*, 2012, **171**, 938–945.
44. R. Shinar, D. Ghosh, B. Choudhury, M. Noack, V. L. Dalal and J. Shinar, *J. Non-Cryst. Solids*, 2006, **352**, 1995–1998.
45. S. Santoro, A. J. Moro, C. Portugal, J. G. Crespo, J. C. Lima and I. M. Coelhoso, *J. Food Eng.*, 2016, **189**, 37–44.
46. J. F. Fernández-Sánchez, T. Roth, R. Cannas, M. K. Nazeeruddin, S. Spichiger, M. Graetzel and U. E. Spichiger-Keller, *Talanta*, 2007, **71**, 242–250.
47. P. Escobedo, M. M. Erenas, N. López-Ruiz, M. A. Carvajal, S. Gonzalez-Chocano, I. De Orbe-Payá, L. F. Capitán-Valley, A. J. Palma and A. Martínez-Olmos, *Anal. Chem.*, 2017, **89**, 1697–1703.
48. K. Koren, L. Hutter, B. Enko, A. Pein, S. M. Borisov and I. Klimant, *Sens. Actuators, B*, 2013, **176**, 344–350.
49. S. M. Borisov and I. Klimant, *Anal. Bioanal. Chem.*, 2012, **404**, 2797–2806.

50. C. S. K. Mak, D. Pentlehner, M. Stich, O. S. Wolfbeis, W. K. Chan and H. Yersin, *Chem. Mater.*, 2009, **21**, 2173–2175.
51. D. E. Achatz, R. J. Meier, L. H. Fischer and O. S. Wolfbeis, *Angew. Chem., Int. Ed.*, 2011, **50**, 260–263.
52. C. R. Schroeder, G. Neurauter and I. Klimant, *Microchim. Acta*, 2007, **158**, 205–218.
53. P. Douglas and K. Eaton, *Sens. Actuators, B*, 2002, **82**, 200–208.
54. S. R. Ricketts and P. Douglas, *Sens. Actuators, B*, 2008, **135**, 46–51.
55. L. Shen, M. Ratterman, D. Klotzkin and I. Papautsky, *Sens. Actuators, B*, 2011, **155**, 430–435.
56. C. Baleiža, S. Nagl, M. Schäferling, M. N. Berberan-Santos and O. S. Wolfbeis, *Anal. Chem.*, 2008, **80**, 6449–6457.
57. K. Eaton and P. Douglas, *Sens. Actuators, B*, 2002, **82**, 94–104.
58. S. Nagl, C. Baleizão, S. M. Borisov, M. Schäferling, M. N. Berberan-Santos and O. S. Wolfbeis, *Angew. Chem., Int. Ed.*, 2007, **46**, 2317–2319.
59. D. Badocco, A. Mondin, P. Pastore, S. Voltolina and S. Gross, *Anal. Chim. Acta*, 2008, **627**, 239–246.
60. S. Banerjee, R. T. Kuznetsova and D. B. Papkovsky, *Sens. Actuators, B*, 2015, **212**, 229–234.
61. D. Badocco, A. Mondin and P. Pastore, *Sens. Actuators, B*, 2012, **163**, 165–170.
62. D. Badocco, A. Mondin and P. Pastore, *Sens. Actuators, B*, 2011, **158**, 54–61.
63. D. Badocco, A. Mondin, A. Fusar and P. Pastore, *J. Phys. Chem. C*, 2009, 20467–20475.
64. H. S. Voraberger, H. Kreimaier, K. Biebernik and W. Kern, *Sens. Actuators, B*, 2001, **74**, 179–185.
65. N. B. Borchert, G. V Ponomarev, J. P. Kerry, D. B. Papkovsky, B. Chemistry and C. Pharmacy, *Anal. Chem.*, 2011, **83**, 18–22.
66. K. Stubenrauch, M. Sandholzer, F. Niedermair, K. Waich, T. Mayr, I. Klimant, G. Trimmel and C. Slugovc, *Eur. Polym. J.*, 2008, **44**, 2558–2566.
67. Z. Pang, X. H. Gu, A. Yekta, Z. Masoumi, J. B. Coll, M. A. Winnik and I. Manners, *Adv. Mater.*, 1996, **8**, 768.
68. L. H. Fischer, M. I. J. Stich, O. S. Wolfbeis, N. Tian, E. Holder and M. Schäferling, *Chem.–Eur. J.*, 2009, **15**, 10857–10863.
69. Y. Xing, C. Liu, X. Song and J. Li, *J. Mater. Chem. C*, 2015, **3**, 2166–2174.
70. Y. Amao, K. Asai, I. Okura, H. Shinohara and H. Nishide, *Analyst*, 2000, **125**, 1911–1914.
71. Y. Matsuda, F. Nagashima, H. Yamaguchi, Y. Egami and T. Niimi, *Sens. Actuators, B*, 2011, **160**, 1464–1467.
72. Y. Amao, T. Miyashita and I. Okura, *React. Funct. Polym.*, 2001, **47**, 49–54.
73. M. E. Köse, B. F. Carroll and K. S. Schanze, *Langmuir*, 2005, **21**, 9121–9129.
74. G. E. Khalil, C. Coston, J. Crafton, G. Jones, S. Grenoble, M. Gouterman, J. B. Callis and L. R. Dalton, *Sens. Actuators, B*, 2004, **97**, 13–21.

75. M. Obata, R. Asato, K. Mitsuo and S. Hirohara, *J. Polym. Sci., Part A: Polym. Chem.*, 2014, **52**, 963–972.
76. J. S. Kwak and Y. G. Choi, *Electron. Mater. Lett.*, 2014, **10**, 991–995.
77. K. Zhang, H. Zhang, Y. Wang, Y. Tian, J. Zhao and Y. Li, *Spectrochim. Acta, Part A*, 2017, **170**, 242–246.
78. N. K. Zaitsev, P. V. Melnikov, V. A. Alferov, A. V. Kopytin and K. E. German, *Procedia Eng.*, 2016, **168**, 309–312.
79. R. N. Gillanders, M. C. Tedford, P. J. Crilly and R. T. Bailey, *J. Photochem. Photobiol., A*, 2004, **162**, 531–535.
80. R. N. Gillanders, M. C. Tedford, P. J. Crilly and R. T. Bailey, *Anal. Chim. Acta*, 2005, **545**, 189–194.
81. P. Lehner, C. Larndorfer, E. Garcia-Robledo, M. Larsen, S. M. Borisov, N. P. Revsbech, R. N. Glud, D. E. Canfield and I. Klimant, *PLoS One*, 2015, **10**, 1–15.
82. A. S. Kocincova, S. M. Borisov, C. Krause and O. S. Wolfbeis, *Anal. Chem.*, 2007, **79**, 8486–8493.
83. B. Lamprecht, A. Tschepp, M. Čajlaković, M. Sagmeister, V. Ribitsch and S. Köstler, *Analyst*, 2013, **138**, 5875–5878.
84. M. Quaranta, M. Murkovic and I. Klimant, *Analyst*, 2013, **138**, 6243–6245.
85. R. Chen, F. Formenti, H. McPeak, A. N. Obeid, C. Hahn and A. Farmery, *Sens. Actuators, B*, 2016, **222**, 531–535.
86. B. D. MacCraith, C. M. McDonagh, G. O'Keeffe, E. T. Keyes, J. G. Vos, B. O'Kelly and J. F. McGilp, *Analyst*, 1993, **118**, 385–388.
87. F. Code, B. D. M. A. C. Craith, C. Mcdonagh, G. O. Keeffe and T. Butler, *J. Sol-Gel Sci. Technol.*, 1994, **2**, 661–665.
88. P. Kiernan, C. Mcdonagh, B. D. Maccraith and K. Mongey, *J. Sol-Gel Sci. Technol.*, 1994, **2**, 513–517.
89. G. Kuncovà and M. Fiaiovà, *Biotechnol. Tech.*, 1995, **9**, 175–178.
90. S.-K. Lee and I. Okura, *Analyst*, 1997, **122**, 81–84.
91. S.-K. Lee and I. Okura, *Anal. Chim. Acta*, 1997, **342**, 181–188.
92. B. J. Basu and J. Kamble, *J. Sol-Gel Sci. Technol.*, 2009, **52**, 24–30.
93. C. S. Chu, *J. Lumin.*, 2013, **135**, 5–9.
94. C. S. Chu and C. Y. Chuang, *J. Lumin.*, 2014, **154**, 475–478.
95. O. Oter, K. Ertekin and S. Derinkuyu, *Mater. Chem. Phys.*, 2009, **113**, 322–328.
96. J. Zaharieva, M. Milanova and D. Todorovsky, *J. Mater. Chem.*, 2011, **21**, 4893.
97. J. M. Costa-Fernández, M. E. Diaz-García and A. Sanz-Medel, *Anal. Chim. Acta*, 1998, **360**, 17–26.
98. J. M. Costa-Fernández, N. Bordel, J. C. Campo, F. J. Ferrero, M. A. Pérez and A. Sanz-Medel, *Microchim. Acta*, 2000, **134**, 145–152.
99. E. A. García, R. G. Fernández and M. E. Díaz-García, *Microporous Mesoporous Mater.*, 2005, **77**, 235–239.
100. M. Valledor, J. C. Campo, I. Sánchez-Barragán, J. C. Viera, J. M. Costa-Fernández and A. Sanz-Medel, *Sens. Actuators, B*, 2006, **117**, 266–273.

101. A. K. McEvoy, C. McDonagh and B. D. MaCcraith, *J. Sol-Gel Sci. Technol.*, 1997, **8**, 1121–1125.
102. C. McDonagh, P. Bowe, K. Mongey and B. D. MacCraith, *J. Non-Cryst. Solids*, 2002, **306**, 138–148.
103. K. F. Mongey, J. G. Vos, B. D. Maccraith, C. M. Mcdonagh, C. Coatesc and J. Mcgarveyc, *J. Mater. Chem.*, 1997, **7**, 1473–1479.
104. M. A. Chan, J. L. Lawless, S. K. Lam and D. Lo, *Anal. Chim. Acta*, 2000, **408**, 33–37.
105. D. J. Osborn III, G. L. Baker and R. N. Ghosh, *J. Sol-Gel Sci. Technol.*, 2005, 5–10.
106. C. S. Chu, T. W. Sung and Y. L. Lo, *Sens. Actuators, B*, 2013, **185**, 287–292.
107. A. K. McEvoy, C. M. McDonagh and B. D. MacCraith, *Analyst*, 1996, **121**, 785–788.
108. P. Lavin, C. M. Mcdonagh and B. D. Maccraith, *J. Sol-Gel Sci. Technol.*, 1999, **13**, 641–645.
109. C. McDonagh, B. D. MacCraith and A. K. McEvoy, *Anal. Chem.*, 1998, **70**, 45–50.
110. C. M. Mcdonagh, A. M. Shields, A. K. Mcevoy, B. D. Maccraith and J. F. Gouin, *J. Sol-Gel Sci. Technol.*, 1998, **13**, 207–211.
111. Z. Tao, E. C. Tehan, Y. Tang and F. V. Bright, *Anal. Chem.*, 2006, **78**, 1939–1945.
112. I. Klimant, F. Ruckruh, G. Liebsch, A. Stangelmayer and O. S. Wolfbeis, *Microchim. Acta*, 1999, **131**, 35–46.
113. Y. Tang, E. C. Tehan, Z. Tao and F. V. Bright, *Anal. Chem.*, 2003, **75**, 2407–2413.
114. T. S. Yeh, C. S. Chu and Y. L. Lo, *Sens. Actuators, B*, 2006, **119**, 701–707.
115. C. S. Chu and Y. L. Lo, *Sens. Actuators, B*, 2007, **124**, 376–382.
116. Y. Xiong, J. Xu, J. W. Wang and Y. F. Guan, *Anal. Bioanal. Chem.*, 2009, **394**, 919–923.
117. Y. Xiong, D. Zhu, S. Chen, H. Peng and Y. Guan, *J. Fluoresc.*, 2010, **20**, 269–274.
118. C. S. Chu and Y. L. Lo, *Sens. Actuators, B*, 2011, **155**, 53–57.
119. Y. Xiong, Z. Ye, J. Xu, Y. Zhu, C. Chen and Y. Guan, *Analyst*, 2013, **138**, 1819–1827.
120. J. Estella, D. Wencel, J. P. Moore, M. Sourdaine and C. McDonagh, *Anal. Chim. Acta*, 2010, **666**, 83–90.
121. Y. Zhao, T. Ye, H. Chen, D. Huang, T. Zhou, C. He and X. Chena, *Luminescence*, 2011, **26**, 29–34.
122. V. S. Tripathi, G. Lakshminarayana and M. Nogami, *Sens. Actuators, B*, 2010, **147**, 741–747.
123. W. Cao and Y. Duan, *Sens. Actuators, B*, 2006, **119**, 363–369.
124. Y. L. Lo, C. S. Chu, J. P. Yur and Y. C. Chang, *Sens. Actuators, B*, 2008, **131**, 479–488.
125. P. A. S. Jorge, C. Maule, A. J. Silva, R. Benrashid, J. L. Santos and F. Farahi, *Anal. Chim. Acta*, 2008, **606**, 223–229.

126. X. D. Wang, X. Chen, Z. X. Xie and X. R. Wang, *Angew. Chem., Int. Ed.*, 2008, **120**, 7560–7563.
127. C. S. Chu and T. H. Lin, *Sens. Actuators, B*, 2015, **210**, 302–309.
128. C. S. Chu and C. A. Lin, *Sens. Actuators, B*, 2014, **195**, 259–265.
129. C. S. Chu and Y. L. Lo, *Sens. Actuators, B*, 2008, **134**, 711–717.
130. C. S. Chu and C. Y. Chuang, *J. Lumin.*, 2015, **167**, 114–119.
131. R. N. Gillanders, M. C. Tedford, P. J. Crilly and R. T. Bailey, *J. Photochem. Photobiol., A*, 2004, **163**, 193–199.
132. D. L. Plata, Y. J. Briones, R. L. Wolfe, M. K. Carroll, S. D. Bakrania, S. G. Mandel and A. M. Anderson, *J. Non-Cryst. Solids*, 2004, **350**, 326–335.
133. N. Leventis, A. M. M. Rawashdeh, I. A. Elder, J. Yang, A. Dass and C. Sotiriou-Leventis, *Chem. Mater.*, 2004, **16**, 1493–1506.
134. C.-L. Lee, H. L. Chapman, M. E. Cifuentes, K. M. Lee, L. D. Merrill, K. L. Ulman and K. Venkataraman, *J. Membr. Sci.*, 1988, **38**, 55–70.
135. S. A. Stern, V. M. Shah and B. J. Hardy, *J. Polym. Sci., Part B: Polym. Phys.*, 1987, **25**, 1263–1298.
136. B. J. Basu, A. Thirumurugan, A. R. Dinesh, C. Anandan and K. S. Rajam, *Sens. Actuators, B*, 2005, **104**, 15–22.
137. J. R. Bacon and J. N. Demas, *Anal. Chem.*, 1987, **59**, 2780–2785.
138. E. R. Carraway, J. N. Demas, B. A. DeGraff and J. R. Bacon, *Anal. Chem.*, 1991, **63**, 337–342.
139. T. Ishiji and M. Kaneko, *Analyst*, 1995, **120**, 1633–1638.
140. B. J. Basu and K. S. Rajam, *Sens. Actuators, B*, 2004, **99**, 459–467.
141. Z. Zhou, R. Shinar, A. J. Allison and J. Shinar, *Adv. Funct. Mater.*, 2007, **17**, 3530–3537.
142. B. Meier, T. Werner, I. Klimant and O. S. Wolfbeis, *Sens. Actuators, B*, 1995, **29**, 240–245.
143. C.-M. Chan, M.-Y. Chan, M. Zhang, W. Lo and K.-Y. Wong, *Analyst*, 1999, **124**, 691–694.
144. M. Kameda, H. Seki, T. Makoshi, Y. Amao and K. Nakakita, *Sens. Actuators, B*, 2012, **171–172**, 343–349.
145. B. Wang, Y. Liu, B. Li, S. Yue and W. Li, *J. Lumin.*, 2008, **128**, 341–347.
146. B. Lei, L. Wang, H. Zhang, Y. Liu, H. Dong, M. Zheng and X. Zhou, *Sens. Actuators, B*, 2016, **230**, 101–108.
147. N. Feng, J. Xie and D. Zhang, *Spectrochim. Acta, Part A*, 2010, **77**, 292–296.
148. J. Haitao, Y. Huilin, L. Fan and L. Yang, *J. Lumin.*, 2012, **132**, 198–204.
149. L. Guangying and F. Jun, *Sens. Actuators, B*, 2016, **233**, 347–354.
150. Y. Liu, B. Li, Y. Cong, L. Zhang, D. Fan and L. Shi, *J. Lumin.*, 2011, **131**, 781–785.
151. X. Y. Xu and H. N. Xiao, *J. Lumin.*, 2012, **132**, 2251–2258.
152. H. Zhang, Y. Sun, K. Ye, P. Zhang and Y. Wang, *J. Mater. Chem.*, 2005, **15**, 3181–3186.
153. J. W. Ye, H. L. Zhou, S. Y. Liu, X. N. Cheng, R. B. Lin, X. L. Qi, J. P. Zhang and X. M. Chen, *Chem. Mater.*, 2015, **27**, 8255–8260.
154. S. M. Borisov, C. Larndorfer and I. Klimant, *Adv. Funct. Mater.*, 2012, **22**, 4360–4368.

155. C. S. Chu, Y. L. Lo and T. W. Sung, *Talanta*, 2010, **82**, 1044–1051.
156. T. W. Sung and Y. L. Lo, *Sens. Actuators, B*, 2012, **173**, 406–413.
157. J. F. Fernández-Sánchez, R. Cannas, S. Spichiger, R. Steiger and U. E. Spichiger-Keller, *Anal. Chim. Acta*, 2006, **566**, 271–282.
158. S. Medina-Rodríguez, S. A. Denisov, Y. Cudré, L. Male, M. Marín-Suarez, A. Fernandez-Gutiérrez, J. F. Fernandez-Sanchez, A. Tron, G. Jonusauskas, N. D. McClenaghan and E. Baranoff, *Analyst*, 2016, **141**, 3090–3097.
159. S. Medina-Rodriguez, M. Marin-Suarez, J. F. Fernandez-Sanchez, A. de la Torre-Vega, E. Baranoff and A. Fernandez-Gutierrez, *Analyst*, 2013, **138**, 4607–4617.
160. M. M. S. Toro, J. F. Fernandez-Sanchez, E. Baranoff, M. K. Nazeeruddin, M. Graetzel and A. Fernandez-Gutierrez, *Talanta*, 2010, **82**, 620–626.
161. S. Medina-Rodrí Guez, F. J. Orriach-Fermá Ndez, C. Poole, P. Kumar, N. De La Torre-Vega, J. F. Ferná Ndez-Sá, E. Baranoff and A. Ferná Ndez-Gutié Rrez, *Chem. Commun.*, 2015, **51**, 11401–11404.
162. J. López-Gejo, D. Haigh and G. Orellana, *Langmuir*, 2010, **26**, 2144–2150.
163. D. Haigh-Flórez, C. De la Hera, E. Costas and G. Orellana, *Biosens. Bioelectron.*, 2014, **54**, 484–491.
164. H. Zhao, L. Zang, L. Wang, F. Qin, Z. Zhang and W. Cao, *Sens. Actuators, B*, 2015, **215**, 405–411.
165. C. A. Kelly, C. Toncelli, J. P. Kerry and D. B. Papkovsky, *J. Mater. Chem. C*, 2014, **2**, 2169–2174.
166. G. S. Vasylevska, S. M. Borisov, C. Krause and O. S. Wolfbeis, *Chem. Mater.*, 2006, **18**, 4609–4616.
167. M. I. J. Stich, S. Nagl, O. S. Wolfbeis, U. Henne and M. Schaeferling, *Adv. Funct. Mater.*, 2008, **18**, 1399–1406.
168. M. I. J. Stich, M. Schaeferling and O. S. Wolfbeis, *Adv. Mater.*, 2009, **21**, 2216–2220.
169. L. H. Fischer, C. Karakus, R. J. Meier, N. Risch, O. S. Wolfbeis, E. Holder and M. Schäferling, *Chem.–Eur. J.*, 2012, **18**, 15706–15713.
170. L. H. Fischer, S. M. Borisov, M. Schaeferling, I. Klimant and O. S. Wolfbeis, *Analyst*, 2010, **135**, 1224–1229.
171. R. N. Gillanders, O. V. Arzhakova, A. Hempel, A. Dolgova, J. P. Kerry, L. M. Yarysheva, N. F. Bakeev, A. L. Volynskii and D. B. Papkovsky, *Anal. Chem.*, 2010, **82**, 466–468.
172. C. A. Kelly, C. Toncelli, M. Cruz-Romero, O. V. Arzhakova, J. P. Kerry and D. B. Papkovsky, *Mater. Des.*, 2015, **77**, 110–113.
173. S. Banerjee, O. V. Arzhakova, A. A. Dolgova and D. B. Papkovsky, *Sens. Actuators, B*, 2016, **230**, 434–441.
174. I. A. Okkelman, A. A. Dolgova, S. Banerjee, J. P. Kerry, A. Volynskii, O. V. Arzhakova and D. B. Papkovsky, *ACS Appl. Mater. Interfaces*, 2017, **9**, 13587–13592.
175. A. Greiner and J. H. Wendorff, *Angew. Chem., Int. Ed.*, 2007, **46**, 5670–5703.
176. Y. Wang, B. Li, L. Zhang, Q. Zuo, P. Li, J. Zhang and Z. Su, *Chem. Commun.*, 2011, **12**, 5868–5870.
177. L. Yingkui, *Spectrochim. Acta, Part A*, 2011, **79**, 356–360.

178. L. Songzhu, D. Xiangting, W. Jinxian, L. Guixia, Y. Wenshen and J. Ruokun, *Spectrochim. Acta, Part A*, 2010, **77**, 885–889.
179. L. Y. Wang, Y. Xu, Z. Lin, N. Zhao and Y. Xu, *J. Lumin.*, 2011, **131**, 1277–1282.
180. C. Wolf, M. Tscherner and S. Köstler, *Sens. Actuators, B*, 2015, **209**, 1064–1069.
181. G. Xu, M. Lu, C. Huang, Y. Wang and S. Ge, *Spectrochim. Acta, Part A*, 2014, **123**, 369–375.
182. C. E. Wang, *J. Lumin.*, 2014, **145**, 531–538.
183. M. Z. Ongun, O. Oter, G. Sabanci, K. Ertekin and E. Celik, *Sens. Actuators, B*, 2013, **183**, 11–19.
184. Z. Jiang, X. Yu, S. Zhai and Y. Hao, *Sensors*, 2017, **17**, 548.
185. K. F. Presley, S. Cheong, A. Cochran, R. D. Tilley, J. E. Collins and J. J. Lannutti, *Sens. Actuators, B*, 2016, **235**, 197–205.
186. C. Elosua, N. De Acha, M. Hernaez, I. R. Matias and F. J. Arregui, *Sens. Actuators, B*, 2015, **207**, 683–689.
187. N. de Acha, C. Elosua, D. Martinez, M. Hernaez, I. R. Matias and F. J. Arregui, *Sens. Actuators, B*, 2017, **239**, 1124–1133.
188. N. de Acha, C. Elosúa, I. R. Matías and F. J. Arregui, *Sens. Actuators, B*, 2017, **248**, 836–847.
189. A. Biswas, A. T. Nagaraja and M. J. McShane, *ACS Appl. Mater. Interfaces*, 2014, **6**, 21193–21201.
190. C. Malins, H. G. Glever, B. D. MacCraith, S. Fanni and J. G. Vos, *Anal. Commun.*, 1999, **36**, 3–4.
191. H. Zhang, B. Li, B. Lei, W. Li and S. Lu, *Sens. Actuators, B*, 2007, **123**, 508–515.
192. Z. Wang, A. R. McWilliams, C. E. B. Evans, X. Lu, S. Chung, M. A. Winnik and I. Manners, *Adv. Funct. Mater.*, 2002, **12**, 415–419.
193. A. Habibagahi, Y. Mébarki, Y. Sultan, G. P. A. Yap and R. J. Crutchley, *ACS Appl. Mater. Interfaces*, 2009, **1**, 1785–1792.
194. A. Habibagahi, Y. Mébarki, Y. Sultan and R. J. Crutchley, *J. Photochem. Photobiol., A*, 2011, **225**, 88–94.
195. M. C. DeRosa, P. J. Mosher, C. E. B. Evans and R. J. Crutchley, *Macromol. Symp.*, 2003, **196**, 235–248.
196. M. E. Köse, R. J. Crutchley, M. C. DeRosa, N. Ananthakrishnan, J. R. Reynolds and K. S. Schanze, *Langmuir*, 2005, **21**, 8255–8262.
197. M. C. DeRosa, P. J. Mosher, G. P. A. Yap, K.-S. Focsaneanu, R. J. Crutchley and C. E. B. Evans, *Inorg. Chem.*, 2003, **42**, 4864–4872.
198. R. D. L. Gaspar, J. H. S. K. Monteiro, I. M. Raimundo, I. O. Mazali and F. A. Sigoli, *Chempluschem*, 2015, **80**, 1721–1724.
199. B. J. Müller, T. Burger, S. M. Borisov and I. Klimant, *Sens. Actuators, B*, 2015, **216**, 527–534.
200. Y. Tian, B. R. Shumway and D. R. Meldrum, *Chem. Mater.*, 2010, **22**, 2069–2078.
201. M. Obata, Y. Tanaka, N. Araki, S. Hirohara, S. Yano, K. Mitsuo, K. Asai, M. Harada, T. Kakuchi and C. Ohtsuki, *J. Polym. Sci., Part A: Polym. Chem.*, 2005, **43**, 2997–3006.

202. Y. Tian, B. R. Shumway, A. C. Youngbull, Y. Li, K. Alex, R. H. Johnson and D. R. Meldrum, *Sens. Acuators, B*, 2011, **147**, 714–722.
203. X. Zhou, F. Su, Y. Tian, R. H. Johnson and D. R. Meldrum, *Sens. Actuators, B*, 2011, **159**, 135–141.
204. L. H. Hutter, B. J. Müller, K. Koren, S. M. Borisov and I. Klimant, *J. Mater. Chem. C*, 2014, **2**, 7589–7598.
205. H. Huang, W. Song, G. Chen, J. M. Reynard, T. Y. Ohulchanskyy, P. N. Prasad, F. V. Bright and J. F. Lovell, *Adv. Healthcare Mater.*, 2014, **3**, 891–896.
206. T. Kimura, S. Watanabe, S. I. Sawada, Y. Shibasaki and Y. Oishi, *J. Polym. Sci., Part A: Polym. Chem.*, 2017, **55**, 1086–1094.
207. M. Anni and R. Rella, *J. Phys. Chem. B*, 2010, **114**, 1559–1561.
208. Y. Fujiwara and Y. Amao, *Sens. Actuators, B*, 2003, **89**, 187–191.
209. Y. Amao and I. Okura, *Sens. Actuators, B*, 2003, **88**, 162–167.
210. Y. Fujiwara and Y. Amao, *Talanta*, 2004, **62**, 655–660.
211. Y. Fujiwara and Y. Amao, *Sens. Actuators, B*, 2004, **99**, 130–133.
212. G. Ashkenasy, A. Ivanisevic, R. Cohen, C. E. Felder, D. Cahen, A. B. Ellis and A. Shanzer, *J. Am. Chem. Soc.*, 2000, **122**, 1116–1122.
213. A. Gulino, S. Giuffrida, P. Mineo, M. Purrazzo, E. Scamporrino, G. Ventimiglia, M. E. Van Der Boom and I. Fragalà, *J. Phys. Chem. B*, 2006, **110**, 16781–16786.
214. B. Lei, B. Li, H. Zhang, S. Lu, Z. Zheng, W. Li and Y. Wang, *Adv. Funct. Mater.*, 2006, **16**, 1883–1891.
215. B. Lei, B. Li, H. Zhang, L. Zhang and W. Li, *J. Phys. Chem. C*, 2007, **111**, 11291–11301.
216. H. Zhang, B. Li, B. Lei and W. Li, *J. Lumin.*, 2008, **128**, 1331–1338.
217. H. Zhang, B. Lei, W. Mai and Y. Liu, *Sens. Actuators, B*, 2011, **160**, 677–683.
218. S. Wang, B. Li, L. Zhang, L. Liu and Y. Wang, *Appl. Organometal. Chem.*, 2011, 21–26.
219. L. Liu, B. Li, R. F. Qin, H. F. Zhao, X. G. Ren and Z. M. Su, *Nanotechnology*, 2010, **21**, 285701.
220. S. M. Borisov, P. Lehner and I. Klimant, *Anal. Chim. Acta*, 2011, **690**, 108–115.
221. B. B. Collier, S. Singh and M. McShane, *Analyst*, 2011, **136**, 962–967.
222. Z. Xie, L. Ma, K. E. DeKrafft, A. Jin and W. Lin, *J. Am. Chem. Soc.*, 2010, **132**, 922–923.
223. J. An, C. M. Shade, D. A. Chengelis-Czegan, S. Petoud and N. L. Rosi, *J. Am. Chem. Soc.*, 2011, **133**, 1220–1223.
224. M.-L. Ho, Y.-A. Chen, T.-C. Chen, P.-J. Chang, Y.-P. Yu, K.-Y. Cheng, C.-H. Shih, G.-H. Lee and H.-S. Sheu, *Dalton Trans.*, 2012, **41**, 2592.
225. S. M. Barrett, C. Wang and W. Lin, *J. Mater. Chem.*, 2012, **22**, 10329–10334.
226. X.-L. Qi, S.-Y. Liu, R.-B. Lin, P.-Q. Liao, J.-W. Ye, Z. Lai, Y. Guan, X.-N. Cheng, J.-P. Zhang and X.-M. Chen, *Chem. Commun.*, 2013, **49**, 6864–6866.
227. X. Dong, L. Wei, Y. Su, Z. Li, H. Geng, C. Yang and Y. Zhang, *J. Mater. Chem. C*, 2015, **3**, 2798–2801.

228. Q. Li, M. Zhou, Q. Yang, Q. Wu, J. Shi, A. Gong and M. Yang, *Chem. Mater.*, 2016, **28**, 8221–8227.
229. J. Samonina-Kosicka, C. A. Derosa, W. A. Morris, Z. Fan and C. L. Fraser, *Macromolecules*, 2014, **47**, 3736–3746.
230. C. A. DeRosa, C. Kerr, Z. Fan, M. Kolpaczynska, A. S. Mathew, R. E. Evans, G. Zhang and C. L. Fraser, *ACS Appl. Mater. Interfaces*, 2015, **7**, 23633–23643.
231. S. J. Kim, V. P. Chodavarapu, A. N. Cartwright, M. T. Swihart and T. J. Bunning, *Sens. Actuators, B*, 2008, **130**, 758–764.
232. P. Zhou, D. Zhou, L. Tao, Y. Zhu, W. Xu, S. Xu, S. Cui, L. Xu and H. Song, *Light: Sci. Appl.*, 2014, **3**, e209.

CHAPTER 3

# *Evolution of Cell-penetrating Phosphorescent $O_2$ Probes*

DMITRI B. PAPKOVSKY AND RUSLAN I. DMITRIEV*

School of Biochemistry and Cell Biology, University College Cork, Cavanagh Pharmacy Building, Cork, Ireland
*E-mail: r.dmitriev@ucc.ie

## 3.1 Evolution of Cell-penetrating Phosphorescent $O_2$ probes

### 3.1.1 Introduction

Molecular oxygen ($O_2$) is one of the key molecules in tissue physiology, cell metabolism, signaling and redox biology.[1–10] It is essential for all kingdoms of life – anaerobes and aerobes, from prokaryotic organisms to eukaryotes, including plants, mammalian and human cells. Biochemically, $O_2$ is delivered to cells from the atmosphere or aqueous environment and utilized by numerous cytoplasmic enzymes and the mitochondria (a process often termed as 'cellular respiration'), to produce energy *via* oxidative phosphorylation as well as other important metabolites and biomolecules. This process, giving rise to the reactive oxygen (ROS) and nitrogen (RNS) species, contributes to redox balance and other vital processes in the cell.[11] At multicellular and tissue levels, $O_2$ is supplied by diffusion and dedicated $O_2$ transport systems, which consist of the vasculature, red blood cells, hemoglobin and myoglobin. This allows mammals to maintain $O_2$ concentration in cells and tissues

Detection Science Series No. 11
Quenched-phosphorescence Detection of Molecular Oxygen: Applications in Life Sciences
Edited by Dmitri B. Papkovsky and Ruslan I. Dmitriev

Published by the Royal Society of Chemistry, www.rsc.org

within physiological optima, rapidly respond to fluctuations in cellular $O_2$ demand and adapt to changing physiological conditions.

Oxygen homeostasis is achieved by balanced $O_2$ supply and consumption. Each type of cell and tissue has its characteristic oxygenation state called physiological *normoxia*, which differs significantly between different tissues and organs.[1] *Hypoxic* (below the norm) and *hyperoxic* (higher than norm) states are usually associated with tissue pathology or diseases such as cancer, inflammation or stroke. While average values for these conditions are approximately defined, the exact local tissue oxygenation and its dynamics remain important questions for vascular, cancer and stem cell biology, tissue engineering, drug discovery, regenerative medicine and other biomedical areas.[3,12] Analytical techniques, which can provide quantitative, real-time, continuous readout of cell and tissue $O_2$ concentration and its 3D distribution, are therefore in high demand.

Quenched phosphorescence-based detection is one of the thriving bioanalytical approaches to measure physiological $O_2$ gradients in various biological and disease models: cells, tissues and organs, from *in vitro* to *in vivo*. Compared to the alternative techniques, such as electron paramagnetic resonance, (micro)electrodes, NIRS oximetry, photoacoustics,[3,8,13–15] indirect staining with antibodies and hypoxic markers,[16] or the use of fluorescent proteins,[17–19] this method allows direct, non-invasive optical readout with stable calibration, subcellular resolution and multiplexing capabilities. The coupling of quenched phosphorescence $O_2$ detection with live cell microscopy and customized imaging formats requires custom-tailored phosphorescent materials in the form of solid-state sensors or soluble probes.[20] Solid-state materials, represented by dot sensors, film coatings, planar sensor foils, scaffolds and other types, usually bearing a hydrophobic phosphorescent $O_2$-responsive indicator dye, are intrinsically restricted to point, macroscopic or contact-based (2D) peri- and extra-cellular measurements of $O_2$. In contrast, $O_2$ sensing and imaging probes, represented by hydrophilic small molecule dyes, supramolecular bioconjugates or nanoparticle structures, can be distributed evenly within the tissue or even stain diffusely individual cells, thus facilitating high-resolution, 3D intracellular $O_2$ measurements and high content *in situ* analysis. Probes that can efficiently stain live cells by simple and passive means and thus facilitate $O_2$ measurement with subcellular spatial resolution, are called 'cell-penetrating' $O_2$ probes.[20,21]

Originally, the design of $O_2$ probes was focused on macromolecular conjugates, dendrimers, nano- and microparticles which did not have intrinsic cell-penetrating ability.[20,22–37] With such probes researchers worked to improve their brightness, compatibility with imaging readout modalities, biocompatibility and stability of calibration in complex biological samples. In order to deliver such probes into the cell, a number of cumbersome approaches were proposed, including conjugation with delivery moieties, gene gun, electroporation or facilitated delivery with transfection reagents such as Endo-Porter.[37] While some of these $O_2$ probes were demonstrated in functional intracellular measurements and analysis of cell oxygenation,[38] these studies were rare. As a result of this work, a need for self-loading and easy-to-use cell-penetrating probes has been recognized and prompted active research in this field.

Over the last 10 years, a number of original papers and reviews have been published covering various aspects of preparation and application of extracellular and intracellular $O_2$ probes, solid-state sensors and dedicated detection and imaging instrumentation.[3,8,12,20,21,39–47] The main practical considerations for the design and selection of optimal intracellular $O_2$ probes and their key operational parameters such as cell staining efficacy, brightness, toxicity, cell-specificity and intracellular distribution, compatibility with detection modes, stability of $O_2$ calibration were defined.

In this chapter, we describe the evolution of cell-penetrating $O_2$ probes, highlighting their desirable structural and functional features, analytical performance and advantages, in order to summarize the main requirements and specifications of such probes for high-resolution $O_2$ imaging and physiological studies with complex 3D tissue models.

### 3.1.2 Why and How to Measure Intracellular $O_2$?

Recent advances in imaging instrumentation, fluorescence microscopy and imaging techniques such as super-resolution and light-sheet microscopy, tissue clearing and expansion microscopy, design of advanced fluorescent protein constructs and biosensors promise us deeper insights into cell and tissue function. However, on the quantitative side many of these new techniques (*e.g.* tissue clearing and expansion microscopy, dealing with dead and fixed tissues[48–50]) rather raise new questions and prompt generation of artifacts,[50] relying on semi-quantitative intensity-based measurements (*e.g.* most of fluorescent protein biosensors) at best.[17] In contrast, truly quantitative fluorescence (FLIM) and phosphorescence (PLIM) lifetime imaging microscopies, were somehow overlooked until recently.[51,52]

The two main limitations of optical imaging during *in vivo* and deep tissue measurements are low in-depth light penetration, high scattering and autofluorescence of mammalian tissue, and bio-delivery of probes and nanosensors. Thus, *in vivo* intracellular and vascular $O_2$ imaging was realized only for rodent brain vasculature, tumors and some other tissues such as gastrointestinal tract and bone marrow,[53–57] mostly using the implanted optical windows[58,59] or the spot (1D) imagers.[60,61]

However, various engineered *in vitro* and *ex vivo* 3D micro-tissues such as organotypic explants, organoids, multicellular aggregates, microfluidic biochips and conventional cell culture already benefit from the use of high-resolution live cell microscopy and intracellular $O_2$ probes. Here, real-time $O_2$ detection helps to analyze lateral and in-depth oxygenation profiles, their heterogeneity and dynamics during tissue functioning at rest, in a diseased state and upon pharmacological treatment.[3,12,62] For engineered 3D tissue models in perfused microfluidic systems, analysis of $O_2$ helps better interpret the effects of drugs on cell metabolism, address the problem of cell and tissue heterogeneity, standardization, create, maintain and characterize stem cell niches and micro-regions for rare cell types.[63–67]

In conventional and primary cell cultures, $O_2$ detection with cell-penetrating probes is essential for monitoring cell oxygenation *in situ*, maintaining cells at physiological $O_2$ levels found *in vivo*, evaluating energy production pathways and local $O_2$ micro-gradients.[38,68,69] For some biological models, researchers can create specific oxygenation conditions and patterns, control and manipulate them: for example, large multicellular aggregates or samples of tissue with limited $O_2$ supply deoxygenate themselves.[62] Similarly, actively respiring cells, even in dense 2D monolayers, form prominent $O_2$ gradients under ambient $O_2$ (20.86% in air).[70]

A number of different readouts can be utilized with cell-penetrating $O_2$ probes: commercial time-resolved fluorescence (TR-F) readers, standard or customized intensity-based or phosphorescence lifetime imaging (PLIM) laser-scanning microscopes and camera-based imaging systems (Table 3.1).

Current TR-F microplate readers, which use pulsed excitation with a xenon flash lamp, offer biomedical users a simple and low-cost detection platform for sensing cellular $O_2$ levels, with minimal sample photodamage, low light exposure, high detection sensitivity, and control of atmospheric $O_2$ in sample compartment.[71]

**Table 3.1** Comparison of microplate TR-F reader and laser-scanning PLIM readouts in intracellular $O_2$ measurements.

| | Plate reader | PLIM microscope |
|---|---|---|
| Main application areas | 2D cultures, mean oxygenation | 3D cultures, $O_2$ micro-gradients, single cell analysis |
| Probe photostability requirement | Moderate | High |
| Cell staining efficiency, intensity signals | Moderate | High |
| $O_2$ probe lifetime range | >20 µs (Xe flashlamp) | Not limited (ps/fs laser pulses) |
| Excitation/Emission of $O_2$ probe | Near-UV (340–420 nm) Visible light excitation is rare | Any (available pulsed lasers) |
| Intracellular $O_2$ probe distribution | Not critical, but must be known | Any |
| Applicability to 3D models | Limited | Very useful, resolution limited by diffraction |
| Applicability to animal models | Not applicable (*in vitro* only) | Applicable, but with limitations and requires dedicated measurement setup |
| Two-photon excitation | Not applicable | Compatible |
| Spatial resolution | Low, if any (>100 micron) | Higher than 1 µm |
| Multiplexing potential | Moderate | High |

These instruments are very useful and convenient for $O_2$ analysis in cell populations, 2D and even 3D tissue cultures, when lateral and local $O_2$ gradients are not so pronounced and heterogeneous. Another advantage is high throughput, allowing parallel analysis of dozens of samples in one experiment and frequent repetitive measurements that can be performed in second, minute and hour timescales.

In contrast to the analysis of mean local $O_2$ levels in cell populations on a TR-F reader, widefield and confocal PLIM allows visualization of spatial distribution $O_2$ in X-Y-Z dimensions and much higher information content with single cell resolution and multi-parametric analysis. This is the method of choice for complex and hetero-cellular 3D tissue models.[67] Commercial time-gated, frequency modulation and TCSPC-based PLIM platforms are available from several vendors, and PLIM microscopy, together with FLIM, FLIM-FRET, FLIM-PLIM systems are rapidly gaining popularity in biomedical research. PLIM can operate under one- and two-photon,[39] confocal laser-scanning, widefield (light emitting diodes and gated cameras) and other excitation modes such as light-sheet microscopy. However, for $O_2$ probes, the preferred readout is time-domain PLIM, which ensures highly accurate $O_2$ measurements, stable $O_2$ calibration (one-off or provided by probe developer/vendor), independence of probe concentration, minimal or no interference by autofluorescence and light scattering. In contrast to most fluorescent dyes used in FLIM, phosphorescent $O_2$ probes tend to show mono-exponential decay facilitating simple and robust lifetime calculation and its conversion into $O_2$ values (μM, kPa or Torr units). PLIM also has broad multiplexing capabilities, both in spectral (due to large Stokes shifts for phosphorescent materials) and temporal (microsecond PLIM/nanosecond FLIM) domains. Long emission decay times (>1 μs) of phosphorescent probes also allow for background-free measurements.

Still, inconsistencies and measurement artifacts are not uncommon with $O_2$ sensors and probes, therefore each $O_2$ probe, experimental biological model and measurement setup should be carefully considered and verified with control experiments. Additional complications can include intrinsic toxicity of intracellular probes, production of singlet oxygen, photoconsumption of $O_2$, triplet–triplet annihilation, depletion of the dye ground state and some others.[72]

A number of phosphorescent $O_2$ probes (see Section 3.3) also allow ratiometric-intensity measurements on certain experimental models, mainly 2D and high-content systems. Still, for such probes $O_2$ calibration is instrument-specific and these intensity-based measurements are susceptible to variability and instability, *e.g.* due to the difference in light penetration depths for reference and $O_2$-sensitive spectral channels or with different tissue models.

For whole animal fluorescence imaging, devices that support PLIM detection mode are very rare, and ratiometric-intensity based $O_2$ probe and detection is usually the only option for (semi)quantitative $O_2$ measurements.

Thus, the cell-penetrating phosphorescent $O_2$ probes have significant application scope in live cell and tissue analysis, including a range of different biological and disease models. Probe applicability is largely aided or limited by the available detection instrumentation. In Section 3.3, we will describe several different types of cell-penetrating $O_2$ probes, their characteristic features and application scope.

### 3.1.3 Different Classes of Cell-penetrating $O_2$ Probes and Their Evolution

Phosphorescent $O_2$-sensitive indicator dyes, despite their attractive photophysical and analytical (brightness, quantum yield, photostability, phosphorescence lifetime range) features, are rarely directly used in biological applications. These dyes frequently contain hydrophobic macrocyclic moiety, which provide them with complex behavior due to non-specific interactions with proteins (*e.g.* albumin) and other biomolecules, aggregation, effects of local environment, or other quenchers. All this makes them hard to use in complex biological samples and for accurate measurement and imaging of tissue $O_2$. Some phosphorescent dyes, such as $Ru(bpy)_3$ complexes, are commercially available and inexpensive, prompting some researchers to use them in biological experiments[73,74] without proper initial validation, calibration, analysis of concentration- and temperature-dependence, and bio-distribution. The results of such studies are often difficult to interpret for their meaning for redox biology. On the other hand, the shortcomings of many existing phosphorescent indicator dyes were highlighted nearly three decades ago and researchers are still working hard to address these issues.[23,34,75] Development and demonstration of advanced cell-penetrating $O_2$ probes were important milestones of this work (Table 3.2).

Evolution of cell-permeable $O_2$ probes followed two main directions: (1) optimization of the probe photophysical characteristics, sensitivity to $O_2$ and stability of calibration; (2) improving the cell and tissue staining performance, biocompatibility and addressing the mechanisms of cell internalization. This evolution is seen as a transition from simple dye structures to multi-functional bioconjugates and dye encapsulation in nanosensor structures and their fine-tuning. Below we discuss the evolution of 'classes' of cell-penetrating $O_2$ probes consecutively.

#### *3.1.3.1 Small Molecule Probes and Other Conjugates*

Chemical conjugates, represented by small molecule dyes, branched dendrimers and protein conjugates can be grouped together as well-defined chemical structures in which the phosphorescent indicator moiety is chemically modified with side chain groups, other dyes, short polymers or biomolecules, in order to acquire the desired spectral and physical-chemical

**Table 3.2** Representative cell-penetrating $O_2$ probes of the different classes.

| Class of $O_2$ probe | Name/Structure | Cell and tissue specificity | Properties | Applications | Comments |
|---|---|---|---|---|---|
| Small molecule conjugate | **PEPP0**/PtCPTE-bactenecin 7 peptide fragment (15-24) conjugate[90] | PC12, SH-SY5Y, HeLa, HCT116 cells | Low photostability, hydrophobic, cell-permeable 390 nm exc., 650 nm em | Analysis of intracellular $O_2$ upon metabolic stimulation on a time-resolved fluorescence plate reader[71,90] | Easy to synthesize and use. Cell-specific calibration |
| Small molecule conjugate | **Pt-Glc**/PtTFPP $(Glc)_4$ conjugate[80] | Adherent cells (PC12, Cos-7, SH-SY5Y, HepG2, HCT116, Caco-2, primary neurons), *ex vivo* tissues, multicellular spheroids, intestinal organoids[65,80,84–86,93] | Bright, photostable, 395 nm exc., 650 nm em. Unrivalled cell staining efficacy | Functional imaging of $O_2$ in 3D tissue models | Easy to synthesize and use |
| Nanoparticles (1st generation) | **NanO2 (MitoImage-NanO2)**/RL100 doped with PtTFPP[105] | Adherent cells. Poorly stains neural cells and 3D tissue models | Bright, photostable 395 nm exc./650 nm em. | Analysis of intracellular $O_2$ upon metabolic stimulation on a time-resolved fluorescence plate reader, PLIM of adherent and multicellular spheroids | Easy to synthesize and use. With optimized staining procedure, can be used with some 3D tissue models |

| | | | | | |
|---|---|---|---|---|---|
| Nanoparticles (2nd generation) | **MM2 (MitoImage – MM2)**/Multi-modal nanoparticles: RL100 doped with PtTFPP and PFO[114] | Same as for NanO2 | Bright, photostable. Compatible with different detection modes: intensity-ratiometric and PLIM. One (exc. 395 nm) and two (760 nm) -photon excited. 430 ($O_2$-insensitive) and 650 nm ($O_2$-sensitive) nm em | Same as for NanO2, with better quality of data due to two-photon excitation | Batch-to-batch variability of sensing properties (ratiometric response, two-photon excitability) is possible |
| Nanoparticles (3rd generation) | **SI-0.2$^+$**/Conjugated polymer nanoparticles: PtTFPP-PFO co-polymer[115] | Can be tuned, depending on the charged groups (positive, negative, zwitter-ionic) | Very bright (due to more efficient and controllable FRET between antenna and PtTFPP), photostable. Spectrally similar to MM2 | Imaging of 2D and 3D tissue models | Versatile and tunable design which can be used to overcome any problems with cell specificity<br>Complex preparation procedure |

properties, cell penetration and bio-delivery, biocompatibility and usable $O_2$-sensing properties for intracellular measurements. Typically, they contain one molecule of $O_2$-sensitive phosphor, which is at least partly protected from micro-environmental influences, *e.g.* by dendrimer structures[26] or other groups. In contrast, in nanosensors the phosphorescent dye acquires novel properties due to the physical dimensions and multiple dye molecules present. Some chemical conjugates (*e.g.* the cell-impermeable dendrimer PtP–C343[76]) sometimes are inaccurately referred as 'nanoprobes', although this term does not adequately reflect their structure and properties.

The first truly cell-penetrating phosphorescent conjugates were independently reported by Neugebauer *et al.* (2008)[77] and Dmitriev *et al.* (2009),[78,79] when the poorly cell permeable $[Ru(bpy)_2(pic)]^{2+}$ and PtCP phosphors were conjugated to cell-penetrating peptides. These probes showed high efficacy of intracellular delivery, and were subsequently used in several biological studies.[80–86] While the Ru(II) dyes produced mostly cytotoxic probes which were problematic to use,[87,88] the optimization of peptide type and linkage for metalloporphyrin moiety helped to develop $O_2$ probes with broad cell-specificity, low toxicity and tunable intracellular distribution (lysosomes, cytoplasm and mitochondria).[80,89–91] Cell-penetrating peptides, such as TAT, oligoarginine and Pro-rich sequences, are commonly used to deliver molecular cargo into the cell.[92] The $O_2$ probes on their basis were suitable for microplate reader analysis and PLIM microscopy, providing the efficient intracellular staining of monolayer cell cultures in as little as 6 h.

New chemistries allowed the use of more bright and photostable phosphorescent dyes for intracellular measurements. Thus, the click-modification with thiols of the widely available but hydrophobic PtTFPP dye produced novel peptide-based conjugates. The hydrophobicity of the phosphor negatively affected the photophysical properties of the conjugate, however the modification of PtTFPP with four monosaccharide moieties helped to achieve the right balance between probe hydrophilicity, $O_2$-sensing properties and cell permeability. The resulting PtTFPP conjugates with thioglucose (Pt-Glc) and thiogalactose (Pt-Gal) showed stable calibration with minor tissue-specificity, minimal toxicity, and efficient tissue staining making them as bright as the nanosensors. Currently, Pt-Glc is one of the most used probes in biological applications with 3D tissue models.[65,80,84–86,93] A somewhat similar type of cell-penetrating conjugate based on PdTCTBP dye with poly(amido)amine was reported later,[94] but no biological studies have been conducted with it so far.

Mitochondrial targeting is often desirable for cell-penetrating $O_2$ probes for efficient labeling of the mitochondria, direct measurement of $O_2$ consumption and analysis of intracellular gradients. A number of peptide vectors and targeting groups (*e.g.* triphenylphosphonium) were tested to create 'mitochondria-targeting probes' such as BTP-mito[95] and Ir(III) complex with

5-amino-1,10-phenanthroline.[96] However, such conjugates displayed low brightness (ε in range of 6700–13 500 $M^{-1}$ $cm^{-1}$) and were not characterized in sufficient detail for concentration dependence of their $O_2$ calibration and cell specificity. Furthermore, when evaluating these probes, HeLa cells with intrinsically low respiration were used, which are not expected to provide any differences between mitochondrial and extracellular $pO_2$.[90] PEPP0 conjugate of the uncharged PtCP with bactenecin 7 peptide fragment also showed mitochondrial localization in HCT116 cells, possibly due to the combination of hydrophobic and positively charged amino acid residues. Again, this probe has low brightness and photostability and is poorly suited for confocal PLIM.[90]

Some dye structures have good solubility in biological media and their direct conjugation with other fluorophores resulted in creating cell-penetrating $O_2$ probes. Thus, linking the Ir-based BTP to coumarin C343 produced a ratiometric intracellular $O_2$ probe.[97] Another research team prepared Ru-coumarin complex for a similar purpose.[98] Ir-BTP derivative BTPD1[99] modified with dimethylamino group also showed cell permeability better than $Ru(dpp)_3^{2+}$ and was used for the estimation of tumor oxygenation *in vivo*. Tris-cyclometalated organo-iridium(III) complexes were evaluated as cell-penetrating probes, but they showed aggregation in aqueous media, making their staining and calibration uncontrollable and unusable for quantitative measurements.[100]

Some protein based structures were suggested for intracellular $O_2$ sensing. Thus, PtCP-transferrin conjugate rapidly internalizing in cultured cells can be used for plate reader-based measurements.[69] Similarly, PtCP metallochelate complex with virus-like nanoparticle has promising features for $O_2$ sensing.[101] However, the progress with protein-based probes is limited by the availability of specific labeling chemistries for the common phosphorescent $O_2$-sensitive dyes.

Thus far, the main strategies in preparing cell-penetrating probes based on conjugates were in attaching the various cell-targeting moieties, which enhance cell penetration while keeping and tuning photophysical properties and analytical performance of the phosphor. The main concern with such structures is the non-ideal shielding of the phosphor from the biological environment, resulting in concentration and cell/tissue-specific staining and $O_2$ calibration. However, the relatively simple and flexible design and small size (up to a few nm) facilitate their fine-tuning for different cell types, 3D models and analytical tasks, extending their application potential from simple *in vitro* models to *in vivo* imaging in live animals.

#### *3.1.3.2 Nanoparticles of First Generation*

Nanoparticle-based $O_2$ probes (nanosensors), usually produced by the precipitation method, are compatible with various phosphors and organic polymers.[102] Choosing suitable and biocompatible polymer and phosphorescent dye allows simple preparation of highly efficient, reproducible and bright cell-penetrating $O_2$ nanosensors of 'first generation', where the dye is impregnated in polymer, in a simple and reproducible manner. Thus, a number of nanoprobes based on difluoroboron dibenzoylmethane-polylactide,[103] polyacrylamide/[Ru(dpp($SO_3Na)_2)_3]Cl_2$,[104] ammonio methacrylate copolymer Eudragit RL100/PtTFPP,[105] styrene and acrylonitrile/PtTFPP[106] were developed in 2008–2013. Their advantages were higher specific brightness than for small molecule conjugates, stable calibration (phosphor is shielded by carrier polymer) and often intrinsic cell-staining efficiency, allowing low doses of nanosensor probe to be used for cell staining (1–10 μg $ml^{-1}$) and $O_2$ analysis. Thus, RL100/PtTFPP probe (commercialized as MitoImage-NanO2) is well suited for lifetime-based time-resolved fluorescence (TR-F) detection and PLIM imaging of cell oxygenation upon physiological and metabolic treatments. However, further research indicated cell-specific staining and partial aggregation of NanO2, noted for example with neurosphere model.[62,80,107] Thus, a number of additional charged polymers were screened to create similar nanosensors with improved cell staining. For example, it was found that negatively charged PMMA-AA/PtTFPP demonstrated improved intracellular staining of neural cell models, including spheroids and live brain slices.[107]

The average size of cell-penetrating $O_2$ nanosensors should be kept within 10–100 nm, in order to retain permeability and minimize the potentially damaging and invasive effects for the cell. Larger size particles were also described[108,109] but their compatibility with minimally invasive cell-based measurements is questionable. Overall, first generation nanosensors demonstrated the power of the approach, but also highlighted its limitations such as less flexible chemical design, incompatibility with two-photon excited PLIM or ratiometric intensity imaging, potential leaching of the phosphor from the nanoparticle due to its physical entrapment by the polymer rather than chemical linkage. Further research has addressed these issues.

#### *3.1.3.3 Nanoparticles of Second Generation*

Compared to first generation, the second generation nanosensors are enriched with additional functionalities, *e.g.* a reference dye improving imaging properties or modified nanoparticle shell improving cell staining.

The incorporation of a second ($O_2$-insensitive reference) dye in carrier nanoparticles, in addition to the phosphor, allows ratiometric intensity measurements, helpful when PLIM detection equipment is not available. Secondary fluorophore can also serve as a light harvesting antenna which improves brightness and two-photon excitability of the phosphor *via* FRET mechanism (symmetric metalloporphyrins and other phosphors frequently display low two-photon absorption[110]). Following a number of 'two-dye' nanosensor structures,[111–113] a concept of 'multi-modal' nanoparticle probes was developed.[114] Thus, MM2 probe comprised cell-penetrating

RL100 nanoparticles doped with PtTFPP dye and polyfluorene polymer as light antenna displayed enhanced two-photon excitability, and compatibility with ratiometric-intensity and lifetime (PLIM) detection modes. The combination of such properties ensured compatibility of the MM2 probe with all the main types of fluorescence imaging platforms. It also provided a framework for the development of nanosensors with improved cell permeability.[107,115]

Recently, Feng and co-workers utilized the 'two-dyes' design with an Ir(III) phosphor chemically conjugated with branched fluorene residues[116] and incorporating the resulting 'hyperbranched' phosphor in PSMA polymer to produce the 22 nm-size negatively charged nanosensors. The main advantage of such a probe over MM2 and similar nanoparticles[114] is improved compatibility with ratiometric detection, and potential use in photodynamic therapy (PDT) to kill the cells, although this data also indicates the significant phototoxicity of the probe.

Kopelman and co-workers pursued another design route and produced the so-called PEBBLE sensors.[117] The modular but rather complex design of PEBBLE sensors allowed tuning of their cell targeting properties with a cell-penetrating peptide coat thus modifying the nanoparticle surface.[118] However, these nanosensors were evaluated in quantitative $O_2$ measurements only briefly and never applied for physiological experiments.

In order to improve intracellular distribution the chemical modification of nanoparticle surface with amino[119] and triphenylphosphonium[120] groups was explored. Both studies claimed to achieve mitochondrial localization of the nanosensors, however the published images do not support this, showing only partial co-localization with MitoTracker and predominantly an endosomal pattern of intracellular distribution for the probes.

Most of the described and studied the $O_2$-sensitive intracellular nanosensors enter the cell *via* endocytosis and become entrapped in lysosomes. This makes their intracellular route to the mitochondrial membrane highly unlikely and inefficient. Analysis of the published data on intracellular $O_2$ nanosensors shows that they all tend to have relatively slow cell staining (12–16 h or longer), compared to small molecule conjugates, a more or less typical pattern of intracellular distribution (endosomes and lysosomes) and relatively limited number of cell types suitable for staining with every particular type of nanosensor. Unfortunately, in many reported studies researchers limit evaluation of their new nanosensors only to one-two cell types or tissue models.[21]

Another second generation nanosensor is the Ir-CM where nanoparticles are produced by co-precipitation of Ir(III)-conjugated poly(*N*-vinylpyrrolidone) and poly(ε-caprolactone)-*b*-poly(*N*-vinylpyrrolidone) polymers.[121] The resulting nanoparticles demonstrated an average size of 30–50 nm and a slight negative charge. They provided efficient staining of cells, as well as of deeply deoxygenated tumors and some other tissues *in vivo* upon intravenous injection. However, the suitability of this nanosensor for quantitative $O_2$ measurements was not evaluated in detail, which makes this study incomplete and hard to compare with other *state-of-the-art* $O_2$ nanosensors.

### 3.1.3.4 Nanoparticles of Third Generation

While often versatile and multi-functional, the nanoparticles of second generation frequently harbor a number of combined features and functionalities, which are hard to systemically optimize or improve. Their excellent cell staining ability can be limited to just a few cell lines and spectrally nanosensors are also bound to a limited number of phosphor and light harvesting antenna dye structures, compatible with the precipitation method. Modification of the shell is also a very challenging task, due to poly-disperse nature of the nanoparticles.[122] A number of improved nanoparticle structures were therefore developed, which address these issues or provide an advanced framework to overcome them.

The third generation of intracellular nanosensors are based on conjugated polymer as nanoparticle carrier which is covalently coupled with a phosphor and surface modification.[123] Since Xiang *et al.* reported that conjugated polyfluorene (reference dye) and PtTFPP ($O_2$-sensitive dye) can display improved spectral properties and tunable FRET (in contrast to simply mixing them together as in second generation nanosensors[114]), it was realized that chemical conjugation of reference dye (modified polyfluorene or benzothiadiazole) with phosphor (PtTFPP, Ir(III) complex or PtTPBPF derivatives) can be utilized in making highly efficient intracellular nanoparticles.[115,124,125] The chemical conjugation also allowed synthesis of multimodal sensing materials, compatible with one- and two-photon excitation modes, ratiometric intensity and phosphorescence lifetime-based detection. The ratio between these two emitters can be kept stable (chemical bonding ensures no leaching), controlled and used to tune the photophysical properties of the probe (*e.g.* sensitivity to $O_2$). Such probes are 10–100 times brighter than first and second generation nanosensors produced by doping of non-emitting carrier polymers, making them the brightest intracellular $O_2$ sensor probes available. Such materials do not need encapsulation as they form nanoparticles by themselves. The third advantage is that the polymer backbone can be chemically modified with charged groups or targeting moieties, in order to tune the surface charge and bio-delivery of the nanosensor.[115] The different spectral and charged variants of the nanosensors ensured the high staining efficiency with respect to different cell types, including 3D tumor spheroid models. Such flexible and modular design (tunable optical and staining properties) and working characteristics meant that these structures could be called versatile 'one-for-all' type of $O_2$-sensitive nanosensors. Their negative side is a more complex fabrication procedure (multi-step chemical synthesis), compared to the nanosensors of previous generations.

## 3.2 Conclusions and Future Outlook

Analysis of the flourishing area of intracellular $O_2$ sensors (2008–2017) shows that still only a few dye structures and chemistries fulfill the requirements of 'optimal' probe for imaging cell and tissue $O_2$. Out of many dye structures tested[40,47] only PtTFPP, PtTPBPF, Ir(III) and some Ruthenium complexes were shown to produce efficient nanosensor structures. The choice of reference and antenna dyes is also limited, with polyfluorene (PFO) being most suitable for multi-modal and conjugated polymer nanoparticles structures.[114,115,125] Many other structures have been shown to produce bright one- and two-photon excitable materials with useful sensitivity to $O_2$ and stable calibration, their characterization and demonstration with complex biological samples and tissue models is often insufficient and analytical capabilities and operational performance are questionable. Many reported nanosensors require significant improvements for use in biological applications: they often suffer from inability to stain certain cell and tissue types, need long staining time (12–18 h for cells), and display limited applicability to *in vivo* measurements.

What we know is that the size of nanoparticle probes must be in range of 20–100 nm and their intracellular accumulation and distribution are not directly related to their charge: cell-permeable cationic, zwitter-ionic and anionic nanoparticles were all reported as efficient nanosensors. It is still unclear how to further improve intracellular staining and distribution of existing materials. With a few targeting moieties tested (PEG, Arg-rich peptides, triphenylphosphonium), the knowledge of how chemical and structural features determine nanoparticle interactions at biological interface is very limited.[126] Here progress can be made with the new polymers and nanoparticle-forming structures, combined with new targeting moieties. Some new directions include the use of metal–organic frameworks[127] and combining phosphors with self-assembling protein nanoparticles,[101] however such probes reported so far do not look superior to existing $O_2$ nanosensors.

The growing demand from biological applications necessitates the deployment of best $O_2$ probes, either small molecule conjugates or sophisticated nanosensors.[62,80,86] One of the areas critically demanding efficient (bright, photostable, allowing quick staining and retaining reliable calibration) intracellular $O_2$ probes is engineered live tissue, which is actively used in regenerative medicine, cancer biology and stem cell research. Knowledge of $O_2$ micro-gradients and even pericellular (at cell surface) oxygenation in micro-tissue samples having sizes 100–500 μm is important. $O_2$ monitoring is also useful in studying drug responses[63,85] and metabolic heterogeneity of engineered organoids.[64,65] It can be valuable addition to the existing pH, redox, ROS, $Ca^{2+}$ and other probes[93,128,129] and other key parameters for the assessment of function and viability of live tissue.

In a 3D context, extracellular $O_2$ can also be sensed within the tissue, *e.g.* with hybrid phosphorescent scaffolds[130,131] or spray-paint delivered $O_2$-sensitive materials.[132] Hopefully, the knowledge generated in the area of design of cell-penetrating $O_2$ probes can be highly useful in the design of new probes and methodologies for deep tissue imaging.

## Acknowledgement

This work was supported by Science Foundation Ireland (SFI) grants 13/SIRG/2144 (RID) and 12/RC/2276 (DBP).

## References

1. M. Erecinska and I. A. Silver, *Respir. Physiol.*, 2001, **128**, 263–276.
2. A. Melvin and S. Rocha, *Cell. Signalling*, 2012, **24**, 35–43.
3. D. B. Papkovsky and R. I. Dmitriev, *Chem. Soc. Rev.*, 2013, **42**, 8700–8732.
4. R. Pittman, *Acta Physiol.*, 2011, **202**, 311–322.
5. S. Sakadzic, E. Roussakis, M. A. Yaseen, E. T. Mandeville, V. J. Srinivasan, K. Arai, S. Ruvinskaya, W. Wu, A. Devor, E. H. Lo, S. A. Vinogradov and D. A. Boas, *J. Visualized Exp.*, 2011, e1694.
6. G. L. Semenza, *Wiley Interdiscip. Rev.: Syst. Biol. Med.*, 2010, **2**, 336–361.
7. M. C. Simon and B. Keith, *Nat. Rev. Mol. Cell Biol.*, 2008, **9**, 285–296.
8. O. S. Wolfbeis, *BioEssays*, 2015, **37**, 921–928.
9. M. D. Brand and D. G. Nicholls, *Biochem. J.*, 2011, **435**, 297.
10. T. Clanton, M. Hogan and L. Gladden, *Compr. Physiol.*, 2013, DOI: 10.1002/cphy.c120030.
11. D. S. Bilan and V. V. Belousov, *Free Radicals Biol. Med.*, 2017, **109**, 167–188.
12. D. B. Papkovsky and A. V. Zhdanov, *Free Radicals Biol. Med.*, 2016, **101**, 202–210.
13. C. Diepart, J. Verrax, P. B. Calderon, O. Feron, B. F. Jordan and B. Gallez, *Anal. Biochem.*, 2010, **396**, 250–256.
14. S. Ashkenazi, S.-W. Huang, T. Horvath, Y.-E. L. Koo and R. Kopelman, *J. Biomed. Opt.*, 2008, **13**, 034023–034024.
15. M. Ferrari and V. Quaresima, *Neuroimage*, 2012, **63**, 921–935.
16. R. B. Elmes, *Chem. Commun.*, 2016, **52**, 8935–8956.
17. R. Erapaneedi, V. V. Belousov, M. Schäfers and F. Kiefer, *EMBO J.*, 2016, **35**, 102–113.
18. M. L. Skiles, R. Fancy, P. Topiwala, S. Sahai and J. O. Blanchette, *J. Biomed. Mater. Res., Part B*, 2011, **97B**, 148–155.
19. E. Takahashi, T. Takano, Y. Nomura, S. Okano, O. Nakajima and M. Sato, *Am. J. Physiol.: Cell Physiol.*, 2006, **291**, C781–C787.
20. R. I. Dmitriev and D. B. Papkovsky, *CMLS, Cell. Mol. Life Sci.*, 2012, **69**, 2025–2039.
21. R. I. Dmitriev and D. B. Papkovsky, *Methods Appl. Fluoresc.*, 2015, **3**, 034001.

22. L.-W. Lo, C. J. Koch and D. F. Wilson, *Anal. Biochem.*, 1996, **236**, 153–160.
23. W. L. Rumsey, J. M. Vanderkooi and D. F. Wilson, *Science*, 1988, **241**, 1649–1651.
24. L. Zheng, A. S. Golub and R. N. Pittman, *Am. J. Physiol.: Heart Circ. Physiol.*, 1996, **271**, H365–H372.
25. M. C. Hogan, *J. Appl. Physiol.*, 1999, **86**, 720–724.
26. I. Dunphy, S. A. Vinogradov and D. F. Wilson, *Anal. Biochem.*, 2002, **310**, 191–198.
27. O. S. Finikova, A. Y. Lebedev, A. Aprelev, T. Troxler, F. Gao, C. Garnacho, S. Muro, R. M. Hochstrasser and S. A. Vinogradov, *ChemPhysChem*, 2008, **9**, 1673–1679.
28. A. Y. Lebedev, A. V. Cheprakov, S. Sakadzic, D. A. Boas, D. F. Wilson and S. A. Vinogradov, *ACS Appl. Mater. Interfaces*, 2009, **1**, 1292–1304.
29. D. F. Wilson, O. S. Finikova, A. Y. Lebedev, S. Apreleva, A. Pastuszko, W. M. F. Lee and S. A. Vinogradov, *Oxygen Transport to Tissue XXXII*, Springer US, 2011, vol. 701, pp. 53–59.
30. S. M. Borisov, T. Mayr and I. Klimant, *Anal. Chem.*, 2008, **80**, 573–582.
31. P. Chojnacki, G. Mistlberger and I. Klimant, *Angew. Chem.*, 2007, **119**, 9006–9009.
32. Y. E. Koo, Y. Cao, R. Kopelman, S. M. Koo, M. Brasuel and M. A. Philbert, *Anal. Chem.*, 2004, **76**, 2498–2505.
33. H. Xu, J. W. Aylott, R. Kopelman, T. J. Miller and M. A. Philbert, *Anal. Chem.*, 2001, **73**, 4124–4133.
34. A. Fercher, G. Ponomarev, D. Yashunski and D. Papkovsky, *Anal. Bioanal. Chem.*, 2010, **396**, 1793–1803.
35. J. Hynes, S. Floyd, A. E. Soini, R. O'Connor and D. B. Papkovsky, *J. Biomol. Screening*, 2003, **8**, 264–272.
36. T. C. O'Riordan, K. Fitzgerald, G. V. Ponomarev, J. Mackrill, J. Hynes, C. Taylor and D. B. Papkovsky, *Am. J. Physiol.: Regul., Integr. Comp. Physiol.*, 2007, **292**, R1613–R1620.
37. T. C. O'Riordan, A. V. Zhdanov, G. V. Ponomarev and D. B. Papkovsky, *Anal. Chem.*, 2007, **79**, 9414–9419.
38. A. V. Zhdanov, V. I. Ogurtsov, C. T. Taylor and D. B. Papkovsky, *Integr. Biol.*, 2010, **2**, 443–451.
39. Y. Chen, R. Guan, C. Zhang, J. Huang, L. Ji and H. Chao, *Coord. Chem. Rev.*, 2016, **310**, 16–40.
40. M. Quaranta, S. M. Borisov and I. Klimant, *Bioanal. Rev.*, 2012, **4**, 115–157.
41. Y. Feng, J. Cheng, L. Zhou, X. Zhou and H. Xiang, *Analyst*, 2012, **137**, 4885–4901.
42. M. A. Filatov, S. Baluschev and K. Landfester, *Chem. Soc. Rev.*, 2016, **45**, 4668–4689.
43. R. Paolesse, S. Nardis, D. Monti, M. Stefanelli and C. Di Natale, *Chem. Rev.*, 2016, **117**, 2517–2583.
44. E. Roussakis, Z. Li, A. J. Nichols and C. L. Evans, *Angew. Chem., Int. Ed.*, 2015, **54**, 8340–8362.

45. T. Yoshihara, Y. Hirakawa, M. Hosaka, M. Nangaku and S. Tobita, *J. Photochem. Photobiol., C*, 2017, **30**, 71–95.
46. X.-D. Wang and O. S. Wolfbeis, *Anal. Chem.*, 2012, **85**, 487–508.
47. J.-n. Liu, W. Bu and J. Shi, *Chem. Rev.*, 2017, **117**, 6160–6224.
48. F. Chen, P. W. Tillberg and E. S. Boyden, *Science*, 2015, **347**, 543–548.
49. R. Tomer, L. Ye, B. Hsueh and K. Deisseroth, *Nat. Protoc.*, 2014, **9**, 1682.
50. U. Schnell, F. Dijk, K. A. Sjollema and B. N. Giepmans, *Nat. Methods*, 2012, **9**, 152–158.
51. W. Becker, *J. Microsc.*, 2012, **247**, 119–136.
52. K. Suhling, L. M. Hirvonen, J. A. Levitt, P.-H. Chung, C. Tregidgo, A. Le Marois, D. A. Rusakov, K. Zheng, S. Ameer-Beg and S. Poland, *Med. Photonics*, 2015, **27**, 3–40.
53. A. Devor, S. Sakadzic, V. J. Srinivasan, M. A. Yaseen, K. Nizar, P. A. Saisan, P. Tian, A. M. Dale, S. A. Vinogradov, M. A. Franceschini and D. A. Boas, *J. Cereb. Blood Flow Metab.*, 2012, **32**, 1259–1276.
54. L. Albenberg, T. V. Esipova, C. P. Judge, K. Bittinger, J. Chen, A. Laughlin, S. Grunberg, R. N. Baldassano, J. D. Lewis and H. Li, *Gastroenterology*, 2014, **147**, 1055–1063.e1058.
55. J. A. Spencer, F. Ferraro, E. Roussakis, A. Klein, J. Wu, J. M. Runnels, W. Zaher, L. J. Mortensen, C. Alt and R. Turcotte, *Nature*, 2014, **508**, 269–273.
56. T. V. Esipova, A. Karagodov, J. Miller, D. F. Wilson, T. M. Busch and S. A. Vinogradov, *Anal. Chem.*, 2011, **83**, 8756–8765.
57. S. Zhang, M. Hosaka, T. Yoshihara, K. Negishi, Y. Iida, S. Tobita and T. Takeuchi, *Cancer Res.*, 2010, **70**, 4490–4498.
58. L. Bruens, S. I. J. Ellenbroek, J. van Rheenen and H. J. Snippert, *Gastroenterology*, 2017, **153**, 674–677.e3.
59. J. R. W. Conway, S. C. Warren and P. Timpson, *Methods*, 2017, **128**, 78–94.
60. V. Tsytsarev, E. Pumbo, S. Borisov, H. Arakawa, R. S. Erzurumlu and D. B. Papkovsky, *J. Neurosci. Methods*, 2013, **216**, 146–151.
61. M. Lukina, A. Orlova, M. Shirmanova, D. Shirokov, A. Pavlikov, A. Neubauer, H. Studier, W. Becker, E. Zagaynova and T. Yoshihara, *Opt. Lett.*, 2017, **42**, 731–734.
62. R. I. Dmitriev, A. V. Zhdanov, Y. M. Nolan and D. B. Papkovsky, *Biomaterials*, 2013, **34**, 9307–9317.
63. R. I. Dmitriev, S. M. Borisov, J. Jenkins and D. B. Papkovsky, *Proc. SPIE*, 2015, 9328.
64. R. I. Dmitriev, I. A. Okkelman, T. Foley and D. B. Papkovsky, *FASEB J.*, 2017, **31**, 590.591.
65. I. A. Okkelman, T. Foley, D. B. Papkovsky and R. I. Dmitriev, *Biomaterials*, 2017, **146**, 86–96.
66. I. A. Okkelman, T. Foley, D. B. Papkovsky and R. I. Dmitriev, *Multi-parametric Live Cell Microscopy of 3D Tissue Models*, Springer, 2017, pp. 85–103, DOI: 10.1007/978-3-319-67358-5_6.
67. N. O'Donnell and R. I. Dmitriev, *Multi-parametric Live Cell Microscopy of 3D Tissue Models*, Springer, 2017, pp. 49–67, DOI: 10.1007/978-3-319-67358-5_4.

68. A. Zhdanov, C. Favre, L. O'Flaherty, J. Adam, R. O'Connor, P. Pollard and D. Papkovsky, *Integr. Biol.*, 2011, **3**, 1135–1142.
69. R. I. Dmitriev, A. V. Zhdanov, G. Jasionek and D. B. Papkovsky, *Anal. Chem.*, 2012, **84**, 2930–2938.
70. A. V. Zhdanov, M. W. Ward, J. H. Prehn and D. B. Papkovsky, *J. Biol. Chem.*, 2008, **283**, 5650–5661.
71. A. V. Zhdanov, R. I. Dmitriev, J. Hynes and D. B. Papkovsky, *Methods Enzymol.*, 2014, **542**, 183–207.
72. P. Lehner, C. Staudinger, S. M. Borisov, J. Regensburger and I. Klimant, *Chem.–Eur. J.*, 2015, **21**, 3978–3986.
73. H. Kurokawa, H. Ito, M. Inoue, K. Tabata, Y. Sato, K. Yamagata, S. Kizaka-Kondoh, T. Kadonosono, S. Yano and M. Inoue, *Sci. Rep.*, 2015, **5**, 10657.
74. S. Kalinina, J. Breymayer, P. Schäfer, E. Calzia, V. Shcheslavskiy, W. Becker and A. Rück, *J. Biophotonics*, 2016, **9**, 800–811.
75. T. Saito, N. Asakura, T. Kamachi and I. Okura, *J. Porphyrins Phthalocyanines*, 2007, **11**, 160–164.
76. S. Sakadzic, E. Roussakis, M. A. Yaseen, E. T. Mandeville, V. J. Srinivasan, K. Arai, S. Ruvinskaya, A. Devor, E. H. Lo, S. A. Vinogradov and D. A. Boas, *Nat. Methods*, 2010, **7**, 755–759.
77. U. Neugebauer, Y. Pellegrin, M. Devocelle, R. J. Forster, W. Signac, N. Moran and T. E. Keyes, *Chem. Commun.*, 2008, 5307–5309.
78. R. I. Dmitriev, A. V. Zhdanov, G. V. Ponomarev, D. V. Yashunski and D. B. Papkovsky, *Anal. Biochem.*, 2010, **398**, 24–33.
79. D. Papkovsky and R. Dmitriev, US Patent App. 13/143,309, 2009.
80. R. I. Dmitriev, A. V. Kondrashina, K. Koren, I. Klimant, A. V. Zhdanov, J. M. Pakan, K. W. McDermott and D. B. Papkovsky, *Biomater. Sci.*, 2014, **2**, 853–866.
81. A. Zhdanov, J. Hynes, R. Dmitriev and D. Papkovsky, *Phosphorescent Oxygen-sensitive Probes*, Springer Basel, 2012, pp. 29–69.
82. A. V. Zhdanov, R. I. Dmitriev, A. V. Golubeva, S. A. Gavrilova and D. B. Papkovsky, *BBA, Biochim. Biophys. Acta, Gen. Subj.*, 2013, **1830**, 3553–3569.
83. A. V. Zhdanov, R. I. Dmitriev and D. B. Papkovsky, *Cell. Mol. Life Sci.*, 2011, **68**, 903–917.
84. A. V. Zhdanov, A. V. Golubeva, I. A. Okkelman, J. F. Cryan and D. Papkovsky, *Am. J. Physiol.: Cell Physiol.*, 2015, **309**, C501–C509.
85. A. V. Zhdanov, I. A. Okkelman, F. W. Collins, S. Melgar and D. B. Papkovsky, *BBA, Biochim. Biophys. Acta, Bioenerg.*, 2015, **1847**, 1254–1266.
86. A. V. Zhdanov, I. A. Okkelman, A. V. Golubeva, B. Doerr, N. P. Hyland, S. Melgar, F. Shanahan, J. F. Cryan and D. B. Papkovsky, *CMLS, Cell. Mol. Life Sci.*, 2017, **74**, 141–151.
87. A. Martin, A. Byrne, C. S. Burke, R. J. Forster and T. E. Keyes, *J. Am. Chem. Soc.*, 2014, **136**, 15300–15309.
88. L. Blackmore, R. Moriarty, C. Dolan, K. Adamson, R. J. Forster, M. Devocelle and T. E. Keyes, *Chem. Commun.*, 2013, **49**, 2658–2660.

89. R. I. Dmitriev, H. Ropiak, G. Ponomarev, D. V. Yashunsky and D. B. Papkovsky, *Bioconjugate Chem.*, 2011, **22**, 2507–2518.
90. R. I. Dmitriev, H. M. Ropiak, D. V. Yashunsky, G. V. Ponomarev, A. V. Zhdanov and D. B. Papkovsky, *FEBS J.*, 2010, **277**, 4651–4661.
91. K. Koren, R. I. Dmitriev, S. M. Borisov, D. B. Papkovsky and I. Klimant, *ChemBioChem*, 2012, **13**, 1184–1190.
92. I. Nakase, T. Takeuchi, G. Tanaka and S. Futaki, *Adv. Drug Delivery Rev.*, 2008, **60**, 598–607.
93. I. A. Okkelman, R. I. Dmitriev, T. Foley and D. B. Papkovsky, *PLoS One*, 2016, **11**, e0167385.
94. A. J. Nichols, E. Roussakis, O. J. Klein and C. L. Evans, *Angew. Chem., Int. Ed.*, 2014, **53**, 3671–3674.
95. T. Murase, T. Yoshihara and S. Tobita, *Chem. Lett.*, 2012, **41**, 262–263.
96. T. Yoshihara, S. Murayama, T. Masuda, T. Kikuchi, K. Yoshida, M. Hosaka and S. Tobita, *J. Photochem. Photobiol., A*, 2015, **299**, 172–182.
97. T. Yoshihara, Y. Yamaguchi, M. Hosaka, T. Takeuchi and S. Tobita, *Angew. Chem., Int. Ed.*, 2012, **51**, 4148–4151.
98. D. Hara, H. Komatsu, A. Son, S.-i. Nishimoto and K. Tanabe, *Bioconjugate Chem.*, 2015, **26**, 645–649.
99. Y. Hirakawa, T. Yoshihara, M. Kamiya, I. Mimura, D. Fujikura, T. Masuda, R. Kikuchi, I. Takahashi, Y. Urano and S. Tobita, *Sci. Rep.*, 2015, **5**, 17838.
100. Y. Fan, J. Zhao, Q. Yan, P. R. Chen and D. Zhao, *ACS Appl. Mater. Interfaces*, 2014, **6**, 3122–3131.
101. R. I. Dmitriev, N. O'Donnell and D. B. Papkovsky, *Bioconjugate Chem.*, 2016, **27**, 439–445.
102. S. M. Borisov, T. Mayr, G. Mistlberger, K. Waich, K. Koren, P. Chojnacki and I. Klimant, *Talanta*, 2009, **79**, 1322–1330.
103. A. Pfister, G. Zhang, J. Zareno, A. F. Horwitz and C. L. Fraser, *ACS Nano*, 2008, **2**, 1252–1258.
104. M. P. Coogan, J. B. Court, V. L. Gray, A. J. Hayes, S. H. Lloyd, C. O. Millet, S. J. A. Pope and D. Lloyd, *Photochem. Photobiol. Sci.*, 2010, **9**, 103–109.
105. A. Fercher, S. M. Borisov, A. V. Zhdanov, I. Klimant and D. B. Papkovsky, *ACS Nano*, 2011, **5**, 5499–5508.
106. H. Liu, H. Yang, X. Hao, H. Xu, Y. Lv, D. Xiao, H. Wang and Z. Tian, *Small*, 2013, **9**, 2639–2648.
107. R. I. Dmitriev, S. M. Borisov, A. V. Kondrashina, M. Janelle, P. Pakan, U. Anilkumar, H. Jochen, M. Prehn, A. V. Zhdanov and K. W. Mcdermott, *Cell. Mol. Life Sci.*, 2015, 72, 367.
108. X.-H. Wang, H.-S. Peng, Z. Chang, L.-L. Hou, F.-T. You, F. Teng, H.-W. Song and B. Dong, *Microchim. Acta*, 2012, 1–6.
109. X.-d. Wang, D. E. Achatz, C. Hupf, M. Sperber, J. Wegener, S. Bange, J. M. Lupton and O. S. Wolfbeis, *Sens. Actuators, B*, 2013, **188**, 257–262.
110. T. V. Esipova, H. J. Rivera-Jacquez, B. Weber, A. E. Masunov and S. A. Vinogradov, *J. Phys. Chem. A*, 2017, **121**, 6243–6255.

111. C. Wu, B. Bull, K. Christensen and J. McNeill, *Angew. Chem., Int. Ed.*, 2009, **48**, 2741–2745.
112. X.-d. Wang, H. H. Gorris, J. A. Stolwijk, R. J. Meier, D. B. M. Groegel, J. Wegener and O. S. Wolfbeis, *Chem. Sci.*, 2011, **2**, 901–906.
113. J. Napp, T. Behnke, L. Fischer, C. Würth, M. Wottawa, D. M. Katschinski, F. Alves, U. Resch-Genger and M. Schäferling, *Anal. Chem.*, 2011, **83**, 9039–9046.
114. A. V. Kondrashina, R. I. Dmitriev, S. M. Borisov, I. Klimant, I. O'Brien, Y. M. Nolan, A. V. Zhdanov and D. B. Papkovsky, *Adv. Funct. Mater.*, 2012, **22**, 4931–4939.
115. R. I. Dmitriev, S. M. Borisov, H. Düssmann, S. Sun, B. J. Müller, J. Prehn, V. P. Baklaushev, I. Klimant and D. B. Papkovsky, *ACS Nano*, 2015, **9**, 5275–5288.
116. Z. Feng, P. Tao, L. Zou, P. Gao, Y. Liu, X. Liu, H. Wang, S. Liu, Q. Dong, J. Li, B. Xu, W. Huang, R. W. Y. Wong and Q. Zhao, *ACS Appl. Mater. Interfaces*, 2017, **9**, 28319–28330.
117. Y.-E. Koo Lee, R. Smith and R. Kopelman, *Annu. Rev. Anal. Chem.*, 2009, **2**, 57–76.
118. Y.-E. Koo Lee, E. E. Ulbrich, G. Kim, H. Hah, C. Strollo, W. Fan, R. Gurjar, S. Koo and R. Kopelman, *Anal. Chem.*, 2010, **82**, 8446–8455.
119. X.-H. Wang, H.-S. Peng, W. Yang, Z.-D. Ren and Y.-A. Liu, *Microchim. Acta*, 2016, **10**, 2723–2731.
120. X. H. Wang, H. S. Peng, L. Yang, F. T. You, F. Teng, L. L. Hou and O. S. Wolfbeis, *Angew. Chem.*, 2014, **126**, 12679–12683.
121. X. Zheng, H. Tang, C. Xie, J. Zhang, W. Wu and X. Jiang, *Angew. Chem., Int. Ed.*, 2015, **54**, 8094–8099.
122. W. R. Algar, D. E. Prasuhn, M. H. Stewart, T. L. Jennings, J. B. Blanco-Canosa, P. E. Dawson and I. L. Medintz, *Bioconjugate Chem.*, 2011, **22**, 825–858.
123. H. Xiang, L. Zhou, Y. Feng, J. Cheng, D. Wu and X. Zhou, *Inorg. Chem.*, 2012, **51**, 5208–5212.
124. H. Shi, X. Ma, Q. Zhao, B. Liu, Q. Qu, Z. An, Y. Zhao and W. Huang, *Adv. Funct. Mater.*, 2014, **24**, 4823–4830.
125. Q. Zhao, X. Zhou, T. Cao, K. Y. Zhang, L. Yang, S. Liu, H. Liang, H. Yang, F. Li and W. Huang, *Chem. Sci.*, 2015, **6**, 1825–1831.
126. A. E. Nel, L. Mädler, D. Velegol, T. Xia, E. M. Hoek, P. Somasundaran, F. Klaessig, V. Castranova and M. Thompson, *Nat. Mater.*, 2009, **8**, 543.
127. R. Xu, Y. Wang, X. Duan, K. Lu, D. Micheroni, A. Hu and W. Lin, *J. Am. Chem. Soc.*, 2016, **138**, 2158–2161.
128. J. Jenkins, S. M. Borisov, D. B. Papkovsky and R. I. Dmitriev, *Anal. Chem.*, 2016, **88**, 10566–10572.
129. J. Jenkins, D. B. Papkovsky and R. I. Dmitriev, *Biochem. J.*, 2016, **473**, 2507–2518.
130. J. Jenkins, R. I. Dmitriev, K. Morten, K. W. McDermott and D. B. Papkovsky, *Acta Biomater.*, 2015, **16**, 126–135.

131. G. Yazgan, R. I. Dmitriev, V. Tyagi, J. Jenkins, G.-M. Rotaru, M. Rottmar, R. M. Rossi, C. Toncelli, D. B. Papkovsky and K. Maniura-Weber, *Sci. Rep.*, 2017, **7**, 158.
132. P. G. Koolen, Z. Li, E. Roussakis, M. A. Paul, A. M. Ibrahim, R. Matyal, T. Huang, C. L. Evans and S. J. Lin, *Plast. Reconstr. Surg.*, 2017, **140**, 89–96.

CHAPTER 4

# *Hydrophilic Ir(III) Complexes for* In vitro *and* In vivo *Oxygen Imaging*

TOSHITADA YOSHIHARA[a], YOSUKE HIRAKAWA[b], MASAOMI NANGAKU[b] AND SEIJI TOBITA*[a]

[a]Gunma University, Department of Chemistry and Chemical Biology, Kiryu, Gunma 376-8515, Japan; [b]The University of Tokyo, Graduate School of Medicine, Hongo, Bunkyo-ku, Tokyo 113-8655, Japan
*E-mail: tobita@gunma-u.ac.jp

## 4.1 Introduction

Luminescence-based imaging technology is developing as a powerful tool for investigating the real-time behaviour of biologically important substances in live cells and animals. The key elements of this technology are suitable luminescent probes, either as small molecules or nanoparticles, as well as detection devices such as various luminescence microscopes and *in vivo* imaging systems. Of the multitude of biologically important materials, oxygen is of particular interest[1–4] because it plays indispensable roles in living organisms, and oxygen deprivation is related to the pathophysiology of various hypoxia-related diseases.[5–7]

For the past ten years we have been developing molecular oxygen probes based on Ir(III) complexes which give phosphorescence with tunable wavelengths and high quantum yield under deoxygenated conditions.[8]

Detection Science Series No. 11
Quenched-phosphorescence Detection of Molecular Oxygen: Applications in Life Sciences
Edited by Dmitri B. Papkovsky and Ruslan I. Dmitriev

Published by the Royal Society of Chemistry, www.rsc.org

Hypoxic tumours of living mice could be imaged by using the red-emitting Ir(III) complex BTP, $(btp)_2Ir(acac)$ (btp = benzothienylpyridine, acac = acetylacetone) which gives brighter emission in hypoxic tissues than in normal tissues.[9,10] An important advantage of using Ir(III) complexes as oxygen probes is the ease of controlling their photophysical properties and cellular properties by appropriately designing the structures of the ligands. The near IR-emitting complex BTPHSA, a COOH derivative of BTPH (Figure 4.1), allowed the visualization of tumours transplanted 6 to 7 mm below the skin surface.[10] Depending on the ligand structure, Ir(III) complexes show different intracellular localizations, which allows organelle-specific oxygen measurements. For example, mitochondria-targeted oxygen probes have been developed by introducing a triphenylphosphonium (TPP) cation into the acac ligand of BTP.[11] Cationic Ir(III) complexes with a 5-amino-1,10-phenanthroline ligand showed mitochondrial accumulation and exhibited oxygen-sensitive emission in living cells.[12] We also synthesized a series of BTP derivatives that possess a modified acac ligand, and found that the cationic BTP derivative BTPDM1, which has a dimethylamino group on the acac ligand (Figure 4.1), shows very high cellular uptake efficiency and

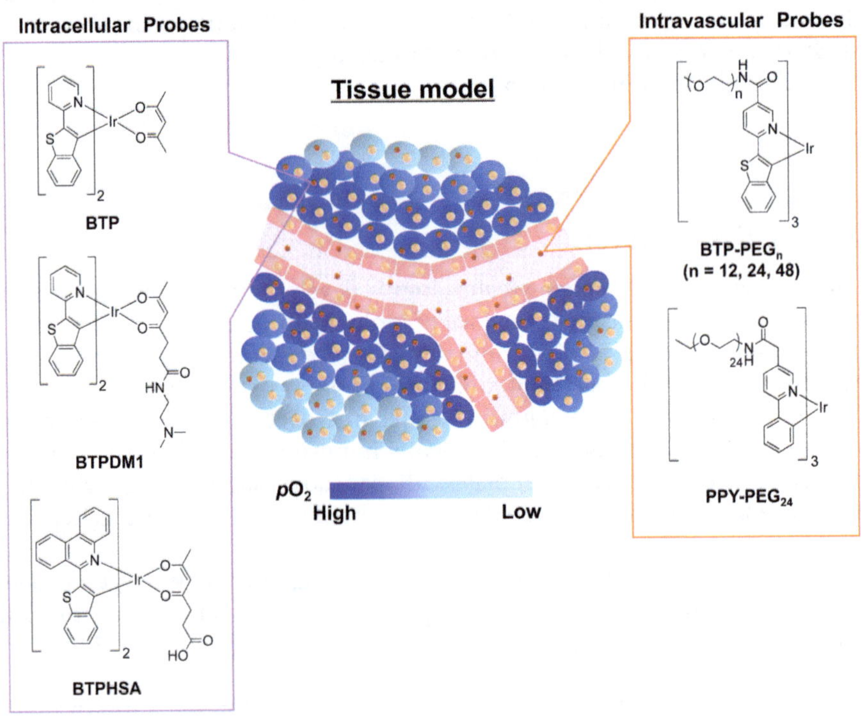


**Figure 4.1** Structures of the Ir(III) complexes developed for intracellular $O_2$ imaging, and of the hydrophilic Ir(III) complexes BTP-PEG$_n$ ($n$ = 12, 24, 48) and PPY-PEG$_{24}$ synthesized in the present work for intravascular $O_2$ imaging.

excellent oxygen sensitivity.[13] Using BTPDM1 as an oxygen probe, we could evaluate the oxygen gradients in monolayer cultured cells by phosphorescence lifetime measurements.[13] Furthermore, we demonstrated that the oxygen status of mouse kidney can be measured by intravenous administration of BTPDM1 followed by lifetime measurements using a bifurcated fibre system.[14] Phosphorescence microscopic images of frozen sections of kidney showed that lipophilic BTPDM1 distributed inside tubular cells after intravenous administration.

Fluorescence and phosphorescence lifetime images of the kidney surface of a BTPDM1-administered mouse showed that BTPDM1 is accumulated in tubules, and its phosphorescence allowed visualization of the oxygen gradient in tubular cells *in vivo*. To reveal the oxygen distribution in specific organs and to clarify hypoxic lesions, it is necessary to stain tissues using oxygen probes that accumulate in different compartments, thereby allowing evaluation of the oxygen status of whole tissues. To this end, here we designed and synthesized hydrophilic BTP analogues bearing a biocompatible hydrophilic unit, polyethylene glycol (PEG), in the benzothienyl pyridinato ligand.

The design concept of our Ir(III) complex probes is schematically shown in Figure 4.1. Following intravenous administration in mice, lipophilic Ir(III) complexes such as BTP, BTPDM1, and BTPHSA can penetrate blood vessel walls and provide information on the intracellular oxygen status of tissues. In contrast, the newly-developed hydrophilic Ir(III) complexes derivatised with PEG chains are expected to remain in vasculature and interstitial space in organs, and also urinary space in kidney, thereby providing insights into the oxygen status of blood and urine. We synthesized Ir(III) complexes with different PEG chain lengths, BTP-PEG$_n$ ($n$ = 12, 24 and 48) in order to understand the effect of PEG length on the hydrophilicity and intravital properties of the complexes. In addition, we synthesized a green light-emitting Ir(III) complex, PPY-PEG$_{24}$, to examine the possibility of multicolour oxygen measurements.

## 4.2 Experimental

### 4.2.1 Synthesis of BTP-PEG$_n$ ($n$ = 12, 24, 48) and PPY-PEG$_{24}$

A mixture of benzo[*b*]thiophene-2-ylboronic acid (1.17 g, 6.6 mmol), 6-bromonicoyinic acid (1.38 g, 6.8 mmol), tetrakis(triphenylphosphine) palladium(0) (0.3 g, 0.26 mmol), toluene (20 ml), ethanol (20 ml), and an aqueous solution of sodium carbonate (2 M, 20 ml) was refluxed under a nitrogen atmosphere for 6 h. After cooling, the solution was acidified by the addition of conc. hydrochloric acid. The formed precipitate was filtered to give a crude white product. The product (1.39 g) was dissolved in anhydrous ethanol (100 ml), then conc. sulphuric acid (2 ml) was added. The solution was refluxed for 5 h. After cooling, some of the ethanol was evaporated under reduced pressure. The concentrated solution was added to chloroform, then washed with distilled water. The organic solution was dried over sodium sulphate and evaporated to dryness under reduced pressure. The residual oil was purified by silica-gel column chromatography using chloroform as eluent. The product, (ethyl 6-(benzo[*b*] thiophen-2-yl)nicotinate, btp-COOEt), was obtained as a white powder (1.45 g, 5.1 mmol, 77%).

A mixture of btp-COOEt (2.42 g, 8.5 mmol), $IrCl_3{\cdot}3H_2O$ (1.47 g, 4.2 mmol), 2-ethoxyethanol (130 ml), and distilled water (40 ml) was refluxed for 15 h. After cooling, the solvent was evaporated to give a chloro-bridged dimer. The chloro-bridged dimer (796 mg, 0.50 mmol) and btp-COOEt (425 mg, 1.5 mmol) were dissolved in 2-ethoxyethanol (30 ml), then the solution was purged with nitrogen gas and silver trifluoroacetate (354 mg, 1.60 mmol) was quickly added. The solution was heated at 110 °C for 24 h, then evaporated under reduced pressure. The brown solid was dissolved in chloroform and excess silver trifluoroacetate was removed. The crude product was purified by silica-gel column chromatography using chloroform as eluent. Further purification was performed by using a recycling preparative HPLC (LC-9225 NEXT, Japan Analytical Industry) to remove btp-COOEt. The product (BTP-COOEt: ethyl ester of BTP-COOH (see Scheme 4.1)) was obtained as a red powder (416 mg, 0.40 mmol, 40%).

BTP-COOEt (108 mg, 0.10 mmol) was dissolved in tetrahydrofuran (9 ml) and ethanol (3 ml), then lithium hydroxide monohydrate (42 mg, 1.0 mmol) was added. The solution was stirred for 2 h, then distilled water (9 ml) was added. The mixed solution was heated at 80 °C for 24 h. After cooling, the solution was acidified by the addition of conc. hydrochloric acid. The solution was added to ethyl acetate, then washed with distilled water. The organic solution was dried over sodium sulphate and evaporated to dryness under

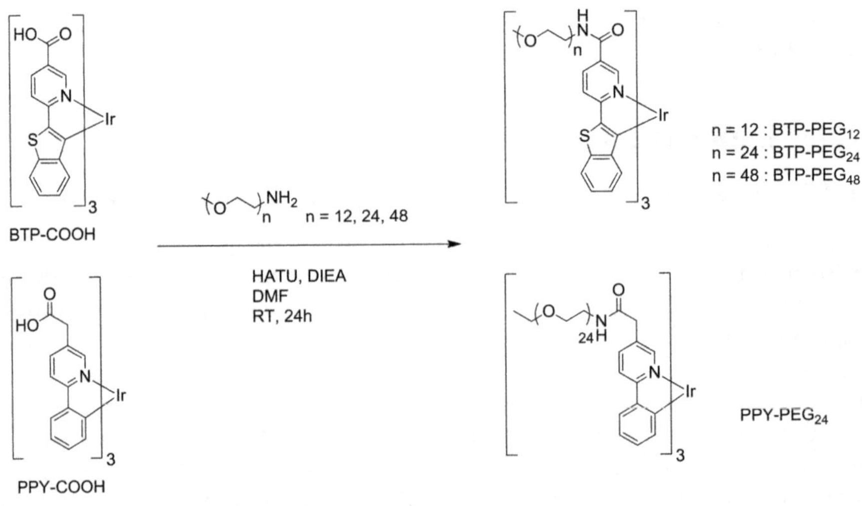


**Scheme 4.1** Condensation reaction between phosphors and PEGylation reagents.

reduced pressure. The product (BTP-COOH) was obtained as a red powder (94 mg, 0.098 mmol, 98%).

2-(6-Chloropyridin-3-yl)acetic acid (5.0 g, 29.1 mmol) was dissolved in anhydrous ethanol (100 ml), then conc. sulphuric acid (3 ml) was added. The solution was refluxed for 5 h. After cooling, some of the ethanol was evaporated under reduced pressure. The concentrated solution was added to chloroform, then washed with distilled water. The organic solution was dried over sodium sulphate and evaporated to dryness under reduced pressure. The residual oil was purified by silica-gel column chromatography using chloroform/methanol (98 : 2, v/v) as eluent. A mixture of the obtained product (5.8 g, 29.1 mmol), phenylboronic acid (3.6 g, 29.5 mmol), tetrakis(triphenylphosphine)palladium(0) (1.1 g, 0.95 mmol), toluene (60 ml), ethanol (20 ml), and an aqueous solution of sodium carbonate (2 M, 40 ml) was refluxed under a nitrogen atmosphere for 6 h. After cooling, the solution was added to chloroform, then washed with distilled water. The organic solution was dried over sodium sulphate and evaporated to dryness under reduced pressure. The residual oil was purified by silica-gel column chromatography using chloroform as eluent. The product (ethyl 2-(6-phenylpyridin-3-yl)acetate, ppy-COOEt) was obtained as a colourless oil (4.9 g, 20.0 mmol, 69%).

A mixture of ppy-COOEt (530 mg, 2.2 mmol), $IrCl_3 \cdot 3H_2O$ (350 mg, 1.0 mmol), 2-ethoxyethanol (30 ml), and distilled water (10 ml) was refluxed for 15 h. After cooling, the solvent was evaporated and chloro-bridged dimers produced by transesterification were obtained. These chloro-bridged dimers (1.13 g) and ppy-COOEt (580 mg, 2.4 mmol) were dissolved in 2-ethoxyethanol (50 ml), then the solution was purged with nitrogen gas and silver trifluoroacetate (354 mg, 1.60 mmol) was quickly added. The solution was heated at 110 °C for 24 h and evaporated under reduced pressure. The brown solid residue was dissolved in chloroform and excess silver trifluoroacetate was removed. The crude products were purified by silica-gel column chromatography using chloroform/methanol (98 : 5, v/v) as eluent. Further purification was performed by using a recycling preparative HPLC (LC-9225 NEXT) to remove ppy-COOEt. The obtained products (400 mg) were dissolved in tetrahydrofuran (30 ml) and ethanol (10 ml), then lithium hydroxide monohydrate (160 mg, 3.8 mmol) was added to the solution. The solution was stirred for 2 h, then distilled water (40 ml) was added. The mixed solution was heated at 80 °C for 24 h. After cooling, the solution was acidified by the addition of conc. hydrochloric acid. The solution was added to ethyl acetate, then washed with distilled water. The organic solution was dried over sodium sulphate and evaporated to dryness under reduced pressure. The product (PPY-COOH) was obtained as a yellow powder (340 mg, 0.41 mmol).

BTP-$PEG_n$ ($n$ = 12, 24, 48) and PPY-$PEG_{24}$ were produced by a condensation reaction, as shown in Scheme 4.1.

BTP-COOH (41 mg, 0.043 mmol) and *O*-(7-azabenzotriazol-1-yl)-*N*,*N*,*N*′,*N*′-tetramethyluronium hexafluorophosphate (HATU, 100 mg, 0.26 mmol) were dissolved in *N*,*N*-dimethylformamide (DMF, 3 ml), then *N*,*N*-diisopropylethylamine (DIEA, 174 µl, 1.0 mmol) and m-dPEG®$_n$-amine ($n = 12$ (100 mg, 0.179 mmol), 24 (200 mg, 0.184 mmol), or 48 (300 mg, 0.140 mmol), Quanta BioDesign) was added to the solution, then the solution was stirred for 24 h at room temperature under a nitrogen atmosphere. The crude product was purified by amine modified silica-gel column chromatography using chloroform:methanol (99 : 1, v/v) as eluent. BTP-PEG$_n$ ($n = 12, 24, 48$)) was obtained as a red oil (BTP-PEG$_{12}$, 70% and BTP-PEG$_{24}$, 65%) or a red solid (BTP-PEG$_{48}$, 80%).

PPY-COOH (50 mg, 0.06 mmol) and HATU (100 mg, 0.26 mmol) were dissolved in DMF (3 ml), then DIEA (174 µl, 1.0 mmol) and m-dPEG®$_{24}$-amine (200 mg, 0.184 mmol) were added and the solution was stirred for 24 h at room temperature. The crude product was purified by amine modified silica-gel column chromatography using chloroform:methanol (99 : 1, v/v) as eluent. PPY-PEG$_{24}$ were obtained as a yellow solid (49%).

### 4.2.2 Photophysical Properties in Aqueous Solutions

Absorption and emission spectra were recorded on a UV/vis spectrophotometer (Jasco, Ubest-V550) and a photonic multichannel analyser (Hamamatsu, C10027-01) equipped with a monochromatized Xe arc lamp, respectively. Each emission spectrum was corrected for spectral sensitivity. Phosphorescence lifetimes of the Ir(III) complexes in solution were measured with a time-correlated single-photon counting fluorometer (Hamamatsu, Quantaurus-Tau C11367). Phosphorescence quantum yield was measured with an absolute photoluminescence quantum yield system (Hamamatsu, C9920-01)[15] consisting of a Xe arc lamp, a monochromator, an integrating sphere, and a multichannel detector. Deionized water was purified using a water purification system (Merck-Millipore, Simplicity UV).

### 4.2.3 Octanol/Water Partition Coefficients ($\log P_{O/W}$)

1-Octanol saturated with water and water saturated with 1-octanol were prepared by vigorous stirring for 24 h. Each Ir(III) complex was dissolved in water saturated with 1-octanol. This stock solution was added to an equal volume of 1-octanol saturated with water, and the mixture was shaken for 20 min. Then, the water layer was carefully separated from the 1-octanol layer. The absorbance of the stock solution ($A_S$) and the aqueous solution ($A_W$) after partitioning was measured. The $\log P_{O/W}$ values were determined by using the following equation:

$$\log P_{O/W} = \log(C_{OC}/C_W) = \log((C_{OC}/(C_S - C_{OC})) = \log((A_S - A_W)/A_W) \qquad (4.1)$$

where $C_{OC}$, $C_W$ and $C_S$ are the concentration of the Ir(III) complex in 1-octanol and in water after partitioning, and the concentration of the Ir(III) complex in the stock solution, respectively.

### 4.2.4 Fluorescence and Phosphorescence Lifetime Imaging Microscopy

Fluorescence and phosphorescence lifetime imaging microscopy (FLIM/PLIM) images were acquired using an inverted fluorescence microscope (Olympus, IX73) equipped with a confocal scanning system (Becker & Hickl GmbH, DCS-120). A ×40 oil immersion objective (Olympus, UPLFLN; NA 1.3) was used for *in vitro* and *in vivo* imaging. FITC-conjugated *Lycopersicon esculentum* lectin, BTP-PEG$_{48}$, and PPY-PEG$_{24}$ were excited at 488 nm using a picosecond diode laser (Becker & Hickl GmbH, BDL-SMC, pulse width: 40–90 ps, repetition rate: 50 MHz). The emission, guided through a 1 mm pinhole, was collected at >590 nm (BTP-PEG$_{48}$) and 510–560 nm (FITC-lectin and PPY-PEG$_{24}$) and detected with a hybrid detector module (Becker & Hickl GmbH, HPM-100-40). Time-resolved emission measurements were performed using a time-correlated single-photon counting (TCSPC) unit (Becker & Hickl GmbH, Simple-Tau 150-DX). FLIM and PLIM images were analysed using SPCImage data analysis software (Becker & Hickl GmbH). The obtained decay data were fitted to a single exponential decay function, unless otherwise stated.

### 4.2.5 Cell Culture and Imaging

Human colorectal adenocarcinoma cells (HT-29) were cultured in McCoy's 5A medium (Gibco). To the medium, 10% foetal bovine serum (FBS), penicillin (50 units ml$^{-1}$), and streptomycin (50 μg ml$^{-1}$) were added. All cells were grown at 37 °C under a 5% $CO_2$ atmosphere. For the intensity imaging of living cells, HT-29 cells were seeded on glass bottom dishes, then incubated with one of the Ir(III) complexes (5 μM) for 2 h. The medium was changed to McCoy's 5A without FBS. An inverted microscope (Olympus IX71) equipped with a ×40 oil-immersion objective lens and an electron multiplying CCD camera (PHOTOMETRICS, Evolve 512) driven by MetaMorph software was used to obtain luminescence microscopic images. For cellular lifetime imaging experiments, HT-29 cells were cultured on glass-base dishes, and the medium was changed to McCoy's 5A with BTP-PEG$_{48}$ (100 μM). After 0.5 h, fluorescence or phosphorescence lifetime imaging was performed using a FLIM/PLIM system equipped with an $O_2$ concentration-changeable multigas incubator (Tokai Hit, INUF-LV200-MT GM-8000). Both imaging experiments were conducted in four fields per dish.

### 4.2.6 *In vivo* Imaging

All protocols for the animal experiments conducted in this study were approved by the Ethical Committee on Animal Experiments of Gunma University (14-053), and all animal experiments were conducted in accordance with the institutional guidelines of Gunma University.

Eight-week-old Balb/c male mice (Japan SLC) were used in this study. General anaesthesia was administered *via* a 40 mg kg$^{-1}$ intraperitoneal injection of pentobarbital, and an additional 8 mg kg$^{-1}$ pentobarbital was added if necessary. Subsequently, a flank incision was made, and the left kidney was exposed. The renal capsule was not removed. FITC-lectin (125 μg/mouse) and BTP-PEG$_{48}$ (250 nmol per mouse) were administered *via* tail vein injection. Then, the mouse was turned with its left kidney on a cover glass chamber (Asahi Glass, IWAKI 5202-001). We checked whether there was any movement on the renal surface caused by respiration. Then, fluorescence or phosphorescence lifetime imaging was performed using the FLIM/PLIM system. Both imaging experiments were conducted in four fields per mouse.

## 4.3 Results and Discussion

### 4.3.1 Electronic Structures of the Ir(III) Complexes

The lowest excited singlet and triplet states of Ir(III) complexes have two electronic characters. One is of the metal-to-ligand charge transfer (MLCT) state, which arises from the promotion of an electron from a metal dπ orbital to a vacant π* orbital. The second is the ligand-centred (LC) state, which arises from the promotion of an electron from the π orbital to the π* orbital in the ligand (Figure 4.2a).[16] In Ir(III) complexes, non-emissive metal-centred (MC) states arising from promotion of an electron from a metal dπ orbital to the dσ* orbital are usually located at higher energies

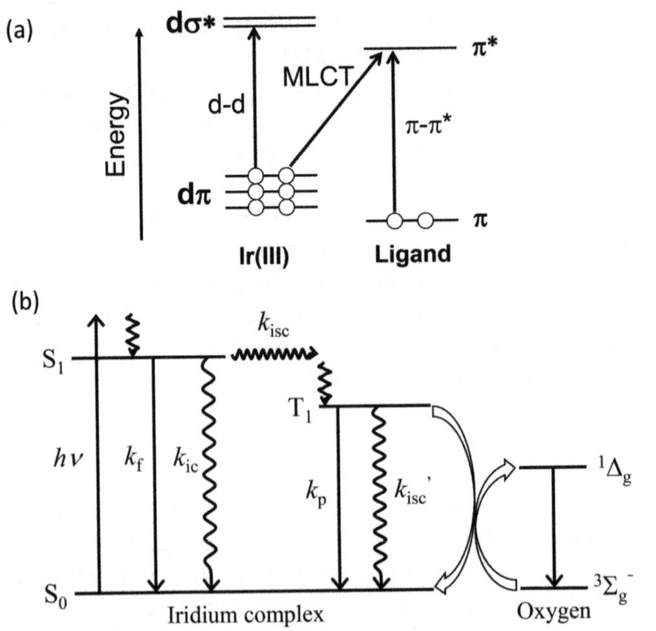


**Figure 4.2** Electronic transitions (a) and relaxation processes (b) of Ir(III) complexes.

relative to the MLCT and MC states. As a result, the lowest excited singlet ($S_1$) and triplet ($T_1$) states of Ir(III) complexes have the characters of the singlet MLCT ($^1$MLCT) and triplet MLCT ($^3$MLCT) states, respectively, with significant contribution from the triplet ligand-centred ($^3$LC) ππ* state.[17] Because of the extremely strong heavy atom effect from the central Ir(III) ion (spin–orbit coupling constant $\zeta = 3909\ cm^{-1}$),[18] the rate constants for the $S_1 \rightarrow T_1$ and $T_1 \rightarrow S_0$ intersystem crossing ($k_{isc}$ and $k_{isc}'$) and the $T_1 \rightarrow S_0$ radiative transition ($k_p$) of Ir(III) complexes are much larger than is typical for organic molecules.

The mixing ratio of the $^3$MLCT and $^3$LC character in the phosphorescent state depends on the ligand structure: BTP has the $T_1$ state with mixed $^3$MLCT and $^3$LC character, while the $T_1$ state of Ir(ppy)$_3$ has predominantly $^3$MLCT character.[17] The lifetime of the $T_1$ state (phosphorescence lifetime), which is related to oxygen sensitivity, tends to become longer with an increased contribution from the $^3$LC state. An increase in the contribution from the $^3$MLCT state results in an increase in the phosphorescence rate constant ($k_p$), as described in the next section.

Phosphorescence quenching by molecular oxygen is mainly caused by bimolecular energy transfer from the $T_1$ state of the Ir(III) complex to the lower-lying excited states ($^1\Sigma_g^+$ and $^1\Delta_g$) of $O_2$ (Figure 4.2b). In the case of Ir(III) complexes which have a relatively low oxidation potential and a high $T_1$ energy, such as (ppy)$_2$Ir(acac), electron (charge) transfer from the $T_1$ state of the Ir(III) complex to $O_2$ can also be involved in the quenching mechanism.[19] In such cases, the bimolecular quenching rate constant ($k_q$) is known to approach the diffusion-controlled rate constant ($k_d$) under the restriction of the spin-conservation rule.[20]

## 4.3.2 Hydrophilicity and Photophysical Properties in Water

In order to impart water solubility to BTP, we introduced PEG groups with different lengths into the btp ligand of Ir(btp)$_3$.[21] The introduction of a substituent into the ligand of an Ir(III) complex is known to shift the absorption and emission wavelengths, with the introduction of an electron-withdrawing substituent at the 5-position of the pyridine ring generally causing red shifts in the emission spectrum. Since longer-wavelength emission is preferable for *in vivo* imaging, we linked the PEG unit with the btp ligand *via* an electron-withdrawing amide group (Figure 4.1). In contrast, in PPY-PEG$_{24}$, the PEG unit was linked *via* a methylene chain to retain its green emission. The hydrophilicities of the synthesized complexes were evaluated by measuring the octanol/water partition coefficient (expressed in log $P_{O/W}$): the log $P_{O/W}$ values of BTP-COOH and PPY-COOH (Scheme 4.1), BTP-PEG$_n$ ($n$ = 12, 24 and 48) and PPY-PEG$_{24}$ were 1.2, 1.3, −0.3, −0.6, −1.2 and −0.5, respectively. The negative log $P_{O/W}$ values of BTP-PEG$_n$ and PPY-PEG$_{24}$ indicate that these complexes have high affinity to water, and that the affinity increases as the length of the PEG chain increases.

The absorption and phosphorescence spectra of BTP-PEG$_n$ are all quite similar, as shown in Figure 4.3, because differences in the PEG chain length have little influence on the electronic properties of the complexes.

The first absorption band of the BTP-PEG$_n$ compounds is around 500 nm, with a molar absorption coefficient of ~$8.0\times10^3$ $M^{-1}cm^{-1}$ at 500 nm, and all exhibit red phosphorescence with a peak at around 645 nm (Table 4.1). In contrast, the first absorption peak of PPY-PEG$_{24}$ is at a much shorter wavelength (374 nm) and the compound exhibits green phosphorescence, with a peak at 540 nm (Table 4.1). Since the absorption bands of PPY-PEG$_{24}$ extend to wavelengths over 500 nm, all the complexes shown in Figure 4.3 can be excited using a 488 nm laser diode. Co-staining of cells or tissues with green-emitting hydrophilic PPY-PEG$_{24}$ and red-emitting lipophilic BTPDM1 should allow multicolour oxygen imaging of cells and tissues.

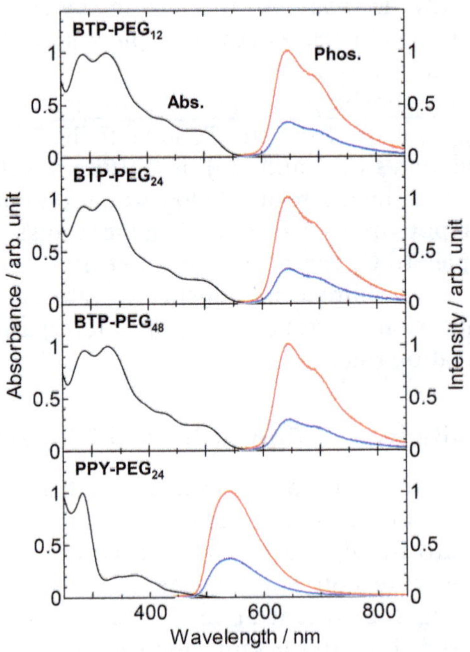


**Figure 4.3** Absorption and phosphorescence spectra of BTP-PEG$_n$ and PPY-PEG$_{24}$ in aerated (blue) and deaerated (red) $H_2O$ at room temperature.

**Table 4.1** Photophysical properties of BTP-PEG$_n$ ($n$ = 12, 24, 48) and PPY-PEG$_{24}$ in $H_2O$ at room temperature (295 K).

| Ir complex | $\lambda_{phos}{}^{max}$/nm | $\tau_p{}^0$/μs | $\tau_p$/μs | $\Phi_p{}^0$ | $\Phi_p$ | $k_p/10^4$ $s^{-1}$ | $k_{isc}'/10^5$ $s^{-1}$ |
|---|---|---|---|---|---|---|---|
| BTP-PEG$_{12}$ | 646 | 3.64 | 1.10 | 0.15 | 0.046 | 4.1 | 2.3 |
| BTP-PEG$_{24}$ | 645 | 3.96 | 1.15 | 0.16 | 0.047 | 4.0 | 2.1 |
| BTP-PEG$_{48}$ | 645 | 4.18 | 1.17 | 0.17 | 0.048 | 4.1 | 2.0 |
| PPY-PEG$_{24}$ | 540 | 1.71 | 0.57 | 0.61 | 0.20 | 35.7 | 2.3 |

The photophysical parameters of BTP-PEG$_n$ and PPY-PEG$_{24}$ in $H_2O$ at 295 K are compiled in Table 4.1. The phosphorescence lifetimes ($\tau_p^0$ and $\tau_p$) and quantum yields ($\Phi_p^0$ and $\Phi_p$) were measured for each complex in deaerated and aerated water.

The phosphorescence lifetimes of BTP-PEG$_n$ in deaerated $H_2O$ tend to decrease as the PEG chain length decreases. The shorter lifetime of BTP-PEG$_{12}$ can be attributed to the enhancement of $k_{isc}'$, which includes water-induced quenching. Indeed, the $\tau_p^0$ value (3.64 μs) of BTP-PEG$_{12}$ in $H_2O$ was significantly shorter than that in acetonitrile (7.98 μs), indicating the induction of non-radiative processes in water. In BTP-PEG$_{24}$ and BTP-PEG$_{48}$, which have longer PEG chains, it is conceivable that interactions of the BTP moiety with water molecules are sterically inhibited by the PEG unit, supressing water-induced quenching. It is also noted from Table 4.1 that the $\tau_p^0$ values of BTP-PEG$_n$ (3.6–4.2 μs) in deaerated $H_2O$ are longer than that of PPY-PEG$_{24}$ (1.71 μs), while the phosphorescence quantum yields (0.15–0.17) in deaerated $H_2O$ are much smaller than that for PPY-PEG$_{24}$ (0.61). Since the quantum yield for the $S_1 \rightarrow T_1$ intersystem crossing of Ir(III) complexes can be assumed to be unity,[22] the $T_1 \rightarrow S_0$ phosphorescence rate constant ($k_p$) and the rate constant for $T_1 \rightarrow S_0$ intersystem crossing ($k_{isc}'$) can be calculated using the relations $\Phi_p^0 = k_p\tau_p^0$ and $k_{isc}' = (\tau_p^0)^{-1} - k_p$. It can be seen from Table 4.1 that the radiative rate constant ($k_p$) of PPY-PEG$_{24}$ ($3.57 \times 10^5$ s$^{-1}$) is one order of magnitude larger than those of BTP-PEG$_n$. This is consistent with the enhanced $^3$MLCT character of Ir(ppy)$_3$.[17]

### 4.3.3 Phosphorescence Quenching by Molecular Oxygen in Solution

The phosphorescence decay curves of BTP-PEG$_n$ and PPY-PEG$_{24}$ in $H_2O$ followed single exponential functions, giving the lifetimes shown in Table 4.1, while those in the presence of human serum albumin (HSA) exhibited different features, depending on the length of the PEG chain. As can be seen from Figure 4.4, the phosphorescence time profile of BTP-PEG$_{12}$ deviated from the single exponential decay, suggesting the interaction of BTP-PEG$_{12}$ with HSA. Lipophilic Ir(III) complexes, such as BTP and BTPDM1, gives bi-exponential decays in the presence of HSA in $H_2O$, due to binding to the protein.[23] The binding of Ir(III) complexes with proteins restrains bimolecular collisions with $O_2$, thereby reducing the phosphorescence decay rate. Depending on the partition of the binding, the phosphorescence time profile becomes a bi-exponential decay. The emission decay profile of BTP-PEG$_{12}$ suggests that similar interactions with HSA are involved. The mono-exponential decays observed for BTP-PEG$_{24}$, BTP-PEG$_{48}$, and PPY-PEG$_{24}$ indicate that in these complexes, interaction with HSA is suppressed because of their highly hydrophilic character. The phosphorescence lifetimes of BTP-PEG$_n$ and PPY-PEG$_{24}$ in the presence and absence of HSA (200 μM) in $H_2O$ are listed in Table 4.2. The $\tau_p$ values of BPT-PEG$_{24}$, BTP-PEG$_{48}$, and PPY-PEG$_{24}$ in the presence of HSA are almost the same as in the absence of HSA in $H_2O$, further indicating that

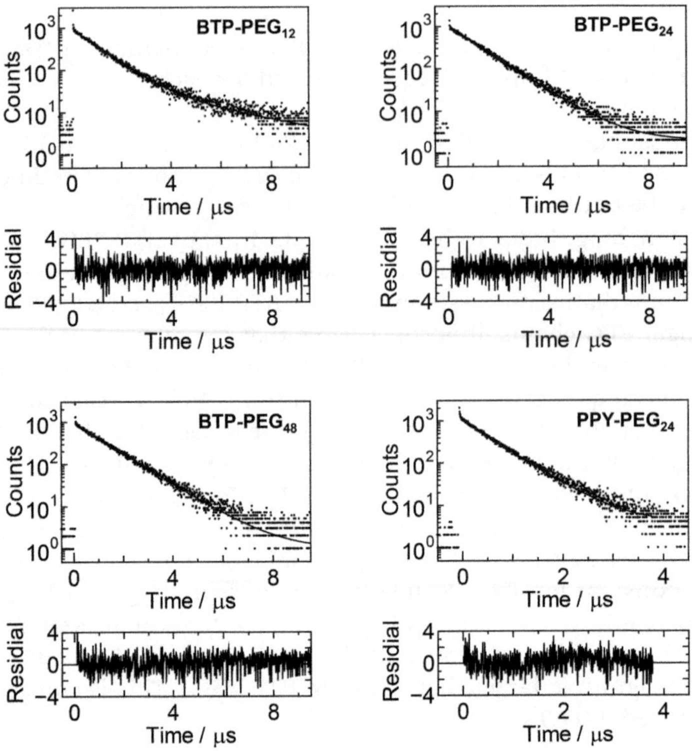


**Figure 4.4** Phosphorescence decay curves of BTP-PEG$_n$ and PPY-PEG$_{24}$ in the presence of HSA in $H_2O$.

**Table 4.2** Phosphorescence lifetimes of BTP-PEG$_n$ and PPY-PEG$_{24}$ in the presence and absence of HSA in $H_2O$.

| | Without HSA | With HSA | |
|---|---|---|---|
| Ir complex | $\tau_p$/μs | $\tau_{p1}$/μs | $\tau_{p2}$/μs |
| BTP-PEG$_{12}$ | 1.10 | 0.96 (93%) | 3.55 (7%) |
| BTP-PEG$_{24}$ | 1.15 | 1.13 | |
| BTP-PEG$_{48}$ | 1.17 | 1.19 | |
| PPY-PEG$_{24}$ | 0.57 | 0.58 | |

interaction with HSA is suppressed in these complexes. We have also found a bi-exponential decay for a BTP derivative with a single PEG$_{24}$ chain ((btp)$_2$Ir-(acac-PEG$_{24}$)) in the presence of HSA in $H_2O$. It is therefore conceivable that the three long PEG chains extending in different directions in BTP-PEG$_n$ and PPY-PEG$_{24}$ effectively prevent interaction of the complexes with HSA.

The Stern–Volmer plots of $\tau_p{}^0/\tau_p$ *vs.* $pO_2$ for BTP-PEG$_{48}$ in McCoy's 5A and in FBS at 37 °C are presented in Figure 4.5. These plots were made on BTP-PEG$_{48}$ in environments similar to those *in vivo*, using the same PLIM

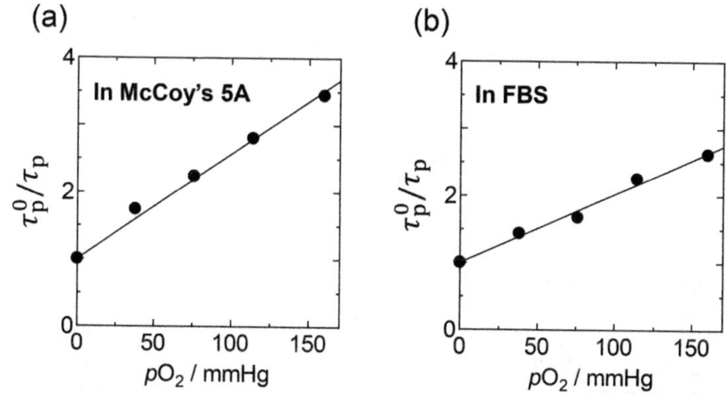


**Figure 4.5** Stern–Volmer plots of $\tau_p{}^0/\tau_p$ *vs.* $pO_2$ for BTP-PEG$_{48}$ in (a) McCoy's 5A and (b) FBS.

apparatus used for *in vivo* measurements. Both plots gave linear relationships, and the quenching rate constant ($k_q$) and $\tau_p{}^0$ were obtained to be $3.86\times10^3$ mmHg$^{-1}$ s$^{-1}$ and 4.09 μs in the McCoy's 5A and $2.57\times10^3$ mmHg$^{-1}$ s$^{-1}$ and 4.00 μs in FBS at 37 °C. The obtained $k_q$ and $\tau_p{}^0$ can be used to evaluate the oxygen partial pressure ($pO_2$) from $\tau_p$ measurements, according to the following equation:

$$pO_2 = \frac{1}{k_q}\left(\frac{1}{\tau_p} - \frac{1}{\tau_p{}^0}\right) \quad (4.2)$$

### 4.3.4 Temperature Effects on Phosphorescence Quenching by O$_2$

The magnitude of the bimolecular quenching rate constant usually varies with temperature because the diffusion rate of $O_2$ depends on the ambient temperature. If the phosphorescence quenching due to oxygen proceeds at a diffusion-controlled rate, the bimolecular quenching rate can be approximated by the diffusion-controlled rate constant ($k_d$), which can be expressed by the following Smoluchowski equation:[24]

$$k_d = 4\pi N_A(D_{Ir} + D_{O_2})(r_{Ir} + r_{O_2}) \quad (4.3)$$

where $k_d$ is the diffusion-controlled rate constant, $D_{Ir}$ and $r_{Ir}$ are the diffusion coefficient and molecular radius of the Ir(III) complex, respectively, $D_{O_2}$ and $r_{O_2}$ are the diffusion coefficient and molecular radius of $O_2$, and $N_A$ is the Avogadro constant. Since the molecular size of $O_2$ is much smaller than that of Ir(III) complexes, $D_{O_2}$ is estimated to be over twenty times larger than $D_{Ir}$.[19] Consequently, the phosphorescence quenching rate by $O_2$ is dominated by $D_{O_2}$,

**Table 4.3** Temperature effects on the phosphorescence lifetimes ($\tau_p{}^0$ and $\tau_p$) of BTP-PEG$_{48}$ in deaerated and aerated saline, and the phosphorescence quenching rate constants ($k_q$) of BTP-PEG$_{48}$ by oxygen.

| T/°C | $\tau_p$/μs | $\tau_p{}^0$/μs | $\tau_p{}^0/\tau_p$ | $k_q/10^3$ mmHg$^{-1}$s$^{-1}$ | $k_q/10^9$ dm$^3$mol$^{-1}$s$^{-1}$ |
|---|---|---|---|---|---|
| 20 | 1.24 | 4.26 | 3.44 | 3.57 | 2.01 |
| 25 | 1.10 | 4.17 | 3.79 | 4.18 | 2.59 |
| 30 | 1.02 | 4.09 | 4.01 | 4.60 | 3.12 |
| 35 | 0.93 | 4.04 | 4.34 | 5.17 | 3.80 |
| 37 | 0.92 | 4.02 | 4.37 | 5.24 | 3.95 |
| 40 | 0.90 | 3.99 | 4.43 | 5.38 | 4.26 |
| 45 | 0.86 | 3.88 | 4.51 | 5.66 | 4.79 |

which depends significantly on the ambient temperature and viscosity of the solvent. Using the Stokes-Einstein equation for the diffusion coefficient, eqn (4.3) can be expressed as eqn (4.4):

$$k_d = \frac{8RT}{3\eta} \quad (4.4)$$

where $R$ is the gas constant, $T$ is the absolute temperature, and $\eta$ is the viscosity of the solvent. Eqn (4.4) predicts that the diffusion-controlled rate, *i.e.*, the quenching rate, depends inversely on the viscosity of the medium. Hence, to use Ir(III) complexes as *in vivo* oxygen probes, it is important to clarify the temperature dependence of the $k_q$ value and of $\tau_p{}^0$. The effect of temperature changes on $\tau_p{}^0$ and $\tau_p$ of BTP-PEG$_{48}$ were investigated in saline using a lifetime measurement system based on the time-correlated single photon counting method. The $\tau_p{}^0$ and $\tau_p$ values were found to decrease with increasing temperature, and the $\tau_p$ values showed more prominent temperature effects than $\tau_p{}^0$ due to the temperature change in the $O_2$ diffusion rate (Table 4.3). The quenching rate constant ($k_q$) at each temperature was estimated from $\tau_p{}^0$ and $\tau_p$, as listed in Table 4.3.

It can be seen that the $k_q$ value increases as the temperature increases. The relatively large $k_q$ and $\tau_p{}^0$ of BTP-PEG$_{48}$ in saline is advantageous for oxygen measurements in the hydrophilic environment of tissues.

### 4.3.5 Cellular Uptake and Oxygen Response

We compared the cellular uptake efficiencies of BTP-PEG$_n$ and PPY-PEG$_{24}$ with that of the lipophilic analogue BTPDM1 using HT-29 cells. Phosphorescence microscopic images of HT-29 cells incubated with each of these complexes (5 μM) for 2 h are depicted in Figure 4.6.

The phosphorescence images clearly demonstrate that BTPDM1 is taken up into cells and gives a phosphorescence image. In contrast, very little BTP-PEG$_n$ or PPY-PEG$_{24}$ is taken up into cells and rather remain in the

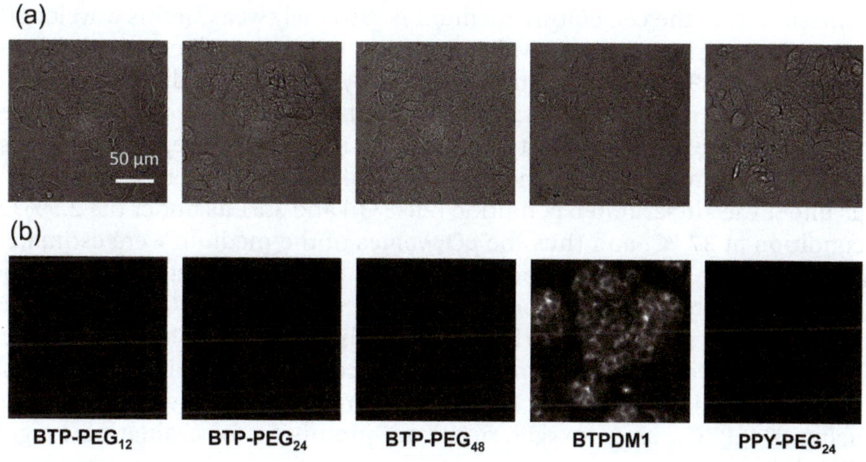


**Figure 4.6** Bright field (a) and phosphorescence (b) images of HT-29 cells incubated with BTP-PEG$_n$ or PPY-PEG$_{24}$ (5 μM) for 2 h.

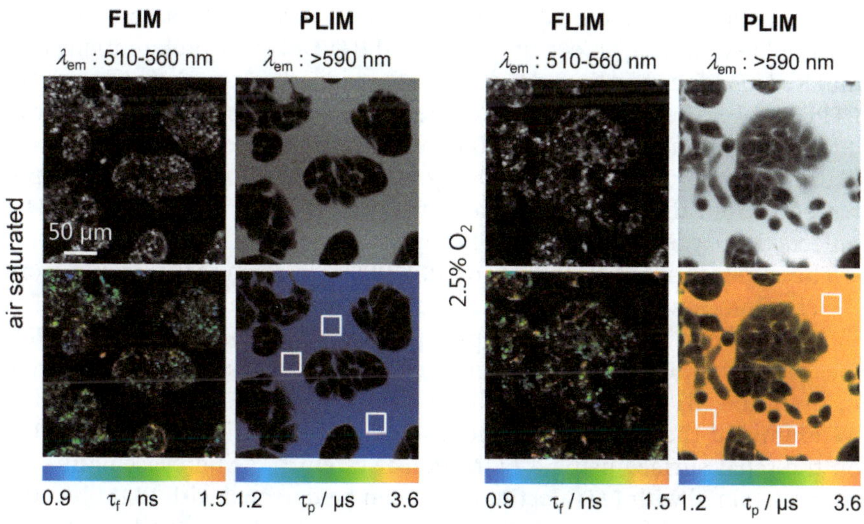


**Figure 4.7** FLIM and PLIM images of HT-29 cells incubated with BTP-PEG$_{48}$ (100 μM) for 30 min under air-saturated and 2.5% $O_2$ conditions at 37 °C.

medium because of their highly hydrophilic character. Fluorescence and phosphorescence intensity and lifetime images of HT-29 cells loaded with BTP-PEG$_{48}$ are shown in Figure 4.7. These images were obtained by confocal fluorescence (FLIM) and phosphorescence (PLIM) lifetime imaging microscopy.

The FLIM images monitored at 510–560 nm under air-saturated and 2.5% $O_2$ conditions are primarily due to the autofluorescence of HT-29 cells, and emission from the cell culture medium is extremely weak in this wavelength range. In contrast, in the PLIM images taken at the emission wavelengths of BTP-PEG$_{48}$ (>590 nm), emission is exclusively from the medium. It is noted that reducing the $O_2$ concentration in the incubator appreciably enhances the phosphorescence intensity and the lifetime. For the region of interest (ROI) shown in Figure 4.7, the mean phosphorescence lifetimes were 1.33 μs under the air-saturated condition (20% $O_2$) and 3.51 μs under the 2.5% $O_2$ condition at 37 °C, and thus the $pO_2$ values of the medium were estimated to be 131 mmHg (17.2%) and 10.5 mmHg (1.4%), respectively, using the calibration data obtained from Figure 4.5a. The derived $pO_2$ values of the medium were slightly lower than those in the incubator. These differences can be ascribed to $O_2$ consumption due to cellular respiration. From these results we could confirm that oxygen measurements based on PLIM images using BTP-PEG$_{48}$ as an oxygen probe are promising for evaluating oxygen distribution *in vivo*.

### 4.3.6 *In vivo* $O_2$ Imaging by PLIM Measurements

Our recent study showed that BTP and BTPDM1 give different intravital distributions after systemic administration into mice: BTP was preferentially accumulated in liver, spleen and lung, and BTPDM1 was preferentially accumulated in kidney and liver.[13] Exploiting the kidney-accumulating property of BTPDM1, we recently studied the oxygen status of mouse kidney.[14] Following intravenous administration, BTPDM1 was transferred from the inside of capillary vessels into tubular cells. Decay measurements of BTPDM1 phosphorescence from the renal cortex demonstrated that the lifetime of BTPDM1 reflects the oxygen level of tubular cells.

The application of BTP-PEG$_n$ or PPY-PEG$_{24}$ to intravital imaging would benefit from lower sensitivity of the phosphorescence lifetime to HSA. We therefore chose BTP-PEG$_{48}$ as a phosphorescent probe for intravital PLIM measurements. Given that low-molecular weight PEGs are rapidly excreted in the urine, BTP-PEG$_{48}$ is also likely excreted in the urine.[25,26] Therefore, we examined the distribution and phosphorescence lifetime of BTP-PEG$_{48}$ on the renal surface using a FLIM/PLIM system.[27–29] Microvasculature was counterstained with FITC-lectin, which can be detected with FLIM by utilizing the different lifetimes of FITC fluorescence (2.0–3.0 ns) and autofluorescence (0.5–2.0 ns). The luminescence images of renal cortex (~10 μm below the renal surface) obtained at 5–6 min after probe injection are shown in Figure 4.8.

The FLIM images taken at 510–560 nm (Figure 4.8, left) correspond to the fluorescence of FITC-lectin from peritubular capillaries (PTC). In contrast, the PLIM images monitored at wavelengths >590 nm (Figure 4.8, right) exhibit phosphorescence due to BTP-PEG$_{48}$, which is emitted not only from PTC but also from inside the tubules, *i.e.*, the urinary space. This indicates that BTP-PEG$_{48}$ indeed passed through the glomeruli and was excreted

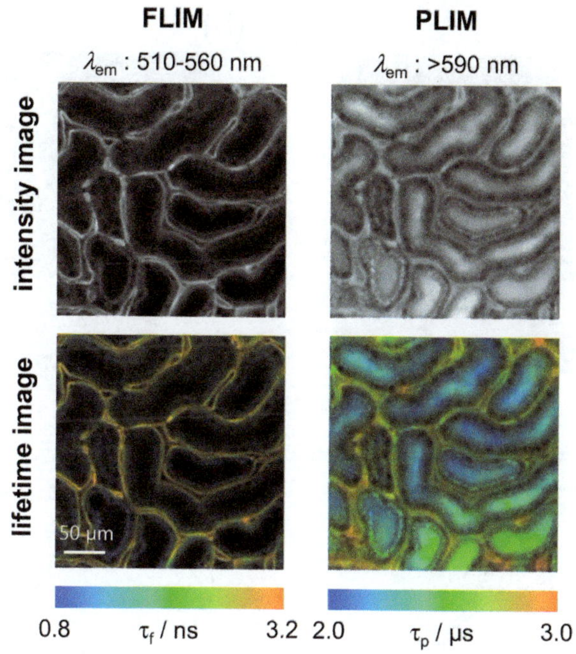


**Figure 4.8** FLIM and PLIM images of the kidney surface of a mouse administered BTP-PEG$_{48}$ (250 nmol) and FITC-lectin (125 μg).

into urine. BTP-PEG$_{48}$ was excreted from the blood in kidney into urine within about ten minutes.

To separately evaluate the oxygen level of the urinary space and PTC, we extracted the phosphorescence lifetime from the ROI as shown in Figure 4.9. The phosphorescence decay profile of the urinary space (Figure 4.9a and c) followed a single exponential function, suggesting that there are no significant interactions between BTP-PEG$_{48}$ and the components of urine.

The average lifetime of BTP-PEG$_{48}$ in urinary space (five different ROI; not shown in Figure 4.9) was 2.31 μs. Using the $k_q$ and $\tau_p{}^0$ values determined in McCoy's 5A at 37 °C, the average $pO_2$ of urinary space was estimated as 49 mmHg. This value of $O_2$ tension in urine inside tubules is in good agreement with the $O_2$ tension inside tubular cells, which we examined previously.[14] The average phosphorescence lifetime obtained for ROI corresponding to PTC (Figure 4.9b and d) was 2.81 μs, from which the oxygen level of PTC was estimated to be 41 mmHg using the calibration data in FBS. The $pO_2$ value (41 mmHg) evaluated for blood in PTC appears to be reasonable. Since the magnitudes of $k_q$ and $\tau_p{}^0$ (especially $k_q$) used in eqn (4.2) depend on the environment of the probe molecules, further studies are necessary to reliably quantify the intravascular $O_2$ level.

Hypoxia, and especially tubule-interstitial hypoxia, are closely related to the progression of renal diseases such as chronic kidney disease.[30,31]

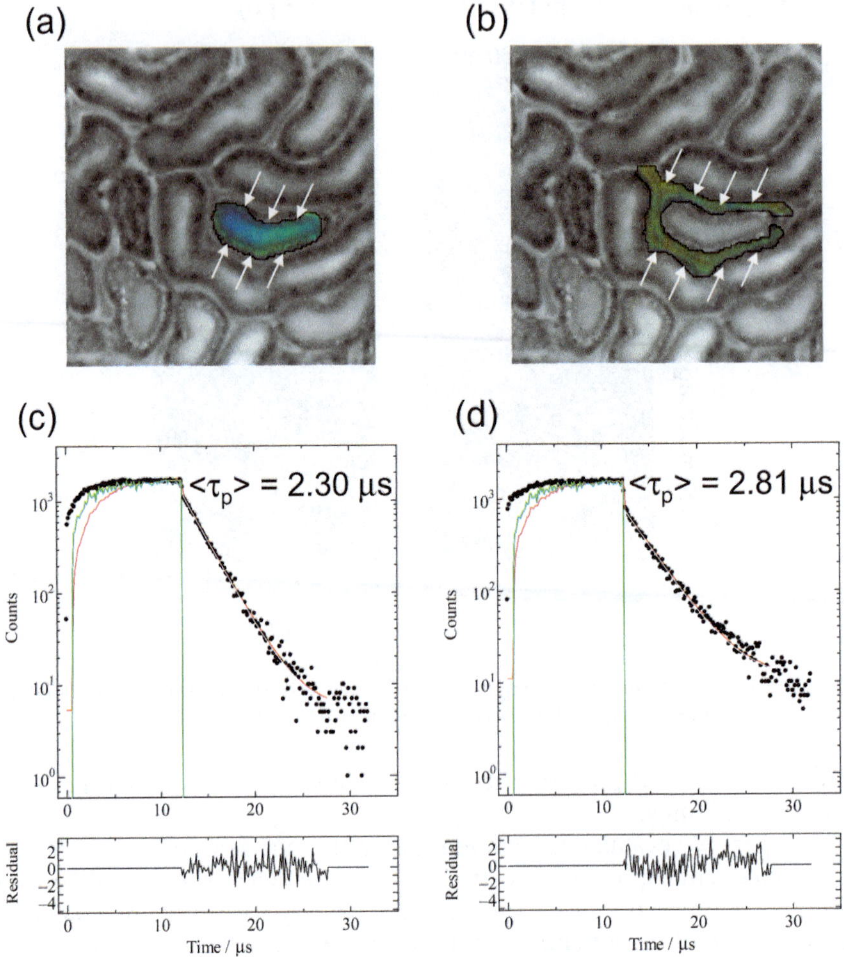


**Figure 4.9** PLIM images of mouse kidney following the administration of BTP-PEG$_{48}$, and phosphorescence decay profiles. The region of interest (ROI) is shown by arrows. (a) Urinary space and (b) peritubular capillaries. (c) Phosphorescence decay profile of BTP-PEG$_{48}$ in the ROI shown in (a) and (d) the phosphorescence decay profile of BTP-PEG$_{48}$ in the ROI shown in (b).

Blood oxygen level dependent-magnetic resonance imaging (BOLD-MRI) is frequently used to assess renal hypoxia,[32] although high resolution images, such as at the cellular level, cannot be obtained using this technique. Urinary oxygen tension is also used as an oxygen window in patients who undergo cardiac surgery and is measured by inserting an oxygen electrode into the bladder.[33] Our technique using BTP-PEG$_{48}$ enables us to examine the oxygen tension in both blood and urine with high spatial resolution, and therefore should be a useful tool for revealing the mechanism of renal hypoxia in renal diseases.

## 4.4 Conclusion

Our aim was to develop molecular oxygen probes operating as intravascular and interstitial $O_2$ sensors in living tissues. We therefore synthesized the hydrophilic Ir(III) complexes BTP-PEG$_n$ ($n$ = 12, 24 and 48) and PPY-PEG$_{24}$ by introducing the PEG group into the btp and ppy ligands in the lipophilic Ir(III) complexes Ir(btp)$_3$ and Ir(ppy)$_3$. BTP-PEG$_n$ and PPY-PEG$_{24}$ showed high hydrophilicity, resulting in very low cellular uptake efficiency. The phosphorescence of these complexes gave high quantum yields ($\Phi_p{}^0$ = 0.15–0.61 in $H_2O$) and relatively long lifetimes ($\tau_p{}^0$ = 1.71–4.18 μs in $H_2O$) under deaerated conditions. The phosphorescence decay profile of BTP-PEG$_{24}$, BTP-PEG$_{48}$, and PPY-PEG$_{24}$ in aerated $H_2O$ followed single exponential functions, even in the presence of HSA. Stern–Volmer plots for the phosphorescence quenching of BTP-PEG$_{48}$ by $O_2$ gave linear relationships both in McCoy's 5A and FBS. Fluorescence and phosphorescence lifetime imaging measurements of kidney were performed on live mice with intravenously administered BTP-PEG$_{48}$. We demonstrated that FLIM images of kidney cortex using FITC-lectin clearly visualize peritubular capillaries, and PLIM images due to hydrophilic BTP-PEG$_{48}$ allowed evaluation of the oxygen status of both the urinary space and peritubular capillaries.

## Acknowledgement

This study was supported by Grants-in-Aid for Scientific Research on Innovative Areas (No. 26111003 to M.N., No. 26111012 to S.T.) from the Ministry of Education, Culture, Sports, Science and Technology, Japan.

## References

1. D. B. Papkovsky and R. I. Dmitriev, *Chem. Soc. Rev.*, 2013, **42**, 8700.
2. X. Wang and O. S. Wolfbeis, *Chem. Soc. Rev.*, 2014, **43**, 3666.
3. J. Liu, W. Bu and J. Shi, *Chem. Rev.*, 2017, **117**, 6160.
4. T. Yoshihara, Y. Hirakawa, M. Hosaka, M. Nangaku and S. Tobita, *J. Photochem. Photobiol., C*, 2017, **30**, 71.
5. E. C. Finger and A. J. Giaccia, *Cancer Metastasis Rev.*, 2010, **29**, 285.
6. I. Mimura and M. Nangaku, *Nat. Rev. Nephrol.*, 2010, **6**, 667.
7. M. C. Manresa, C. Godson and C. T. Taylor, *Am. J. Physiol.: Regul., Integr. Comp. Physiol.*, 2014, **307**, R1369.
8. S. Tobita and T. Yoshihara, *Curr. Opin. Chem. Biol.*, 2016, **33**, 39.
9. T. Yoshihara, Y. Karasawa, S. Zhang, M. Hosaka, T. Takeuchi, Y. Iida, K. Endo, T. Imamura and S. Tobita, *Proc. SPIE*, 2009, **7190**, 71900X1.
10. S. Zhang, M. Hosaka, T. Yoshihara, K. Negishi, Y. Iida, S. Tobita and T. Takeuchi, *Cancer Res.*, 2010, **70**, 4490.
11. T. Murase, T. Yoshihara and S. Tobita, *Chem. Lett.*, 2012, **41**, 262.
12. T. Yoshihara, S. Murayama, T. Masuda, T. Kikuchi, K. Yoshida, M. Hosaka and S. Tobita, *J. Photochem. Photobiol., A*, 2015, **15**, 13503.

13. T. Yoshihara, M. Hosaka, M. Terata, K. Ichikawa, S. Murayama, A. Tanaka, M. Mori, H. Itabashi, T. Takeuchi and S. Tobita, *Anal. Chem.*, 2015, **87**, 2710.
14. Y. Hirakawa, T. Yoshihara, M. Kamiya, I. Mimura, D. Fujikura, T. Masuda, R. Kikuchi, I. Takahashi, Y. Urano, S. Tobita and M. Nangaku, *Sci. Rep.*, 2015, **5**, 17838.
15. K. Suzuki, A. Kobayashi, S. Kaneko, K. Takehira, T. Yoshihara, H. Ishida, S. Oishi and S. Tobita, *Phys. Chem. Chem. Phys.*, 2009, **11**, 9850.
16. V. Balzani, P. Ceroni and A. Juris, *Photochemistry and Photophysics: Concepts, Research, Application*, Wiley-VCH, Weinheim, 2014, p. 47.
17. H. Yersin, *Highly Efficient OLEDs with Phosphorescent Materials*, Wiley-VCH: Weinheim, Germany, 2008.
18. M. Montalti, A. Credi, L. Prodi and M. T. Gandolfi, *Handbook of Photochemistry*, CRC-Taylor & Francis, Boca Raton, 3rd edn, 2006, p. 620.
19. N. Hasebe, Y. Deguchi, S. Murayama, T. Yoshihara, H. Horiuchi, T. Okutsu and S. Tobita, *J. Photochem. Photobiol., A*, 2016, **324**, 134.
20. P. I. Djurovich, D. Murphy, M. E. Thompson, B. Hernandez, R. Gao, P. L. Hunt and M. Selke, *Dalton Trans.*, 2007, 3763.
21. K. M. Jung, K. H. Kim, J.-I. Jin, M. J. Cho and D. H. Choi, *J. Polym. Sci., Part A: Polym. Chem.*, 2008, **46**, 7517.
22. A. Endo, K. Suzuki, T. Yoshihara, S. Tobita, M. Yahiro and C. Adachi, *Chem. Phys. Lett.*, 2008, **460**, 155.
23. I. Dunphy, S. A. Vinogradov and D. F. Wilson, *Anal. Biochem.*, 2002, **310**, 191.
24. A. H. Alwattar, M. D. Lumb and J. B. Birks, in *Organic Molecular Photophysics*, ed. J. B. Birks, Wiley, New York, 1973, vol. 1, p. 403.
25. C. B. Shaffer, F. H. Critchfield and C. P. Carpenter, *Am. J. Physiol.*, 1948, **152**, 93.
26. M. Magnusson, K. E. Magnusson, T. Sundqvist and T. Denneberg, *Nephron*, 1990, **56**, 312.
27. W. Becker and B. S. A. Bergmann, *Proc. SPIE*, 2011, 7903.
28. E. Baggaley, S. W. Botchway, J. W. Haycock, H. Morris, I. V. Sazanovich, J. A. Williams and J. A. Weinstein, *Chem. Sci.*, 2014, **5**, 879.
29. K. Jahn, V. Buschmann and C. Hille, *Sci. Rep.*, 2015, **5**, 14334.
30. L. G. Fine, D. Bandyopadhay and J. T. Norman, *Kidney Int. Suppl.*, 2000, **75**, S22.
31. M. Nangaku, *J. Am. Soc. Nephrol.*, 2006, **17**, 17.
32. M. Pruijm, B. Milani and M. Burnier, *Front. Physiol.*, 2016, **7**, 667.
33. R. G. Evans, *et al.*, *Am. J. Physiol.: Regul., Integr. Comp. Physiol.*, 2014, **306**, R45.

CHAPTER 5

# Protection of Triplet Excited State Materials from Oxygen Quenching and Photooxidation in Optical Sensing Applications

MIKHAIL A. FILATOV

School of Chemical and Pharmaceutical Science, Dublin Institute of Technology, Kevin Street, Dublin 8, Ireland
*E-mail: mikhail.filatov@dit.ie

## 5.1 Introduction

Molecular oxygen is known to interact with triplet excited states in an energy transfer process resulting in non-emissive deactivation of the phosphor.[1–3] The effect of oxygen on the phosphorescence intensity and triplet state lifetimes is used for the quantification of oxygen concentration using the Stern–Volmer equation.[4–6] Biomedical applications of oxygen-dependent phosphorescence quenching are currently under active development.[7–15] Compared to other techniques, this method provides high selectivity and sensitivity, excellent temporal and spatial resolution, while being relatively simple in implementation.

In oxygen-saturated conditions, intensive deactivation of triplet excited states results in very short decay times and weak emission intensity that is

Detection Science Series No. 11
Quenched-phosphorescence Detection of Molecular Oxygen: Applications in Life Sciences
Edited by Dmitri B. Papkovsky and Ruslan I. Dmitriev

Published by the Royal Society of Chemistry, www.rsc.org

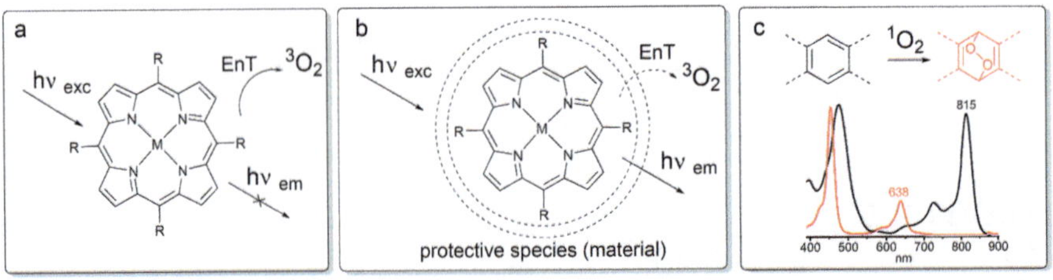


**Figure 5.1** (a) Complete deactivation of triplet excited states in oxygen-saturated conditions. (b) Protection from quenching by encapsulation of the phosphor into a protective material or adding oxygen scavengers. (c) Possible photooxidation reaction mediated by singlet oxygen and corresponding change in the absorption spectra.

deemed to result in incorrect estimation of oxygen concentration (Figure 5.1a). In this case, a physical barrier, preventing oxygen diffusion, or a protective material, reacting with the excess of oxygen, is required to attenuate quenching constants in order to provide reliable response of the sensor (Figure 5.1b). A straightforward solution being applied for most phosphorescent probes is based on encapsulation into a polymer, to achieve the desired sensitivity towards quenching in specific conditions. However, non-exponential phosphorescence decays and nonlinear Stern–Volmer plots are often observed for such materials due to heterogeneity of binding sites in the polymer and differential oxygen quenching between dyes localized in different binding sites.[16] These effects are particularly pronounced for the dyes possessing only limited solubility in a polymer matrix that possibly cause phase separation in a solid state. Moreover, for *in vivo* applications, it is preferable to use water-soluble molecular probes which can be directly delivered into the tissue, rather than polymeric optodes, which can cause mechanical damage and decrease measurement accuracy.

Another important issue in the application of the phosphorescent dyes is the possibility of photooxidation processes in oxygen-saturated conditions. Energy transfer between the triplet excited state and molecular oxygen leads to singlet oxygen ($^1O_2$), which is a high energy state of oxygen and possesses profound reactivity towards organic molecules.[17] As many optical probes rely on molecules with an extended conjugated π-system, the reactions with self-generated $^1O_2$ deteriorate the operating performance of the corresponding devices. In particular, singlet oxygen typically takes part in [4 + 2] cycloaddition reactions with aromatic compounds, breaking the conjugation in the π-systems and resulting in a blue-shift of the absorption and emission wavelengths (Figure 5.1c). Such processes are responsible for the photobleaching of various chromophores, including phosphorescent metal complexes.[18] Thus, in addition to the restriction of oxygen access to triplet excited states, approaches towards scavenging of singlet oxygen are required for the development of robust sensor materials.

Due to the growing interest in applications of phosphorescent dyes as biomedical probes, the specific problem of controlling the sensitivity of the triplet excited states towards oxygen has recently attracted broad attention. Apart from oxygen sensing applications, phosphorescent dyes possess huge potential in biophotonics, due to the large Stokes shift of the phosphorescence emission that allows for imaging in the "optical transparency window" of biological tissues.[19] In these applications, oxygen quenching of the triplet states is an unwanted complication, reducing the efficiency of corresponding probes. To date, a few reviews have been devoted to the protection of triplet excited state materials against quenching and photooxidation.[20,21] This chapter focuses more specifically on the approaches towards phosphorescent compositions with controllable quenching rates and enhanced stability against singlet oxygen. Two general solutions for this challenging problem that have been proposed to date are discussed: (i) active protection of the triplet states, based on the application of oxygen scavenging species and (ii) passive protection, based on barrier materials to reduce physical contact between oxygen and excited states.

## 5.2 Phosphorescent Probes with Appended Protective Groups

### 5.2.1 Phosphorescent Dendrimers

Accuracy of quantification of the dissolved molecular oxygen based on the phosphorescence quenching critically depends on quenching rates $k_q$ and excited-state lifetimes $\tau$. For typical phosphors, such as metalloporphyrins and transition metal complexes, the phosphorescence lifetime and intensity are already rather low at intermediate physiological $O_2$ levels due to high $k_q$. In order to keep the accuracy and dynamic range of the method the $k_q$ values in water must be in the range of $0.5–2.5 \times 10^8$ $M^{-1}$ $s^{-1}$.[22] One way to tune $k_q$ values is based on the introduction of steric congestion around the phosphorescent center, which can be caused by bulky peripheral substituents. A comparison of platinum tetraphenylporphyrin (PtTPP) and porphyrin **1**, containing bulky 2,4,6-triethylphenyl groups (Figure 5.2), demonstrates the role of the substituents in suppressing quenching of the triplet excited states: $k_q$ of **1** and PtTPP were found to be $4.1 \times 10^8$ $M^{-1}$ $s^{-1}$ and $14.8 \times 10^8$ $M^{-1}$ $s^{-1}$, respectively.[23] The difference was attributed to the steric effects of the ethyl groups, which "screen" the core from interacting with oxygen and thus protect the excited states from quenching.

Dendrimers surrounding luminescent centers have been shown to reduce deactivation of the excited states by oxygen and other quenchers due to a diffusion barrier.[24–27] Balzani and co-workers[28,29] studied tris(bipyridine) ruthenium(II) complexes ($[Ru(bpy)_3]^{2+}$) bearing lengthy dendritic branches attached in the 4,4′-positions of the ligands. Dendritic complexes, showed longer triplet state lifetimes in aerated solutions due to the shielding effect of the dendrimer branches on the central core. The rate constants for oxygen quenching were found to depend on the length of the dendritic branches. For higher generation dendrimer **2** (Figure 5.2), $k_q = 0.22 \times 10^9$ $M^{-1}$ $s^{-1}$ in

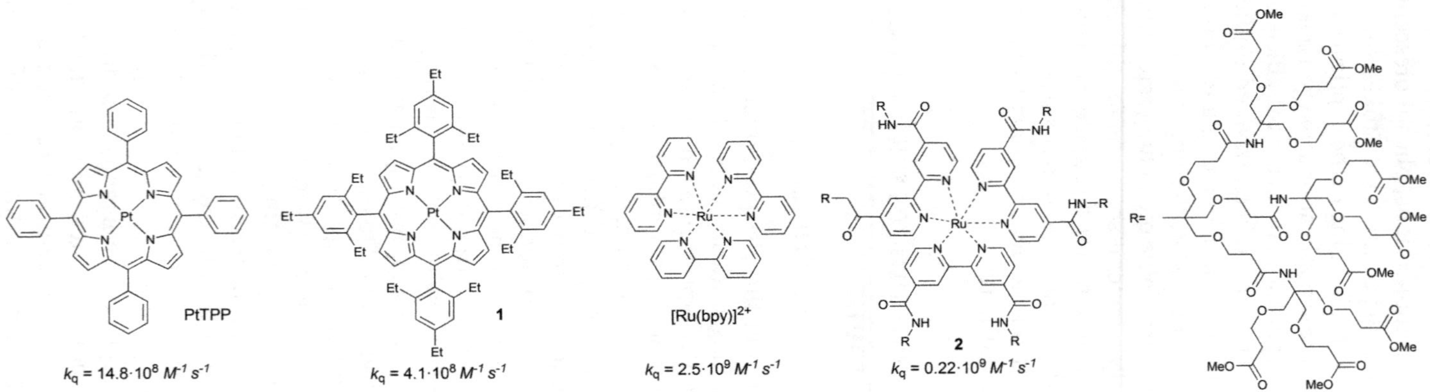


**Figure 5.2** Sterically shielded phosphorescent porphyrin and ruthenium bpy complexes.

air-saturated acetonitrile solution is 12-fold lower, compared to the parent $[Ru(bpy)_3]^{2+}$. This approach for the design of the phosphorescent probes has recently attracted much attention. Dendritic branches provide biological compatibility of the probe, in particular, good solubility in water and protection from interactions with biomolecules. On the other hand, the length and chemical nature of the branches allow control of the values of oxygen quenching rates. Particularly, folding of hydrophobic dendrimers strongly depends on the media polarity and in a water environment enhances the oxygen diffusion barrier.[30] Furthermore, multiple peripheral functionalities on the dendrimer make possible further synthetic modifications to minimize interactions with proteins and other biomolecules.

Biocompatible oxygen probes with variable oxygen quenching rates based on Pd(II) and Pt(II) porphyrin complexes equipped with a protective dendrimer layer have been developed by Vinogradov and co-workers. Poly-*l*-glutamic acid[31] and poly(arylglycine) (AG) dendrons[32] were found to be especially well-suited for the construction of the phosphorescent oxygen probes (Figure 5.3). The resulting molecules closely resemble natural heme proteins ensuring low toxicity and immunoreactivity. In water the values of $k_q$ are decreased 3–4 times due to the lower oxygen diffusion through the dendrimer layer, that is significantly less than the similar effect observed for the dendritic $[Ru(bpy)_3]^{2+}$. This was accounted for by the size of the phosphorescent metalloporphyrin core, which is considerably larger than the $Ru^{2+}$ complexes, leading to much less restricted $O_2$ access in a similar dendrimer. For *in vivo* applications, probes possessing absorption bands in the NIR region were developed based on π-extended porphyrins with external aromatic rings annelated to the central macrocycle, such as tetraaryltetrabenzoporphyrins (TBP) **5–6** and tetraaryltetranaphthoporphyrins (TNP) **7**.[33–36] The absorption bands of these molecules lie in the region of 630–950 nm, where the absorption of natural chromophores is negligible. Peripheral poly(ethylene glycol) (PEG) groups on the dendrimers eliminate interactions with biomacromolecules, while keeping the probes highly hydrophilic. Notably, dendrimers derived from porphyrins with *meso*-3,5-dicarboxyphenyl groups exhibited strongly reduced $k_q$ rates since the diffusion of $O_2$ through such dense dendrimer layers is much more restricted, compared to *meso*-4-carboxyphenyl derivatives. The potential of these probes was demonstrated in the physiological studies, where they allowed for the determination of the absolute blood $pO_2$.[37–40]

A critical limitation for applications of dendrimer-appended phosphorescent probes is the rather complicated synthesis of these materials, which involves stepwise construction of the dendrimer layers. Modification of the porphyrin precursors with pre-organized dendritic building blocks using the "click chemistry" approach represents an attractive alternative route to such protected probes. A new family of "clickable" phosphorescent molecules incorporating dendritic residues bound *via* Huisgen reaction has recently been reported by Evans and co-workers.[41] Following this approach, an alkynyl-substituted platinum(II) tetrabenzoporphyrin precursor was

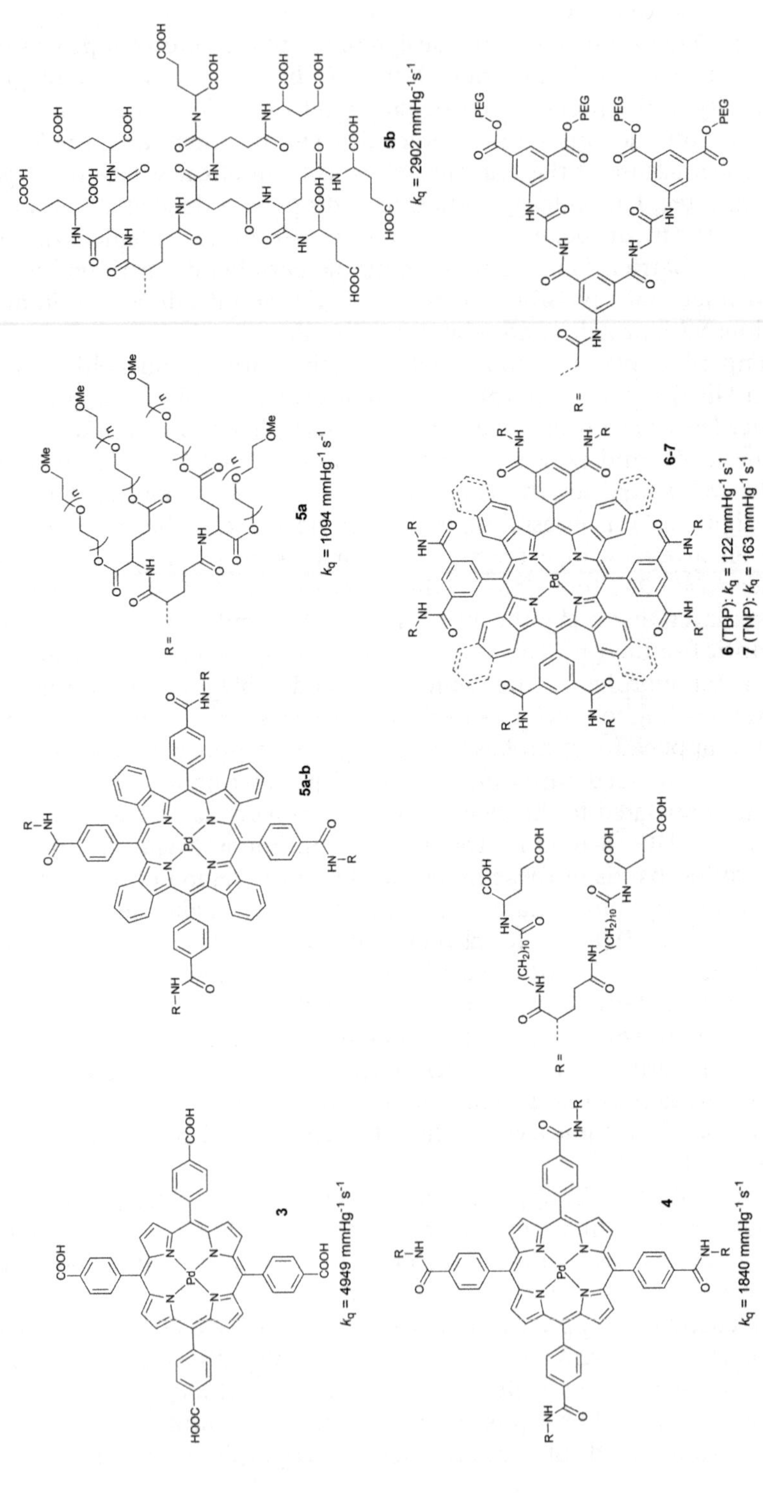


**Figure 5.3** Palladium(II) porphyrin-based phosphorescent dendrimers probes.

converted into a glutamic dendrimer in one synthetic step through the reaction with eight azido-terminated glutamic dendrons. The starting porphyrin was found to be completely non-emissive in air saturated conditions due to the quenching, which was substantially attenuated in the dendrimer. The probe was further incorporated into a wound dressing bandage and applied for visualization and quantification of the skin burns oxygenation by the naked eye under room lighting conditions, providing a clinical diagnostic tool for mapping oxygen consumption in wounds.[42]

### 5.2.2 "Self-healing" Phosphorescent Complexes

In phosphorescent dendrimers, the modulation of quenching rates is achieved by the restriction of physical contact between oxygen molecules and excited state cores. Such an approach can be referred to as "passive protection". An alternative "active protection" is based on surrounding a dye molecule with special subunits able to react with oxygen and thus reduce the quenching rates. This approach has not been explored in the design of the phosphorescent systems until very recently due to a lack of suitable molecular building blocks, which could provide chemical "trapping" of oxygen. In its ground state, which is of triplet character ($^3\Sigma_g$), reactivity of oxygen is rather low. However, the transition of an oxygen molecule into the higher energy singlet state ($^1\Delta_g$) significantly increases its reactivity. Taking into account that such a transition takes place during the interaction between oxygen and triplet excited states of organic molecules, phosphorescence dyes capable of binding oxygen molecules can be developed. Importantly, the processes of oxygen binding can be reversible, as certain types of organic molecules are known to release oxygen back from initially formed adduct *via* thermal dissociation even at ambient temperature.[43,44] Such protection is of sacrificial character and limited in timescale. However, the thermal release of bound oxygen upon heating allows for regeneration of the protective moieties. Anthracene-appended porphyrin **8** was shown to bind up to four molecules of oxygen in solution under irradiation, forming corresponding endoperoxide (Figure 5.4a).[45] Anthracene groups do not affect photophysical properties of the porphyrin either before or after the reaction with oxygen due to a lack of conjugation between the aromatic systems. This process was demonstrated for enhancement of the porphyrin phosphorescence intensity in (i) oxygen contaminated (100 ppm) and (ii) oxygen-saturated solutions. In the first case, increase of the phosphorescence was obtained after irradiation of the whole sample with 633 nm light for 10 min. During the irradiation, dissolved oxygen, which quenches triplet excited states, is being converted into singlet oxygen, reacting with anthracene groups to form endoperoxide that leads to reduction of the overall oxygen concentration in the sample and enhancement of the phosphorescence intensity (Figure 5.4b). Alternatively, the process was performed under oxygen-saturated conditions with excitation of the sample in a local area (400 μm diameter). The excitation pulse of rather low intensity

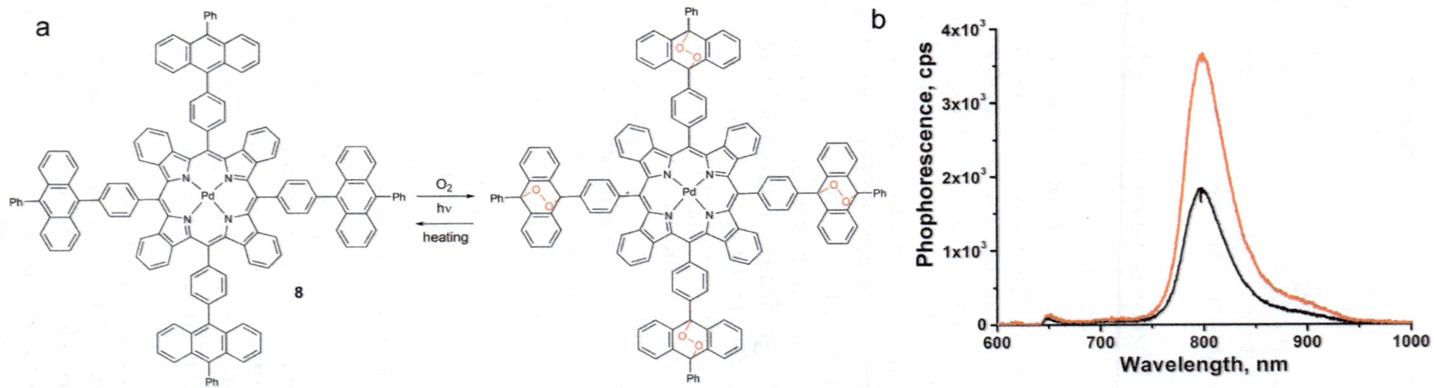


**Figure 5.4** (a) Photosensitized oxygen addition to porphyrin **8.** (b) Increase of the phosphorescence intensity in toluene solution prepared in the atmosphere containing 100 ppm of oxygen (black line) after 10 minutes of continuous irradiation (633 nm, 250 $\mu Wcm^{-2}$) (red line).

(500 μWcm$^{-2}$) caused real-time "deoxygenation" of the local area, resulting in up to 60% intensity increase of the phosphorescence signal within 10–20 s. Endoperoxide product, formed in the process, regenerated parent sensitizer molecules upon heating.

Chemical binding of oxygen to the phosphorescent dyes represents an interesting alternative to currently known approaches towards oxygen sensing based on the phosphorescence quenching. Further development of such probes requires efforts on molecular design, particularly access to water-soluble biocompatible derivatives. Systems based on the anthracene as oxygen trapping subunits are not suitable for *in vivo* studies, due to their hydrophobic character. Porphyrins with four and eight 2-pyridone groups appended at the periphery of the macrocycle have recently been demonstrated for the reversible oxygen binding on *in vitro* cell models.[46] These systems are well suited for biomedical applications due to their facile synthesis and good solubility in water, enhanced by intrinsic polarity of 2-pyridone substituents.

## 5.3 Host-guest Complexes and Aggregates

### 5.3.1 Tryptophan Phosphorescence in Proteins

Incorporation of the phosphorescent molecules into the rigid environment, which confines the molecular diffusion, allows the reduce quenching of the excited states even in oxygen-saturated conditions due to keeping the phosphor and the quencher out of the collision distance. Development of practical methods based on this principle was initially stimulated by the studies of tryptophan phosphorescence in proteins.[47] Normally the phosphorescence of proteins is hard to detect due to high sensitivity towards quenching by low levels of dissolved oxygen in solution. However, in certain cases the quenching constants were found to be reduced due to the steric hindrance of tryptophan residues resulting from the folding of the polypeptide chains. The observations of protein phosphorescence with increasing temperature revealed that the folding of the polypeptide chains in protein molecules is hindering the diffusion of $O_2$ to this region of the macromolecule and reducing the quenching rates of the triplet states. As was shown by Galley and co-workers, tryptophan phosphorescence in horse liver alcohol dehydrogenase, can be observed even in aerated solutions at room temperature due the low quenching values ($6.5 \times 10^4$ 1 M$^{-1}$ s$^{-1}$).[48] It has been shown that certain protein conformations strongly inhibit quenching of tryptophan triplets by dissolved oxygen and that $k_q$ is directly correlated to the rigidity of the protein matrix surrounding the chromophore.[49] Oxygen-dependent quenching of tryptophan phosphorescence in apoazurin, liver alcohol dehydrogenase, and alkaline phosphatase as a function of temperature (0–50 °C) and applied pressure (up to 3 kbar) has been applied by Strambini and Cioni to study structural dynamics of these proteins.[50] This method was further used in the studies of the small molecules migration inside protein structures.[51]

### 5.3.2 Phosphorescence of Cyclodextrin Complexes in the Presence of Oxygen

Turro and co-workers first described inhibition of oxygen quenching upon complexation of the phosphorescent molecules with cyclodextrins (CD). A series of substituted naphthalene derivatives in aqueous solutions containing γ-CD was found to display two distinct excited state decays: the fast decay and the slow decay (Figure 5.5). The oxygen completely quenched the fast decay component, while the slow decay was not influenced. It was further shown that the fast decay originates from the 1:1 complex of the phosphor and γ-CD, while the slow component corresponds to the 1:2 complex.[52] At the same time, no phosphorescence in the presence of oxygen was observed for the complexes of the same naphthalenes with α- and β-CD.

Phosphorescence under oxygen-saturated conditions can be observed if cyclodextrin–phosphor inclusion complex interacts with other molecules, which serve as space-filling components (Figure 5.6).[53] In the resulting complexes, the motion of the phosphor and diffusion of the oxygen into the cavity are largely restricted, inhibiting both the non-radiative decays of the triplet state and oxygen quenching. The enhancement of the phosphorescence can be very strong, in certain cases up to a $10^5$ increase in intensity has been observed, and depends on the fit between the sizes of the phosphor, spacer regulator and CD cavity.[54]

Several types of space regulation in phosphorescent CD inclusion complexes have been described.[55] 6-Bromo-2-naphthol the 1:1 inclusion complex with α-CD is not phosphorescent in aerated solution, while the 2:1 complex showed intense emission under similar conditions due to protection of the chromophore in the cavity formed by two CD molecules.[56,57] A similar effect was observed for β-CD which forms the 2:2 complex with naphthalene stabilized by intramolecular hydrogen bonding between the hydroxy groups of two CDs.[58]

Protic solvent molecules can play the role of space-filling component: due to the formation of H-bonds with hydroxyls of CD, the entrance of the CD cavity can become hindered (Figure 5.7) that prevents oxygen molecule from entering the cavity.[59] Phosphorescence enhancement strongly depends on the fit of the alcohol lid on top of the CD cup. The rate constants for oxygen quenching generally decreases with the increase in the alcohol bulkiness.

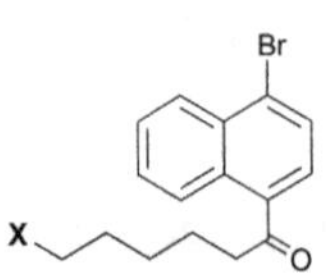


| | Lifetime, τ (ms) | | |
|---|---|---|---|
| | Fast decay | Slow decay ($N_2$) | Slow decay ($O_2$) |
| X = Br | 0.506 | 3.9 | 3.8 |
| X = -$NMe_3$Br | 0.645 | 3.3 | 2.5 |
| X = -O-CO-Ph-CO-Ph | - | 3.4 | 2.8 |

**Figure 5.5** Structures and excited state lifetimes of the phosphorescent naphthalene derivatives in γ–CD aqueous solution.

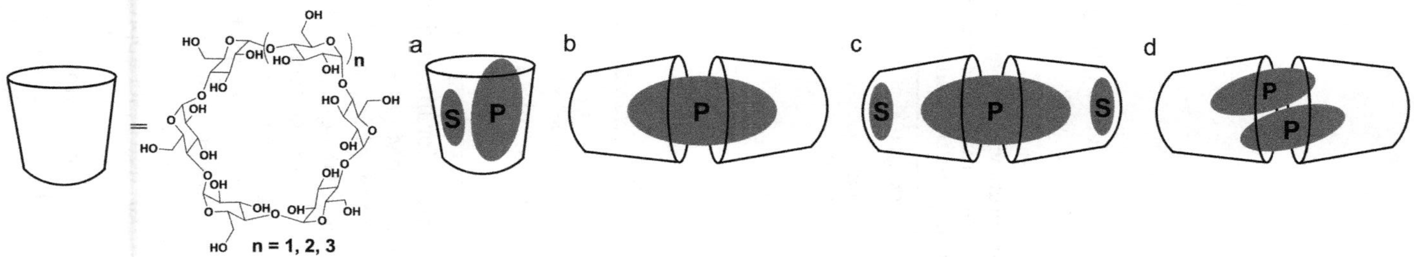


**Figure 5.6** Cyclodextrin inclusion complexes (S is a space regulator and P is a phosphorescent molecule). a – α-CD/P/S, b – α-CD2/P, c – β-CD2/P2/S2, d – β-CD2/P2.

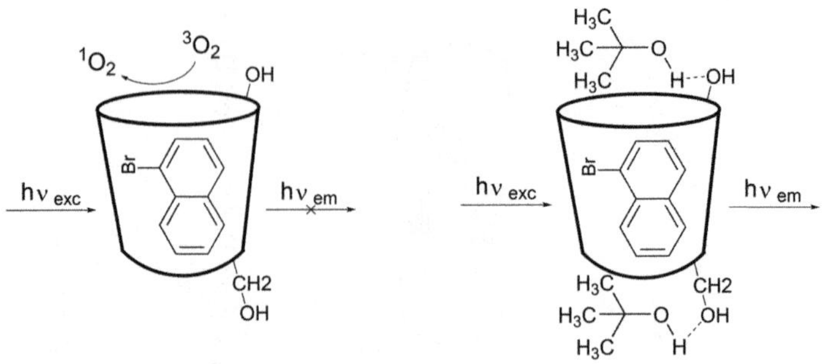


**Figure 5.7** "Covering" effect of the solvent on the quenching of the phosphorescent inclusion complex.

**Table 5.1** Phosphorescence quantum yields in air-saturated solutions and oxygen quenching rates of the complex between 1-bromonaphthalene and glucosyl-β-cyclodextrin (Gβ-CD) in different alcohols.[54]

| Alcohol | $\Phi_{phos}/10^{-4}$ | $k_q/10^5$, $M^{-1}$ $s^{-1}$ |
|---|---|---|
| *tert*-Butanol | 340 | 8.9 |
| Neopentanol | 1149 | 0.3 |
| 3,3-Dimethyl-1-butanol | 789 | 1.6 |
| Cyclohexanol | 350 | 1.2 |
| Cyclohexylmethanol | 771 | 0.2 |
| Cyclohexylethanol | 364 | 0.8 |
| *cis*-1,2-Cyclohexanediol | 883 | 2 |
| 1,3-Cyclohexanediol | 51 | 32 |
| 1,4-Cyclohexanediol | <0.01 | >$10^4$ |

Accordingly, neopentanol and 3,3-dimethyl-1-butanol strongly reduce oxygen quenching rates and provide the highest phosphorescence quantum yields (Table 5.1).

Based on this approach towards phosphorescence protection, Zhang and Johnson developed a technique for measuring over-saturated dissolved oxygen concentrations up to 40.2 mM (14 500% saturation with respect to air-equilibrated water).[60,61] *tert*-Butanol was used as a lid for Gβ-CD complex with 1-bromonaphthalene prepared in aqueous solution. Phosphorescence of the complex decreases with increasing concentrations of oxygen. The phosphorescence lifetime measurements were found to be in good agreement with the Stern–Volmer equation even for very high dissolved oxygen levels.

Phosphorescence protection through complexation with CD can be anticipated to provide new developments in the biological oxygen sensing techniques in the near future. Compared to other approaches towards triplet state protection it has the following advantages: (i) it does not require efforts on the synthesis and purification, as the complexes readily form upon mixing the components and possess high stability constants; (ii) the availability of various CDs and space-filling molecules allows to tune the size of the cavity for a specific phosphor molecule; (iii) oxygen quenching rates can be tuned *e.g. via* changing the lid for the CD complex; (iv) other hosts molecules which have structures similar to cyclodextrins and can act as protective media are available, *e.g.* cucurbit[n]uril[62] and its analogues;[63] (v) complexes with CD can be formed not only in solution but also in solid films or nanoparticles.

### 5.3.3 Steroids as Protective Matrixes

Reduced oxygen quenching of the phosphorescence in molecular aggregates based on sodium deoxycholate (SDC) was first described by Jin and co-workers.[64] 1-Bromo-4-(bromoacetyl)naphthalene (**9**) was found to exhibit phosphorescence in aqueous solution without deoxygenation upon addition of SDS. The oxygen quenching rate constant was measured to be $4.15 \times 10^4$ $M^{-1}$ $s^{-1}$ in air-saturated SDS solution at 1 atm. The effect of SDC on the phosphorescence was interpreted by assuming a formation of "sandwich" type dimers in which the dye is trapped between two steroid scaffolds (Figure 5.8). Although such dimers could not be isolated, molecular modelling study allows the conclusion that the naphthalene molecule is stacked between two SDS molecules by the apolar faces. The effect was much less pronounced for other aggregate forming molecules, such as sodium dodecylsulfate (SDS) and DNA. A similar effect of SDS phosphorescence protection was later reported for water-soluble palladium(II) *meso*-tetraarylporphyrins.[65] The phosphorescence of air-saturated porphyrin solution showed a gradual increase of the intensity upon addition of SDC along with increase of the lifetime. However, at SDC concentrations higher than $4 \times 10^{-3}$ M the emission gradually decays to negligible values. It suggests that the "protective dimer", which is formed at low concentrations of SDC, can further transform into larger aggregates possessing less rigid structure and allowing oxygen diffusion.[66]

Based on this strategy, phosphorescent materials with oxygen-persistent emission and long lifetimes (>1 s) have been recently developed by Adachi and co-workers.[67] β-Estradiol and cholesterol have been applied as host matrixes for a series of organic phosphors based on aromatic hydrocarbons. An advantage of a steroid matrix is the possibility for oxygen removal by repetitive heating above the melting point and cooling. Due to slow diffusion of oxygen in a solid matrix the resulting materials did not show any drop of the phosphorescence intensity over long periods of time.

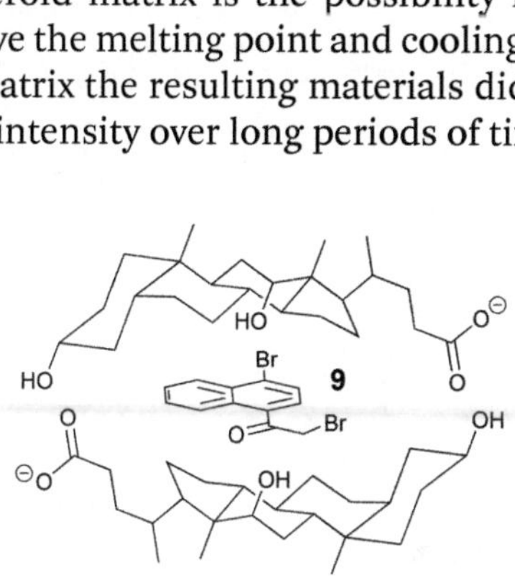


**Figure 5.8** Phosphorescent aggregate of SDC and **9**.

## 5.3.4 Gel Matrixes

The possibility to apply organogels and hydrogels as quenching protective matrixes for phosphorescent materials was recognized very recently and few reports on the incorporation of phosphors into protective gel matrixes have been made. Shirakawa and co-workers developed phosphorescent gels based on copper, palladium and platinum complexes with 8-quinolinol **10** bearing 3,4,5-tris(n-dodecyloxy)benzoylamide substituents (Figure 5.9a).[68,69] Theses complexes were found to form gels in various organic solvents at very low critical gelation concentrations of 0.10 mg $L^{-1}$ (0.05 mM). Due to the formation of a crystal-like structure in the gel state, up to a 3-fold increase of the emission intensity in aerated conditions was observed compared to reference complexes having no gel-forming ligands. It was shown that the packing structure is responsible for reduced oxygen diffusion into the gel phase. De Cola and co-workers reported phosphorescent hydrogels based on host–guest interactions between water-soluble Pt (II) complex **11** (Figure 5.9b) with attached tetraethylene glycol chains and cyclodextrins (α- and β-CD). The materials were found to be strongly emissive and not sensitive towards quenching, although the behavior of the system has not been studied in detail.[70] Further, Yang and co-workers reported 1,3:2,4-di-O-benzylidene-D-sorbitol (DBS) as a gelator for phosphorescent 3-bromoquinoline. The supramolecular gels were prepared by self-assembly of DBS in DMF-water mixture. It was found that deoxygenation of the samples is not required to observe phosphorescence. Moreover, the emission intensities of a pre-deoxygenated sample showed similar values compared to air-saturated.[71]

Organogel matrixes offer several advantages for oxygen sensor techniques. Various phosphorescent molecules can be modified with gel-forming functionalities or entrapped into pre-formed organogels. The matrix does not affect optical properties of the dye and restricts the access of the quencher. The resulting materials can be obtained in any shape (films, fibers, beads). Oxygen sensing *via* phosphorescence quenching of $[Ru(bpy)_3]^{2+}$, entrapped in organogels films, has been demonstrated by Díaz García and co-workers.[72]

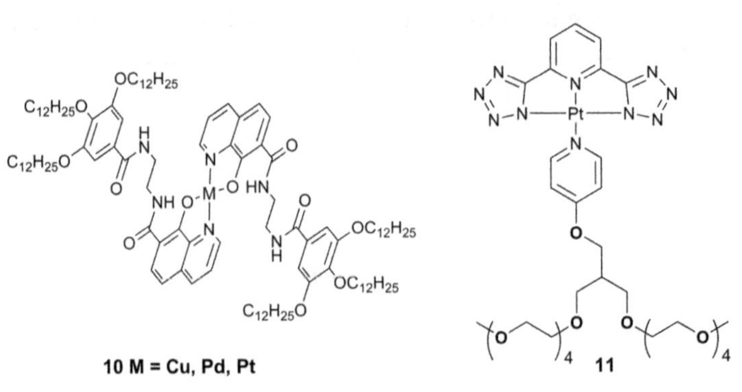


**10** M = Cu, Pd, Pt　　**11**

**Figure 5.9** Gel-forming phosphorescent complexes.

Gelatine gels prepared from reverse micellar solutions were employed for the immobilization of the complex. The resulting films possess long-term stability and showed no swelling or mechanical stability problems if contacted with organic solvent. Phosphorescence emission of the entrapped complex at 580 nm is quenched by oxygen, allowing oxygen detection with a detection limit of 0.2 ppm. An attractive feature of the developed organogel is its high photochemical stability: no substantial fading of the phosphorescence intensity was observed after more than 20 h of continuous illumination.

## 5.4 Application of Oxygen Scavengers

In optical oxygen sensing methods, a calibration of the sensor response in a range of concentrations is usually required. In particular, depletion of oxygen, for oxygen-free measurements has to be performed. Deoxygenation of liquid photoactive compositions *via* bubbling of inert gases or freeze–thaw cycles leaves residual amounts of dissolved oxygen (nM concentration range), which can still affect the phosphorescence intensity. Moreover, in certain cases, such methods lead to serious complications. For instance, with colloidal or micellar systems, gas bubbling results in mechanical damaging of the material and in the formation of bubbles, which can scatter excitation and emission light.[73] For this reason, chemical deoxygenation of the phosphorescent samples *via* the addition of oxygen scavengers is preferable.

On the other hand, formation of singlet oxygen during triplet excited state quenching can lead to fast oxidation of dye molecules incorporated into oxygen sensor material. Additions of compounds, which selectively react with singlet oxygen in the samples or their incorporation into the probe, provide extended lifetimes and reliable operation of the corresponding materials. This can be achieved only if scavenger and products of its oxidation are inert with respect to photoactive components of the sample.

### 5.4.1 Inorganic Oxygen Scavengers

Sodium sulfite was the first demonstrated suitable oxygen scavenger for solution deoxygenation based on the redox reaction (5.1).

$$2SO_3^{2-} + O_2 \rightarrow 2SO_4^{2-} \tag{5.1}$$

Díaz-García and Sanz-Medel first applied sodium sulfite as an alternative to nitrogen bubbling for the phosphorescence measurements in the SDS micellar solutions.[74] Addition of sodium sulfite leads to a gradual increase of the phosphorescence intensity, since a diffusion rate of oxygen limits the efficiency of the phosphorescence quenching coming from the solubilized phosphor molecules. Due to the dynamic nature of the micellar equilibrium, the molar concentration of negatively charged sulfite ion is higher in the bulk solution compared to the micelle surface. This accounts for more rapid oxygen consumption in the bulk phase than inside the micelle. At high concentration of the surfactant, it takes more time to reduce the dissolved oxygen as it is more soluble in the micellar phase than in water and quenches the phosphorescence emission. Increase of sulfite concentration progressively heightens phosphorescence intensity and shortens signal stabilization time to <1 s at concentration of 1.0 mmol.[75]

This technique allows for the protection of corresponding samples on a timescale of hours in the case the sample is open to air or for a longer time, if the sample is sealed. Sulfite-based $O_2$ scavenging to observe phosphorescence emission in aqueous solutions was applied for the development of analytical methods for the detection of a variety of organic compounds (including agricultural, pharmaceutical, petroleum, and biological-related samples) and extended the technique to facile metal ion determinations in solution using room temperature phosphorescence (RTP).[76–80]

In phosphorescence quenching oxygen sensing, measurement of the phosphorescence and decay times in sodium sulfite solution is commonly applied for the calibrations of the phosphor response in anoxic conditions.[81]

### 5.4.2 Application of Natural Antioxidants

Although the addition of inorganic salts into the solutions provides convenient protection of the phosphorescence, it causes certain side effects, such as a change of pH and a formation of solid deposits resulting from oxidation products. In addition, application of sodium sulfite is limited only to non-biological samples. To reduce oxygen concentration in the photoactive materials for application in bio-imaging and sensing, many groups applied the enzymatic oxygen-scavenging system,[82] which utilizes a mixture of glucose oxidase and catalase. Glucose oxidase catalyzes the oxidation of D-glucose according to the reaction (5.2) that results in the net loss of $O_2$ in solution.

$$\text{D-glucose} + O_2 + H_2O \rightarrow H_2O_2 + \text{D-gluconic acid} \tag{5.2}$$

In the samples without glucose oxidase, the $O_2$ level is in equilibrium with the medium. Upon addition of glucose oxidase, $O_2$ level is decreased, depending on glucose concentration. The relationship between glucose and oxygen concentration is not linear because the reaction follows typical enzyme kinetics. This approach allows for "gradual" deoxygenation of the sample by varying the amount of added glucose.[83]

Liu and co-workers showed that natural lipid compounds could be applied as oxygen scavengers.[84] The main components of soybean oil are linoleic acid and oleic acid, which contain double bonds, capable of undergoing oxidation reaction with singlet oxygen (Figure 5.10), providing photo-initiated deoxygenation of the samples. The particular reaction depends on the nature of the double bonds and different types of reaction products (epoxides, dioxetanes, alcohols, *etc.*) can be generated.[85] The effect was studied for platinum(II) tetraphenyltetrabenzoporphyrin (PtTPBP) in aerated soybean oil. After 635 nm irradiation for 30 s, the phosphorescent lifetime of PtTPBP in soybean oil in the

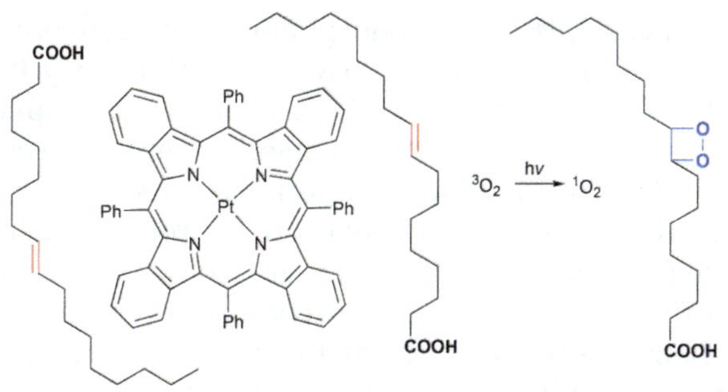


**Figure 5.10** Platinum tetrabenzoporphyrin phosphorescence protection by a component of soybean oil.

presence of oxygen was found to be ~39 μs, similar to that of PtTPBP in toluene under an argon environment (40 μs). For comparison, the effect of ascorbyl palmitate, an efficient reducing agent, was studied. The phosphorescence of PtTPBP in toluene with prolonged exposure to 635 nm laser for 240 s was increased to ~32 μs at $2.0 \times 10^{-3}$ M concentration of the scavenger. These results indicate that soybean oil can act as a protective solvent to prevent oxygen quenching and photooxidation. Linoleic acid and oleic acid of high purity showed similar effect.

### 5.4.3 Scavengers of Singlet Oxygen

Singlet oxygen generation by a phosphorescent probe can significantly influence the measurements and result in incorrect values due to degradation of dye molecules or matrix material, especially if UV radiation or powerful lasers are used as excitation sources. For instance, long-term aging of Pd(II) tetraphenylporphyrin probes encapsulated into poly(methylmethacrylate) (PMMA) films revealed 4-fold decrease of the quenching rates and oxygen diffusion constants.[86] Similarly, Ru(II) complexes with phenanthroline ligands exhibited a significant photodegradation, manifested by decay time decrease and specific absorption spectral changes.[87,88]

Photodegradation process can be reduced by addition of certain amines. Singlet oxygen is known to be susceptible to physical quenching through collisions with solvents or other molecules, causing decay back to the ground triplet state ($^3\Sigma_g$) on a timescale of few microseconds.[89] Quenching of singlet oxygen by amines involves the formation of a charge–transfer complex *via* electron transfer from the amine. It could be expected that phosphorescent materials containing amine additives possess higher photostability. However, not any type of amine is suitable for this purpose, as triplets excited states of the phosphor can also be quenched *via* electron transfer from the amine. Hartman studied the effect of different amines on the phosphorescence of Ru(II) complexes and showed and 1,4-diazabicyclo[2.2.2]octane (DABCO) is one of the most suitable scavengers as it is photochemically stable, does not significantly quench triplets of the phosphor in contrast to other amines and does not react chemically with singlet oxygen, requiring only 1% (w/w) of the additive for efficient protection.[87] Ricketts and Douglas applied DABCO as a stabiliser for oxygen-sensitive Pt (II) octaethylporphyrin (PtOEP) in ethyl cellulose containing $TiO_2$ or ZnO as scattering agents.[90] The resulting material showed stability of the emission signal in the presence of oxygen during continuous use for over an hour. Klimant and co-workers studied the effect of DABCO on the photodegradation processes of Pd(II) and Pt(II) tetra-(pentafluorophenyl)porphyrin and PtOEP in different polymers.[91] It has been shown that in addition to dye photooxidation, singlet oxygen is capable of reacting with polymer matrix that ultimately alters its physical properties and influences mechanical stability, hydrophobicity, and solubility. Polystyrene and poly(phenylsilesquioxane) were found to have the highest stability in the presence of singlet oxygen, while different poly(methylmethacrylates) have the lowest stability. Addition of DABCO into the matrix inhibited photooxidation in most polymers, except for PMMA and its derivatives. *N*-Alkylated derivative of DABCO and the polystyrene-containing covalently grafted DABCO have been prepared in order to overcome the volatility and water solubility of the quencher. However, these derivatives did not show any positive effect on the photostability of the photosensitizer and polymer matrix.

## 5.5 Encapsulation of the Phosphorescent Molecules into Polymers

Polymers are widely used as carriers for phosphorescent dyes and selection of the specific material depends on the target application. The polymer provides the necessary protection for the phosphorescent dye from the environment and helps to tune oxygen sensitivity. The use of various polymers in oxygen sensing and phosphorescence imaging applications has been discussed in several reviews.[92–94]

Permeation rate of oxygen into a polymer film is characterized by the oxygen permeability coefficient $P$. Since the permeation of oxygen into the polymer layer includes the steps of its dissolution and further diffusion, the permeability coefficient is given by $P = D{\cdot}S$, where $D$ and $S$ are the diffusion constant and the oxygen solubility coefficient, respectively. Corresponding parameters for various polymers are given in Table 5.2.

Polysiloxanes possess exceptionally high oxygen solubility and diffusion constants compared to most other polymers, as reflected in Table 5.2. Matrices based on polysilicons were among the first applied in the development of optical oxygen sensors and are still used quite often.[95–99] Organic glassy polymers such as polystyrene, poly(methyl methacrylate), poly(isobutyl methacrylate), poly(vinyl chloride) have lower permeability but secure mechanical strength to thin-films. Among these polymers, polystyrene has been one of the most commonly used polymer matrixes for the preparation of oxygen

**Table 5.2** Permeation rate, solubility and diffusion constants of oxygen in different polymers at 20 °C.[124]

| Polymer | P($\times10^{13}$) $cm^3$(STP) $cm^{-2}$ s cmHg | D($\times10^6$) $cm^2\ s^{-1}$ | S ($\times10^6$) $cm^3$(STP) $cm^{-3}$ $cmHg^{-1}$ |
|---|---|---|---|
| Poly(dimethylsiloxane) | 695 | 40 | 24 |
| Poly(1-trimethylsily-1-propyne) | 7700 | 47 | 170 |
| Polyethylene (low density) | 0.7–0.9 | 46–60 | 21 |
| Polyethylene (high density) | 0.13–2.3 | 17 | 8 |
| Polystyrene | 2.63 | – | – |
| Poly(methylmethacrylate) | 9 | 10 | 8.5 |
| Poly(vinylchloride) | 0.34 | 1.2 | 2.9 |
| Poly(vinylacetate) | 0.4 | 5.5 | 6.2 |
| Poly(isobutylmethacrylate) | 20 | – | – |
| Poly(2,2,2-trifluoroethyl-methacrylate) | 32 | 15 | 0.27 |
| Ethylcellulose | 11.0 | 0.639 | 1.73 |
| Teflon | 1.65 | 15 | 1.3–2.5 |
| Cellulose acetobutyrate | 3.56 | – | – |
| Cellulose acetate | 5.85 | – | – |
| Water | – | 0.025 | – |

sensors due its good oxygen permeability, but complete impermeability to ions, low cost, and easiness in material manufacturing.[100–102] The dynamic range and sensitivity of the probes can be fine-tuned by employing differently substituted polystyrenes, *e.g.* poly(2,6-dichlorostyrene), poly(4-*tert*-butylstyrene) or poly(2,6-fluorostyrene).[103]

Introduction of fluorine atoms into the polymer backbone improves the sensor stability against photooxidation upon prolonged light irradiation. The C–F chemical bond is short and strong (binding energy of 116 kcal $mol^{-1}$), providing high resistance towards photooxidation. These polymers have high $D$ and $S$ values and are useful matrixes of optical oxygen sensor based on the phosphorescent metalloporphyrins.[104–107]

In the applications where oxygen quenching is an unwanted complication, *e.g.* in phosphorescence bio-imaging, low oxygen permeability matrixes are required. Wolfbeis and co-workers showed that ruthenium(II) complexes in films of poly(acrylonitrile) (PAN) are not sensitive to quenching by oxygen and applied this to develop luminescent temperature sensors.[108] Such behavior was observed for other phosphorescent complexes in beads made of PAN and various copolymers due to the low oxygen permeability. Their phosphorescence intensity and lifetime have been reported to be similar to the corresponding phosphorescent molecules in an oxygen-free environment.[109] PAN nanoparticles loaded with Ru(II) phosphorescent complexes were obtained by precipitating them from solutions in organic solvents.[110] Such particles were found to be impermeable to oxygen and negligible quenching of the phosphorescence was observed, making them suitable as phosphorescent labels or as oxygen-insensitive probes for temperature.

However, many phosphorescent dyes possess only limited solubility in PAN and related co-polymers. Halogen-containing polymers encapsulation matrix for organic and organometallic phosphorescent molecules, based on poly(vinylfluoride) (PVF), poly(vinylchloride) (PVC) and their co-polymers, were developed by Song and co-workers[111] to produce phosphorescent materials which are substantially unaffected by oxygen at ambient conditions. In contrast to PAN, these polymers provide a particularly high loading of the phosphorescent molecules. The phosphorescence intensity of Pt(II) tetra-*meso*-fluorophenylporphine (PtTMPFP) encapsulated into PAN particles was observed to drop at >1% loading due to the dye aggregation. For PVF and PVC the phosphorescence increases with loading of PtTMPFT up to 3%. With the same loading (1%) of the dye, PVF and PVC showed 4 and 2 times stronger phosphorescence than PAN particles, respectively. Nanoparticles based on these halogen-containing polymer matrixes and PtTMPFT have been prepared and covalently bound to antibodies to perform C-reactive protein phosphorescent assay.

## 5.6 Inorganic Matrix Materials

Inorganic materials for the fabrication of optical oxygen sensors attracted much attention due their high mechanical strength, thermal and oxidative stability. Various glasses, silica, sol–gels, organically modified silicates and metal oxides have been employed for this purpose.[94] The development of specific types of nanosensors for imaging applications, such as silica nanoparticles and hydrogel nanosensors, has been the subject of comprehensive recent reviews.[112,113] Specific aspects of such materials design related to the control of the phosphorescence quenching rates are discussed here.

It has been demonstrated that silicon oxide possesses excellent gas barrier properties due to very tight interstitial spaces of the Si–O lattice, resulting in oxygen diffusivities in the range $10^{-5}$–$10^{-9}$ $cm^2$ $s^{-1}$. Even nanometer-thick silica layers can act as an oxygen barrier in silica-coated polymer films.[114]

MacCraith and co-workers reported that oxygen quenching rates of the phosphorescent Ru(II) complexes in sol–gel silica films can be tuned to fit different oxygen concentration ranges in different environments.[115] These materials were prepared through hydrolysis of tetraethylorthosilicate precursor in acidic solution.[116] Quenching rates of the immobilized phosphor strongly depend on the porosity of the material and average pore size which, in turn, is determined by such parameters as the dip speed, water-to-precursor ratio, pH value and sol aging time. By adjusting these parameters, film properties can be tailored to optimise oxygen quenching sensitivity. Using these matrixes, optical oxygen sensors with controllable sensitivity towards quenching were fabricated based on PtOEP as a phosphor.[117–119]

The use of silanes modified with hydrophobic groups (*i.e.* compound of general formula R–Si$(OR)_3$, where R is alkyl or phenyl) as sol–gel precursors gives so-called ormosils, which were found to provide certain advantages as matrixes, particularly long-term quenching stability and good reproducibility within a batch of films.[120] Xerogel ormosils doped with platinum(II) porphyrins provide widely adjustable sensitivity towards quenching, which can be readily tuned by adjusting the xerogel composition and the phosphor.[121]

Winnik *et al.* have studied oxygen quenching of PtOEP in the hybrid polymers obtained from polydimethylsiloxane (PDMS), poly(n-butylaminothionylphosphazene) (C4PATP) and 10 nm silica particles.[122] Silica particles were shown to improve the mechanical properties of the matrix and affect oxygen diffusion and permeation. The quenching constants depended on the fraction of silica particles, which are acting as obstacles to oxygen diffusion.

Mirenda and co-workers reported phosphorescence properties of tris (bipyridine)ruthenium(II)-doped silica nanoparticles.[123] While the emission of the complex in solution is strongly quenched by oxygen, the emission of immobilized complex remains the same in $O_2$-free and $O_2$-saturated conditions. A similar effect has been observed by Zhang and co-workers for encapsulated $[Ru(phen)_3]^{2+}$, that was attributed to the low oxygen permeability of the silica shell.[124,125] In addition, encapsulated complexes showed higher photostability compared to free solutions. Leakage of dye molecules from the silica particles is negligible, due to strong electrostatic attractions between the positively charged ruthenium complex and the negatively charged silica allowing the use of such materials in bioanalytical assays.

## 5.7 Conclusions and Perspectives

A significant advance in applications of triplet excited state chromophores has been achieved during the last two decades. In particular, oxygen sensing techniques based on the phosphorescence quenching attracted much attention and efforts which included multistep synthesis of new probe molecules, design of nano-carriers, development of specific instrumentation and biological studies. These activities did not only enable impressive applications, such as 2D or 3D oxygen imaging in tissue, but also delivered new fundamental knowledge of the interactions between optically excited molecules and oxygen.

An ultimate requirement for accurate quantification of oxygen *via* phosphorescence quenching method is the possibility for tuning quenching rates and preventing photooxidation of sensor components. In this chapter, various approaches to control these processes are described. These can be generally divided into two categories referred to as “active” and “passive” protection. Attenuation of triplet state quenching in most of the probes reported so far is achieved through incorporation of the photoactive components into organic polymers or inorganic materials with defined oxygen permeability. However, applicability of this method in certain cases is limited, especially for sensing in living objects. Alternatively, chemical modification of the phosphorescent molecules with special groups allows both limiting oxygen diffusion and ensuring the desired localization of the probe or its interactions with biomolecules. Due to the fact that access to such tailor-made probes usually requires significant synthetic efforts, supramolecular chemistry approaches for the phosphor’s design are currently attracting attention.

On the other hand, the limited lifetime of sensor probes due to singlet oxygen-mediated photooxidation is another important issue in practical applications. Further efforts are needed to better understand the consequences of such deteriorative processes and to improve current protection approaches. An appealing solution is based on the introduction of "self-healing" phosphorescent molecules and matrix materials capable of reversible binding of singlet oxygen or its deactivation into ground state. Chemical toolkits that enable such transformations are under active development.

## Acknowledgements

The support from the European Commission (CONSORT, Grant No. 655142) is acknowledged.

## References


1. H. Kautsky, A. Hirsch and W. Baumeister, *Ber. Dtsch. Chem. Ges.*, 1931, **64**, 2053.
2. D. R. Kearns and A. U. Khan, *Photochem. Photobiol.*, 1969, **10**, 193.
3. C. Grewer and H.-D. Brauer, *J. Phys. Chem.*, 1994, **98**, 4230.
4. J. M. Vanderkooi, G. Maniara, T. J. Green and D. F. Wilson, *J. Biol. Chem.*, 1987, **262**, 5476.
5. D. F. Wilson, W. L. Rumsey, T. J. Green and J. M. Vanderkooi, *J. Biol. Chem.*, 1988, **263**, 2712.
6. D. B. Papkovsky and R. I. Dmitriev, *Chem. Soc. Rev.*, 2013, **42**, 8700.
7. T. V. Esipova, A. Karagodov, J. Miller, D. F. Wilson, T. M. Busch and S. A. Vinogradov, *Anal. Chem.*, 2011, **83**, 8756.
8. O. S. Finikova, A. Galkin, V. Rozhkov, M. Cordero, C. Hagerhall and S. A. Vinogradov, *J. Am. Chem. Soc.*, 2003, **125**, 4882.
9. D. F. Wilson, W. M. F. Lee, S. Makonnen, O. S. Finikova, S. Apreleva and S. A. Vinogradov, *J. Appl. Physiol.*, 2006, **101**, 1648.
10. Q. Zhao, M. X. Yu, L. X. Shi, S. J. Liu, C. Y. Li, M. Shi, Z. G. Zhou, C. H. Huang and F. Y. Li, *Organometallics*, 2010, **29**, 1085.
11. Q. Liu, B. R. Yin, T. S. Yang, Y. C. Yang, Z. Shen, P. Yao and F. Y. Li, *J. Am. Chem. Soc.*, 2013, **135**, 5029.
12. M. Chen, Z. Lei, W. Feng, C. Y. Li, Q. M. Wang and F. Y. Li, *Biomaterials*, 2013, **34**, 4284.
13. K. Koren, R. Dmitriev, S. Borisov, D. Papkovsky and I. Klimant, *ChemBioChem*, 2012, **13**, 1184.
14. A. V. Kondrashina, R. I. Dmitriev, S. Borisov, I. Klimant, I. O'Brien, Y. M. Nolan, A. Zdanov and D. B. Papkovsky, *Adv. Funct. Mater.*, 2012, **22**, 4931.
15. A. Fercher, S. Borisov, A. Zdanov, I. Klimant and D. B. Papkovsky, *ACS Nano*, 2011, **5**, 5499.
16. S. Draxler, in *Advanced Concepts in Fluorescence Sensing*, ed. C. D. Geddes and J. R. Lakowicz, Springer, Norwell, 2005, pp. 241–274.

17. P. R. Ogilby, *Chem. Soc. Rev.*, 2010, **39**, 3181.
18. M. A. Filatov, S. Baluschev, I. Z. Ilieva, V. Enkelman, T. Miteva, K. Landfester, S. E. Aleshchenkov and A. V. Cheprakov, *J. Org. Chem.*, 2012, **77**, 11119.
19. P. N. Prasad, *Introduction to Biophotonics*, Wiley-Interscience, New York, 2004, p. 162.
20. M. A. Filatov, S. Baluschev and K. Landfester, *Chem. Soc. Rev.*, 2016, **45**, 4668.
21. S. Baluschev, K. Katta, Y. Avlasevich and K. Landfester, *Mater. Horiz.*, 2016, **3**, 478.
22. I. B. Rietveld, E. Kim and S. A. Vinogradov, *Tetrahedron*, 2003, **59**, 3821.
23. A. G. Moiseev, E. A. Margulies, J. A. Schneider, F. Beanger-Garieyb and D. F. Perepichka, *Dalton Trans.*, 2014, **43**, 2676.
24. R. H. Jin, T. Aida and S. Inoue, *Chem. Commun.*, 1993, 1260.
25. R. Sadamoto, N. Tomioka and T. Aida, *J. Am. Chem. Soc.*, 1996, **118**, 3978.
26. M. Kimura, K. Nakada, Y. Yamaguchi, K. Hanabusa, H. Shirai and N. Kobayashi, *Chem. Commun.*, 1997, 1215.
27. J. M. Riley, S. Alkan, A. D. Chen, M. Shapiro, W. A. Khan, W. R. Murphy and J. E. Hanson, *Macromolecules*, 2001, **34**, 1797.
28. J. Issberner, F. Vogtle, L. DeCola and V. Balzani, *Chem.–Eur. J.*, 1997, **3**, 706.
29. F. Vogtle, M. Plevoets, M. Nieger, G. C. Azzellini, A. Credi, L. De Cola, V. De Marchis, M. Venturi and V. Balzani, *J. Am. Chem. Soc.*, 1999, **121**, 6290.
30. V. Rozhkov, D. Wilson and S. Vinogradov, *Macromolecules*, 2002, **35**, 1991.
31. S. A. Vinogradov, L.-W. Lo and D. F. Wilson, *Chem.–Eur. J.*, 1999, **5**, 1338.
32. S. A. Vinogradov, *Org. Lett.*, 2005, **7**, 1761.
33. O. S. Finikova, S. Y. Chernov, A. V. Cheprakov, M. A. Filatov, S. A. Vinogradov and I. P. Beletskaya, *Dokl. Chem.*, 2003, **391**, 222.
34. A. Y. Lebedev, A. V. Cheprakov, S. Sakadzic, D. A. Boas, D. F. Wilson and S. A. Vinogradov, *ACS Appl. Mater. Interfaces*, 2009, **1**, 1292.
35. A. V. Cheprakov and M. A. Filatov, *J. Porphyrins Phthalocyanines*, 2009, **13**, 291.
36. M. A. Filatov and A. V. Cheprakov, *Tetrahedron*, 2011, **67**, 3559.
37. I. Dunphy, S. A. Vinogradov and D. F. Wilson, *Anal. Biochem.*, 2002, **310**, 191.
38. M. A. Yaseen, V. J. Srinivasan, S. Sakadžić, W. Wu, S. Ruvinskaya, S. A. Vinogradov and D. A. Boas, *Opt. Express*, 2009, **17**, 22341.
39. D. F. Wilson, W. M. Lee, S. Makonnen, O. Finikova, S. Apreleva and S. A. Vinogradov, *J. Appl. Physiol.*, 2006, **101**, 1648.
40. S. V. Apreleva, D. F. Wilson and S. A. Vinogradov, *Appl. Opt.*, 2006, **45**, 8547.
41. E. Roussakis, Z. Li, N. H. Nowell, A. J. Nichols and C. L. Evans, *Angew. Chem., Int. Ed.*, 2015, **54**, 14728.
42. P. G. L. Koolen, Z. Li, E. Roussakis, M. A. Paul, A. M. S. Ibrahim, R. Matyal, T. Huang, C. L. Evans and S. J. Lin, *Plast. Reconstr. Surg.*, 2017, **140**, 89.

43. J.-M. Aubry, C. Pierlot, J. Rigaudy and R. Schmidt, *Acc. Chem. Res.*, 2003, **36**, 668.
44. M. A. Filatov and M. O. Senge, *Mol. Syst. Des. Eng.*, 2016, **1**, 258.
45. M. A. Filatov, E. Heinrich, D. Busko, I. Z. Ilieva, K. Landfester and S. Baluschev, *Phys. Chem. Chem. Phys.*, 2015, **17**, 6501.
46. S. Callaghan, M. A. Filatov, E. Sitte, H. Savoie, R. W. Boyle, K. J. Flanagan and M. O. Senge, *Photochem. Photobiol. Sci.*, 2017, **16**, 1371.
47. Y. Kai and K. Imakubo, *Photochem. Photobiol.*, 1979, **29**, 261.
48. M. L. Saviotti and W. C. Galley, *Proc. Natl. Acad. Sci. U. S. A.*, 1974, **71**, 4154.
49. M. W. Geiger and N. J. Turro, *Photochem. Photobiol.*, 1975, **22**, 273.
50. G. B. Strambini and P. Cioni, *J. Am. Chem. Soc.*, 1999, **121**, 8337.
51. P. Cioni, *Biophys. Chem.*, 2000, **87**, 15.
52. N. J. Turro, G. Sidney Cox and X. Li, *Photochem. Photobiol.*, 1983, **37**, 149.
53. Y. L. Peng, Y. T. Wang, Y. Wang and W. J. Jin, *J. Photochem. Photobiol., A*, 2005, **173**, 301.
54. W. K. Hartmann, M. H. B. Gray, A. Ponce, D. G. Nocera and P. A. Wong, *Inorg. Chim. Acta*, 1996, **243**, 239.
55. J. Kuijt, F. Ariese, U. A. Th. Brinkman and C. Gooijer, *Anal. Chim. Acta*, 2003, **488**, 135.
56. S. Hamai, *J. Phys. Chem.*, 1995, **99**, 12109.
57. S. Hamai, *Chem. Commun.*, 1994, **19**, 2243.
58. G. Grabner, K. Rechthaler, B. Mayer and G. Kohler, *J. Phys. Chem. A*, 2000, **104**, 1365.
59. Y. Zhao, L. D. Li and A. J. Tong, *Chin. J. Anal. Chem.*, 1997, **25**, 1016.
60. Y. Zhang and D. Johnson, *Sens. Actuators, B*, 2003, **96**, 379.
61. Y. Zhang and D. Johnson, *Anal. Chim. Acta*, 2004, **511**, 333.
62. Z.-W. Gao, X. Feng, L. Mu, X.-L. Ni, L.-L. Liang, S.-F. Xue, Z. Tao, X. Zeng, B. E. Chapman, P. W. Kuchel, L. F. Lindoy and G. Wei, *Dalton Trans.*, 2013, **42**, 2608.
63. N. Jayaraj, M. V. Maddipatla, R. Prabhakar, S. Jockusch, N. J. Turro and V. Ramamurthy, *J. Phys. Chem. B*, 2010, **114**, 14320.
64. Y. Wang, W.-J. Jin, J.-B. Chao and L.-P. Qin, *Supramol. Chem.*, 2003, **15**, 459.
65. G.-R. Li, J.-J. Wu, W.-J. Jin and J.-W. Xie, *Talanta*, 2003, **60**, 555.
66. Y.-T. Wang, X.-W. Wang and Y. Zhang, *Photochem. Photobiol.*, 2011, **87**, 772.
67. S. Hirata, K. Totani, J. Zhang, T. Yamashita, H. Kaji, S. R. Marder, T. Watanabe and C. Adachi, *Adv. Funct. Mater.*, 2013, **23**, 3386.
68. M. Shirakawa, N. Fujita, T. Tani, K. Kanekob and S. Shinkai, *Chem. Commun.*, 2005, 4149.
69. M. Shirakawa, N. Fujita, T. Tani, K. Kaneko, M. Ojima, A. Fujii, M. Ozaki and S. Shinkai, *Chem.–Eur. J.*, 2007, **13**, 4155.
70. N. Kumar Allampally, M. Bredol, C. A. Strassert and L. De Cola, *Chem.–Eur. J.*, 2014, **20**, 16863.
71. H. Wang, H. Wang, X. Yang, Q. Wang and Y. Yang, *Langmuir*, 2015, **31**, 486.

72. N. Velasco-Garcia, M. J. Valencia-Gonzalez and M. E. Díaz García, *Analyst*, 1997, **122**, 1405.
73. S. H. C. Askes, N. Lopez Mora, R. Harkes, R. I. Koning, B. Koster, T. Schmidt and S. Bonnet, *Chem. Commun.*, 2015, **51**, 9137.
74. M. E. Díaz García and A. Sanz-Medel, *Anal. Chem.*, 1986, **58**, 1436.
75. B. Cañabate-Díaz, A. Segura Carretero, C. Cruces-Blanco and A. Fernández-Gutiérrez, *Analyst*, 2003, **128**, 411.
76. M. R. Fernández de la Campa, Y. Ming Liu, M. E. Díaz García and A. Sanz Medel, *Anal. Chim. Acta*, 1990, **238**, 297.
77. A. Segura-Carretero, C. Cruces-Blanco and A. Fernández-Gutiérrez, *Talanta*, 1996, **43**, 1001.
78. A. Segura Carretero, C. Cruces Blanco and A. Fernández Gutiérrez, *Anal. Chim. Acta*, 1997, **353**, 337.
79. A. Segura-Carretero, C. Cruces-Blanco, B. Cañabate-Diaz, J. F. Fernández-Sánchez and A. Fernández-Gutiérrez, *Anal. Chim. Acta*, 2000, **417**, 19.
80. J. A. Arancibia and G. M. Escandar, *Analyst*, 2001, **126**, 917.
81. C. O'Donovan, J. Hynes, D. Yashunski and D. B. Papkovsky, *J. Mater. Chem.*, 2005, **15**, 2946.
82. R. E. Benesch and R. Benesch, *Science*, 1953, **118**, 447.
83. R. L. Plant and D. H. Burns, *Appl. Spectrosc.*, 1993, **47**, 1594.
84. Q. Liu, B. R. Yin, T. S. Yang, Y. Yang, Z. Shen, P. Yao and F. Y. Li, *J. Am. Chem. Soc.*, 2013, **135**, 5029.
85. D. P. Jason, G. O. Brian, P. B. Justin, K. H. Mohammad, R. Jo Ann, S. W. Jeffrey, W. R. James and N. Sergei, *High-oxygen Barrier Materials Based on Hyperbranched Aliphatic Polyesters. Polymer Degradation and Performance*, American Chemical Society, Washington, 2009, pp. 17–30.
86. K. Õige, T. Avarmaa, A. Suisalu and R. Jaaniso, *Sens. Actuators, B*, 2005, **106**, 424.
87. P. Hartmann, M. J. P. Leiner and P. Kohlbacher, *Sens. Actuators, B*, 1998, **51**, 196.
88. P. Hartmann, *Anal. Chem.*, 2000, **72**, 2828.
89. C. Schweitzer and R. Schmidt, *Chem. Rev.*, 2003, **103**, 1685.
90. S. R. Ricketts and P. Douglas, *Sens. Actuators, B*, 2008, **135**, 46.
91. B. Enko, S. M. Borisov, J. Regensburger, W. Bäumler, G. Gescheidt and I. Klimant, *J. Phys. Chem. A*, 2013, **117**, 8873.
92. B. Adhikari and S. Majumdar, *Prog. Polym. Sci.*, 2004, **29**, 699.
93. G. Korotcenkov, *Handbook of Gas Sensor Materials*, Springer, New York, 2014, pp. 249–263.
94. X.-D. Wang and O. S. Wolfbeis, *Chem. Soc. Rev.*, 2014, **43**, 3666.
95. J. R. Bacon and J. N. Demas, *Anal. Chem.*, 1987, **59**, 2780.
96. M. E. Cox and B. Dunn, *Appl. Opt.*, 1985, **24**, 2114.
97. T. Ishiji and M. Kaneko, *Analyst*, 1995, **120**, 1633.
98. E. R. Carraway, J. N. Demas and B. A. DeGraff, *Langmuir*, 1991, **7**, 2991.
99. S. B. Bambot, R. Holavanahali, J. R. Lakowicz, G. M. Carter and G. Rao, *Biotechnol. Bioeng.*, 1994, **43**, 1139.
100. S.-K. Lee and I. Okura, *Anal. Commun.*, 1997, **34**, 185.

101. E. Schmälzlin, J. T. van Dongen, I. Klimant, B. Marmodee, M. Steup, J. Fisahn, P. Geigenberger and H. G. Löhmannsröben, *Biophys. J.*, 2005, **89**, 1339.
102. A. Bizzarri, H. Koehler, M. Cajlakovic, A. Pasic, L. Schaupp, I. Klimant and V. Ribitsch, *Anal. Chim. Acta*, 2006, **573**, 48.
103. K. Koren, L. Hutter, B. Enko, A. Pein, S. M. Borisov and I. Klimant, *Sens. Actuators, B*, 2013, **176**, 344.
104. Y. Amao, I. Okura, H. Shinohara and H. Nishide, *Polym. J.*, 2002, **34**, 411.
105. Y. Amao, K. Asai, I. Okura, H. Shinohara and H. Nishide, *Analyst*, 2000, **125**, 1911.
106. N. Yi-Yan, R. M. Felder and W. J. Koros, *J. Appl. Polym. Sci.*, 1980, **25**, 1755.
107. W. A. Zisma, *Adv. Chem. Ser.*, 1964, **43**, 1.
108. G. Liebsch, I. Klimant and O. S. Wolfbeis, *Adv. Mater.*, 1999, **11**, 1296.
109. J. M. Kürner, I. Klimant, C. Krause, H. Preu, W. Kunz and O. S. Wolfbeis, *Bioconjugate Chem.*, 2001, **12**, 883.
110. O. S. Wolfbeis, *Chem. Soc. Rev.*, 2015, **44**, 4743.
111. X. Song, L. Huang and B. Wu, *Anal. Chem.*, 2008, **80**, 5501.
112. B. Korzeniowska, R. Nooney, D. Wencel and C. McDonagh, *Nanotechnology*, 2013, **24**, 442002.
113. A. Ray and R. Kopelman, *Nanomedicine*, 2013, **8**, 1829.
114. Y. Leterrier, *Prog. Mater. Sci.*, 2003, **48**, 1.
115. A. K. McEvoy, C. McDonagh and B. D. MacCraith, *J. Sol–Gel Sci. Technol.*, 1997, **8**, 1121.
116. B. D. MacCraith, C. M. McDonagh, G. O'Keeffe, E. T. Keyes, J. G. Vos, B. O'Kelly and J. F. McGilp, *Analyst*, 1993, **118**, 385.
117. S. K. Lee and I. Okura, *Analyst*, 1997, **122**, 81.
118. T.-S. Yeh, C.-S. Chu and Y.-L. Lo, *Sens. Actuators, B*, 2006, **119**, 701.
119. C.-S. Chu and Y.-L. Lo, *Sens. Actuators, B*, 2007, **124**, 376.
120. C. McDonagh, B. D. MacCraith and A. K. McEvoy, *Anal. Chem.*, 1998, **70**, 45.
121. V. S. Tripathi, G. Lakshminarayana and M. Nogami, *Sens. Actuators, B*, 2010, **147**, 741.
122. X. Lu, I. Manners and M. A. Winnik, *Macromolecules*, 2001, **34**, 1917.
123. M. Mirenda, V. Levi, M. L. Bossi, L. Bruno, A. V. Bordoni, A. E. Regazzoni and A. Wolosiuk, *J. Colloid Interface Sci.*, 2013, **392**, 96.
124. D. Zhang, Z. Wu, J. Xu, J. Liang, J. Li and W. Yang, *Langmuir*, 2010, **26**, 6657.
125. L. M. Rossi, L. Shi, F. H. Quina and Z. Rosenzweig, *Langmuir*, 2005, **21**, 4277.

CHAPTER 6

# Progress in Phosphorescence Lifetime Measurement Instrumentation for Oxygen Sensing

ANGEL DE LA TORRE-VEGA*[a], SANTIAGO MEDINA-RODRÍGUEZ[b], CARLOS MEDINA-RODRÍGUEZ[b] AND JORGE F. FERNÁNDEZ-SÁNCHEZ[c]

[a]Department of Signal Theory, Telematics and Communications, University of Granada, 18071, Granada, Spain; [b]MODOTIC S.L., C/ Jardines 7, 18195, Cúllar Vega (Granada), Spain; [c]Department of Analytical Chemistry, University of Granada, 18071, Granada, Spain
*E-mail: atv@ugr.es

## 6.1 Introduction

The phosphorescence of a sensing phase is associated to the radiative deactivation of excited states in a phosphorescent material. In oxygen sensitive materials, the phosphorescence is altered by the presence of oxygen because the oxygen provides non-radiative deactivation mechanisms from the excited to the fundamental states. The non-radiative deactivation quenches the phosphorescence and reduces both the luminescent intensity and the luminescence lifetime. Since the magnitude of the quenching is a function of the

Detection Science Series No. 11
Quenched-phosphorescence Detection of Molecular Oxygen: Applications in Life Sciences
Edited by Dmitri B. Papkovsky and Ruslan I. Dmitriev

Published by the Royal Society of Chemistry, www.rsc.org

oxygen concentration, an appropriate measurement of the phosphorescence intensity and/or lifetime, together with an appropriate calibration, is the principle of phosphorescence instrumentation for oxygen sensing.[1–4]

The phosphorescent sensing phase can be considered as a linear system, excited with a light source and generating a phosphorescence response according to its corresponding linear differential equation. The relationship between the input and the output of the sensing phase can be equivalently described with the linear differential equation (*i.e.* the coefficients of the equation), its impulsive response (*i.e.* the phosphorescence response to an impulsive excitation) or its frequency response (*i.e.* the attenuation and the phase-shift of the response when it is excited with a sinusoidally modulated source). As in any linear system, the frequency response $H(j\omega)$ is the Fourier's transform of the impulsive response $h(t)$ (where $\omega = 2\pi f$ is the angular frequency in radians/s and $j$ is the imaginary unit of complex numbers), and the emission response $x_{em}(t)$ to an arbitrary excitation $x_{exc}(t)$ can be calculated either in the time domain as the convolution of the excitation and the impulsive response, $x_{em}(t) = x_{exc}(t) * h(t)$, or in the frequency domain as the product of the Fourier's transform of the excitation signal with the frequency response, $X_{em}(j\omega) = X_{exc}(j\omega) \cdot H(j\omega)$.[5,6]

The simplest model for describing phosphorescent sensing phases is a first order differential equation, with a time constant (or lifetime) as unique relevant coefficient, which is a function of the analyte concentration. The solution of this equation describing the phosphorescence when the system is excited with a pulsed light is a mono-exponential decay characterized by the lifetime constant, while the frequency response is a first order low-pass filter with a cut-off frequency inversely proportional to the lifetime.[7] The change of the lifetime with the quencher concentration is reasonably described by the Stern–Volmer equation as a first approach.[1,8] Even though the Stern–Volmer equation and the first order differential equation qualitatively describe a wide range of sensing phases (and also provide accurate modelling for some of them), a higher order differential equation and a more complex description of the quenching effect are necessary to provide an accurate description of many sensing phases.[9]

From the point of view of instrumentation for oxygen sensing, the phosphorescence lifetime can be determined in the time domain or in the frequency domain. In the first case, a pulsed illumination source is required and the oxygen determination is obtained by analyzing the evolution in time of the phosphorescence response (which is approximately a mono-exponential decay). The frequency domain measurements require a modulated illumination source and the analysis of the amplitude and/or phase of the phosphorescence response relative to those of the excitation signal (which approximately correspond to a first order low-pass filter). Even though both are mathematically equivalent, there are some differences to be considered for designing a phosphorescence instrument:

- Usually the phosphorescent system is not a first order model (*i.e.* the mono-exponential decay or the first order low-pass filter are not accurate enough). In this case a lifetime constant is not well defined, and in time domain measurements, the observed lifetime depends on the time interval used for measuring it.[1,4]

- The measurements are always affected by background noise due to the noise generated by the optoelectronic devices, amplifiers and background illumination. Additionally, the signal-to-noise ratio (SNR) is particularly low in some applications, such as in the case of small samples (due to the small amount of light emitted by the sensing phase).[10,11] Dealing with noise (and therefore optimizing the accuracy in the measurements) is significantly easier in the frequency domain than in the time domain.[12,13]
- Time domain measurements are based on pulsed illumination sources and require expensive and complex instrumentation using high-speed photodetectors.[1] On the other hand, frequency domain lifetime determination can be implemented with cheaper light sources and electronic components.[14,15]

In recent years, optoelectronic devices appropriate for frequency domain measurements (LEDs for medium power at short wavelengths, photodiodes, preamplified photodiodes, phototransistors, *etc.*) have become cheaper. Additionally, small computers (Raspberry-Pi) and microcontrollers have improved their capabilities (in terms of available memory, computational efficiency, velocity, analog-to-digital (AD) and digital-to-analog (DA) conversion, interconnectivity with other systems, power consumption, *etc.*) at a reasonable cost. These components and devices provide a new context in which flexible and non-expensive phosphorescence instruments based on frequency domain measurements can be designed.[16–20] A modular architecture like that proposed in ref. 21, 22 considers the phosphorescent system as a linear system, modelled with a linear differential equation (or a frequency response), which is excited with a modulated light source and responds with a phosphorescence recorded by a photodetector device. A computer or a microcontroller can be used to prepare a digital signal to modulate the illumination source (the digital signal is DA-converted and amplified to excite a LED source). The electrical signal provided by the photodetector can be amplified and AD-converted, and the digital signal can be processed by a microcontroller or a computer.

Under this perspective, the flexibility for exciting the sensing phase and processing the phosphorescence response can be exploited in order to optimize the information provided by the phosphorescence instrument. Several obvious advantages are derived from the proposed architecture. Since both the excitation and the recorded signals are digital, signal processing can be performed with numerical algorithms programmed in a microcontroller or a computer, and the complexity of the algorithms is not a strong limitation in the design of the phosphorescence instrument. Since the phosphorescent system is assumed to be linear, all the signal theory related to linear systems (and commonly applied in modern communication systems) can be applied and exploited.[5,7] And finally, since the configuration of the measuring procedure requires just programming an algorithm, the design of the instrument can be easily adapted to specific requirements or measuring scenarios in order to optimize the information provided by the sensing phase. In this chapter we describe how signal processing applied to instruments based on this architecture can be exploited to provide an optimal characterization of the sensing phase, to improve the accuracy of the oxygen determination, to provide a confidence interval of the oxygen determination (based on the estimated background noise), or to optimally integrate information from several harmonics in a single accurate oxygen determination.

## 6.2 The Phosphorescence Emission as a Linear System

This section describes the fundamental equations of a phosphorescent system, and how it can be considered as a linear system described by its corresponding differential equation or its frequency response.

### 6.2.1 First Order Phosphorescent Systems

In the simplest phosphorescence model, the population of the excited state $N(t)$ increases linearly with the illumination $x_{exc}(t)$ and decreases linearly with the population according to the constants $\Gamma_r$ and $\Gamma_{nr}$ describing, respectively, the radiative and the non-radiative decays from the excited to the fundamental states:

$$\frac{dN(t)}{dt} = a_0 x_{exc}(t) - \left(\Gamma_r + \Gamma_{nr}\right) N(t) \tag{6.1}$$

where $a_0$ is a constant that relates the illumination used as excitation with the increase of the excited state population in the phosphorescent material. This first order linear differential equation is characterized by a time constant or lifetime that is inversely proportional to the sum of the decay constants. The simplest quenching model considers that the non-radiative decay constant increases linearly with the quencher concentration $C$ and, in this case, the dependence of the lifetime with $C$ is described by the Stern–Volmer equation:

$$\tau_q = \frac{1}{\Gamma_r + \Gamma_{nr}} = \frac{1/\Gamma_r}{1 + \Gamma_{nr}/\Gamma_r} = \frac{\tau_0}{1 + kC} \tag{6.2}$$

where $k$ is the Stern–Volmer constant and $\tau_0$ is the lifetime in absence of quencher.[1,8]

The phosphorescence emission $x_{em}(t)$ is proportional to the radiative decay ($\Gamma_r N(t)$) and, therefore, the phosphorescence can be described with a similar differential equation:

$$\frac{dx_{em}(t)}{dt} = a_1 x_{exc}(t) - \frac{x_{em}(t)}{\tau_q} \tag{6.3}$$

where $a_1$ is a constant that relates excitation with emission. This equation can be rewritten in the Fourier domain:[6]

$$j\omega\tau_q X_{em}(j\omega) = \tau_q\, a_1\, X_{exc}(j\omega) - X_{em}(j\omega) \tag{6.4}$$

where the derivative operator has been transformed into $j\omega$ and $X_e(j\omega)$ is the Fourier transform of $x_e(t)$ (for both the excitation or the emission signals). The frequency response of the phosphorescent system is therefore:

$$H(j\omega) = \frac{X_{em}(j\omega)}{X_{exc}(j\omega)} = \frac{\tau_q a_1}{1 + j\omega\tau_q} \tag{6.5}$$

or equivalently, using frequency instead of angular frequency:

$$H(C,f) = M_0 \frac{\tau_q}{\tau_0} \frac{1}{1 + j2\pi f\tau_q} \tag{6.6}$$

where $M_0$ is the modulation-factor (ratio between emission and excitation) at null quencher concentration and low frequencies, and the dependence of the lifetime with the quencher concentration is given by eqn (6.2). This equation describes that an increment of the quencher concentration reduces both the lifetime and the luminescence intensity. Additionally, the frequency response of the phosphorescent system is a low-pass filter with a cut-off frequency that increases with the quencher concentration.

The frequency response is a complex function, *i.e.*, it contains real and imaginary parts, or, equivalently, magnitude and phase. The magnitude or modulation-factor describes the ratio between the amplitudes of the emission and the excitation when the phosphorescent system is excited with a sinusoidally modulated illumination:[1]

$$m(C,f) = \sqrt{|H(C,f)|^2} = M_0 \frac{\tau_q}{\tau_0} \frac{1}{\sqrt{1 + (2\pi f\tau_q)^2}} \tag{6.7}$$

and the phase-shift represents the delay between both sinusoidal waves:[1]

$$\phi(C,f) = -\arctan(2\pi f\tau_q) \tag{6.8}$$

### 6.2.2 Multi-exponential Phosphorescent Systems

A usual model for describing higher order phosphorescent systems includes several phosphorescence processes. This multi-site phosphorescence model is described by a frequency response including the additive contribution of each individual process:

$$H(C,f) = M_{01} \frac{\tau_{q1}}{\tau_{01}} \frac{1}{1 + j2\pi f\tau_{q1}} + M_{02} \frac{\tau_{q2}}{\tau_{02}} \frac{1}{1 + j2\pi f\tau_{q2}} + \ldots \tag{6.9}$$

where $M_{0i}$, $\tau_{0i}$ and $\tau_{qi}$ are associated with the *i*th process and $\tau_{qi}$ describes the reduction of the *i*th lifetime according to a Stern–Volmer constant $k_i$. The frequency response is again a complex function with real and imaginary parts, or with magnitude and phase describing the modulation-factor and the phase-shift of the emission with respect to the excitation signals, for each frequency component *f*.

The time domain version of the multi-site model is an impulsive response (obtained as the inverse Fourier transform of the frequency response) consisting of a multi-exponential decay (additive contribution of several exponential processes), dominated by the shortest lifetimes for the early part of the decay and by the longest lifetimes for the late part of the decay. Obviously, there is a lifetime $\tau_{qi}$ defined for each subprocess but the global process cannot be described as a mono-exponential decay. For this reason, the estimation of $\tau_q$ in a multi-exponential system is an apparent lifetime (rather than a proper lifetime) whose value depends on the procedure defined for measuring it.[1]

### 6.2.3 Lifetime Derived from Modulation Factor and Phase-shift

From eqn (6.7) and (6.8), modulation-factor and phase-shift based lifetimes can be derived for mono-exponential phosphorescent systems:[2]

$$\tau_\phi(f) = \frac{-\tan(\phi(f))}{2\pi f} \tag{6.10}$$

$$\tau_m(f) = \tau_0 \frac{m(f)/m_0(f)}{\sqrt{1 + (2\pi f \tau_0)^2 \left(1 - (m(f)/m_0(f))^2\right)}} \tag{6.11}$$

where $m_0(f)$ is the modulation-factor at frequency *f* and at null quencher concentration†. Both lifetime estimations should be equal in mono-exponential systems, but they should be considered apparent lifetimes in a general phosphorescent system and both estimations are not expected to be equal. Similarly, the estimated lifetimes at different modulation frequencies should be identical in a mono-exponential system, but apparent lifetimes could change with the frequency in a general phosphorescent system.[1] From the instrumentation point of view, phase-shift lifetime estimations are usually preferred to those based on modulation-factors because the degradation of the phosphorescent material[23] or sub-optimal optical

†Note that for low modulation frequencies (*i.e.* much smaller than the cut-off frequency) $\tau_m(f) \approx \tau_0 \frac{m(f)}{m_0(f)} \approx \tau_0 \frac{I}{I_0}$, and the modulation-factor based lifetime can be estimated with the phosphorescence intensity relative to that at null quencher concentration.

coupling (when some optical element is replaced) reduces the constant $m_0(f)$ and an accurate determination of $\tau_m(f)$ usually requires a frequent recalibration of $m_0(f)$.

### 6.2.4 Modelling and Calibration of Phosphorescent Systems

Assuming that the phosphorescent system is linear, the description of the phosphorescence at a given quencher concentration is provided by the frequency response, and therefore a complete description of the phosphorescence requires the estimation of the frequency response at different quencher concentrations. A consistent description of the phosphorescence, appropriate for multi-frequency measurements, should avoid the use of apparent lifetimes and would fit the multi-exponential frequency response proposed in eqn (6.9) for different frequencies and quencher concentrations.

Since many phosphorescence instruments work at a single frequency, an alternative description consists of the estimation of the apparent lifetime (either modulation-factor or phase-shift based) at the selected modulation frequency as a function of the quencher concentration.[2,23–25] Usually, the Demas model (derived from the multi-exponential model) is applied to describe the dependence of the apparent lifetime with the quencher concentration:[1,26]

$$\tau(C,f) = \tau_0(f)\left(\frac{x_1(f)}{1 + k_1(f)C} + \frac{x_2(f)}{1 + k_2(f)C} + \dots\right) \tag{6.12}$$

where the dependence with the frequency is indicated to remark that the Stern–Volmer constants $k_i$, the apparent lifetimes (at null and at $C$ concentrations) and the constants $x_i$ can change with the frequency due to the inconsistency inherent to the use of apparent lifetimes.

For modelling purposes, a global modelling using eqn (6.9) (providing consistent information about individual subprocesses) is preferable. For oxygen determination purposes, the Demas model applied at each individual frequency to each apparent lifetime provides more flexibility in the modelling (more constants to be fitted) and a more accurate determination of the oxygen concentration can be obtained if the measuring instrument is appropriately calibrated.[2,9]

## 6.3 Architecture for Measuring the Frequency Response

As previously discussed, a phosphorescent system appropriate for oxygen determination can be modeled as a linear system characterized by its frequency response (which significantly changes according to the oxygen concentration). A first order model (given by eqn (6.6) and (6.2)) qualitatively describes most phosphorescent systems, even though an accurate enough description requires the multi-exponential model in eqn (6.9). In any case, the phosphorescence instrumentation for oxygen determination can be conceived as an instrument for estimating the frequency response (*i.e.* the modulation-factor and the phase-shift) at one or several modulation frequencies, and more specifically, to accurately measure the changes in the frequency response associated to variations in the oxygen concentration.

### 6.3.1 Modular Architecture of the Phosphorescence Instrument

Medina *et al.* have proposed a modular architecture for measuring the frequency response of the phosphorescent system.[2,21,27,28] As described in Figure 6.1, a central system (a computer or a microcontroller) configures the signal to be used as excitation (a digital signal at a sampling frequency, with a specific duration, containing one or several frequency components). This signal is digital-to-analog converted and amplified in order to provide a modulated illumination with a LED at the excitation wavelength. The sensing phase, excited by the LED illumination, generates a phosphorescence response at the emitting wavelength conditioned by its frequency response. The phosphorescence response is converted into an electrical signal by the photodetector. This signal is amplified and analog-to-digital converted. Both the excitation and the emission digital signals are then processed by the central system in order to estimate the frequency response (modulation-factor and phase-shift) of the sensing phase at the modulation frequency (or at several modulation frequencies if the excitation signal contains several frequency components). This frequency response is used either for the characterization of the sensing phase or for the determination of the oxygen concentration (after appropriate calibration).

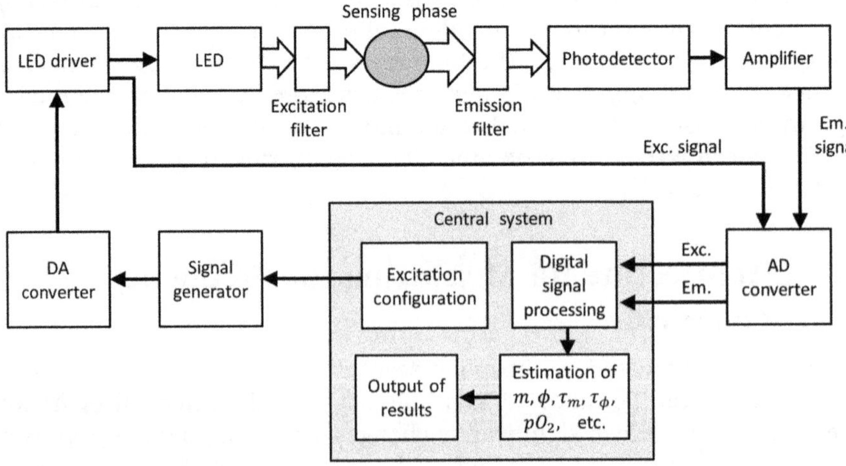


**Figure 6.1** Architecture of the phosphorescence instrument for estimating the frequency response.

This modular architecture allows an easy adaptation to specific requirements. For example, the excitation can be prepared with a computer and a DA board (for an easier and faster configuration) or with a microcontroller (for a cheaper and more portable instrument). The modulation frequency, amplitude and duration of the excitation signal can be programmed (in order to acquire faster measurements or more accurate measurements). The LED driver and the LED can be chosen in order to accommodate the illumination power and excitation wavelength requirements. Similarly, the photodetector and its amplifier can be chosen in order to accommodate the emission wavelength, the phosphorescence intensity and the detector sensitivity. The phosphorescence response can be recorded with a digital oscilloscope, with an AD board or with a microcontroller (with the corresponding implications regarding costs and flexibility). A two-channel AD converter is recommendable to simultaneously record both the excitation and the emission signals, in order to guarantee synchronization between both signals and accurately measure the phase-shift. Digital signal processing of the samples as well as the calibration and determination of the oxygen concentration can be performed with a microcontroller or with a computer.

### 6.3.2 Estimation of the Frequency Response

In order to estimate the frequency response of the phosphorescent system at a given modulation frequency $f_m$, the excitation signal should contain a component of this frequency. The emission signal will then contain a component of this modulation frequency. The digital versions of the excitation and emission signals, digitized at a sampling frequency $f_s$, are:

$$x_e[l] = A_e \cos(2\pi f_m l/f_s + \phi_e) + N_e[l] \tag{6.13}$$

where $l$ is the sample index, subindex e stands for excitation or emission signals, $A_e$ and $\phi_e$ are the amplitude and phase of the signals, and $N_e[l]$ is the noise affecting the signals (including the DC[‡] component, other frequency components and random noise). The correlation of the input signal with a pair of sinusoidal signals in quadrature (a cosine and a sine of frequency $f_m$) provide the in-phase and the quadrature components:[2,7]

$$I_e(f_m) = \frac{1}{K}\sum_l x_e[l]\cos(2\pi f_m l/f_s) \tag{6.14}$$

$$Q_e(f_m) = \frac{1}{K}\sum_l x_e[l]\sin(2\pi f_m l/f_s) \tag{6.15}$$

‡DC (from "Direct Current"): level of continuous signal or, equivalently, component at null frequency.

where $K$ is a normalization constant. The I/Q components provide an estimation of the amplitude and phase of the $f_m$ component of $x_e[l]$:

$$A_e(f_m) = \sqrt{I_e^2(f_m) + Q_e^2(f_m)} \tag{6.16}$$

$$\phi_e(f_m) = -\arctan(Q_e(f_m)/I_e(f_m)) \tag{6.17}$$

and from the amplitude and phase of the excitation and emission signals, the modulation-factor and the phase-shift of the frequency response can be estimated at the frequency $f_m$:

$$m(f_m) = A_{em}(f_m)/A_{exc}(f_m) \qquad \phi(f_m) = \phi_{em}(f_m) - \phi_{exc}(f_m) \tag{6.18}$$

This procedure, known as In-phase/Quadrature estimation (or I/Q),[2,7] is also the principle of lock-in amplifiers (LIA)[29,30] commonly used in phosphorescence instrumentation. From a reference input (usually the excitation signal), the LIA instrument generates a pair of sinusoidal signals in quadrature (a cosine wave, in phase with the reference signal, and a sine wave, in quadrature with it). Both signals are multiplied with the signals to be analyzed ($x_{exc}(t)$ and $x_{em}(t)$) and the resulting signals are then low-pass filtered to obtain the in-phase and the quadrature components for each signal. They are then used to estimate the amplitudes and phases of the input signals, and to estimate the modulation-factor and the phase-shift at the modulation frequency.

I/Q detection is optimal in the sense that the noise minimally affects the estimations at the modulation frequency (the estimations are affected only by the component of the noise at the frequency $f_m$). Even though the measurement principles used in I/Q and LIA are very similar, there are some practical details to take into consideration. LIA instruments work at a single frequency, while I/Q method can independently be applied to each frequency component of a signal containing several harmonics. Additionally, LIA instruments include predefined low-pass filters (to select configurable measuring times) and the estimations of the in-phase and quadrature components are affected by the impulsive response of the selected low-pass filter. I/Q and LIA based instruments have been compared in ref. 27 at different SNR conditions and using both, synthetic signals and signals recorded from a phosphorescent system for measuring the oxygen concentration. Performances for both instruments are similar even though I/Q provides slightly better accuracy (mainly due to the restriction of the low-pass filters in the LIA instruments) and more flexibility in the configuration of the measuring process.

The modulation-factor and the phase-shift can be also estimated from the Fast Fourier Transform (FFT) of the excitation and emission signals.[13,31] The in-phase and quadrature components for the modulation frequency $f_m$ are equivalently estimated as the real and imaginary parts of the FFT component corresponding to $f_m$. This requires that all the energy of the modulation frequency is concentrated in just one frequency component of the FFT. This condition is verified if the duration of the signal to be processed contains an integer number of cycles at the modulation frequency. While using the I/Q method with pre-stored cosine and sine waves is more efficient for single-frequency estimations or multi-frequency estimations with few frequency components, the FFT method is computationally more efficient when the excitation signal contains more frequency components.

## 6.4 Multifrequency Measurements and Applications

The proposed architecture of the phosphorescence instrument allows the use of multi-frequency signals.[2,21,31] The excitation signal can be designed to contain several frequency components and independent modulation-factors and phase-shifts can be estimated for each frequency component by applying the I/Q or the FFT based procedures to the $x_{exc}[l]$ and $x_{em}[l]$ digital signals.

There are several advantages derived from multi-frequency measurements. In the context of the characterization of the phosphorescent system, since it cannot be assumed to be a first order system, an accurate characterization should include the estimation of the frequency response at several modulation frequencies (and, of course, for several quencher concentrations). So, data acquisition using a single-frequency phosphorescence instrument requires repeating the measurements at each single frequency,[23] while a multi-frequency instrument can simultaneously obtain measurements at several modulation frequencies from a single excitation-emission recording.[2,24,25]

In the context of oxygen determination, selecting the optimal modulation frequency for a single-frequency instrument is not trivial: the modulation frequency providing the largest change in the frequency response depends on the quencher concentration. So, a critical decision in the design of a single-frequency phosphorescence instrument is the selection of the modulation frequency, which should be chosen taking into account the sensing phase and the concentration range to be measured.[1,19] The use of multi-frequency measurements allows selecting the optimal modulation frequency at each measuring condition.[21] Finally, since all the modulation frequencies provide information about the oxygen concentration, the determinations provided by each harmonic can be combined in order to obtain a more accurate oxygen determination.[2,31]

Additionally, multi-frequency measurements provide a significantly more efficient management of the illumination intensity. A pure single-frequency modulation is not possible with a light source because sinusoidal signals are negative half of the time (negative illumination is not possible) and light modulation always includes a DC level. This DC level is useless for the frequency response estimation, but increases the power dissipation of the LED (degrading the device and limiting the modulation amplitude) and the photodegradation of the phosphorescent sensing phase. Appropriate design with multi-frequency signals improves the useful/useless ratio of the illumination intensity.[2,11] A particularly interesting multi-frequency excitation is the short duty-cycle rectangular signal:[2,24,25] it provides sufficient energy at several harmonics with low energy at the DC component. Additionally, rectangular signals are easier (and cheaper) generated since switching electronic components can be used (instead of analog amplifiers).

The next sections describe some applications of the multi-frequency spectroscopy for (a) an accurate characterization of the phosphorescent system, (b) selecting the optimal single modulation frequency and (c) improving the accuracy of the oxygen determination by combining the information provided by the different harmonics.

### 6.4.1 Characterization of the Sensing Phase

The accurate characterization of a sensing phase is an important step in the design of a phosphorescence instrument. The frequency response measured at each oxygen concentration provides information about the concentration range in which the instrument can optimally operate (*i.e.* the range in which a small change in concentration produces a large change in the frequency response) or the optimal modulation frequency (*i.e.* frequency providing the largest measurable change in the modulation-factor or phase-shift). An appropriate characterization also helps to choose the most appropriate sensing phase to determine oxygen under specific conditions.

Conventional single-frequency instruments can be used for the characterization of the sensing phase, but measurements must be repeated for each modulation frequency. In ref. 23, for example, optical sensing nanocomposites are evaluated with single-frequency measurements for selecting the optimal modulation frequency and evaluate the phase-shift changes at different concentration ranges. This characterization required frequency-by-frequency measurements. A similar analysis was performed in ref. 24 (using sensing phases based on copper complexes) but the frequency response was measured with a multi-frequency I/Q scheme. In this case, using 10% duty-cycle rectangular signals, multi-frequency analysis was applied to measure the frequency response simultaneously at several harmonics, with a substantial reduction of the measuring effort. The comparison of the frequency response curves at different oxygen concentrations for two sensing phases presented in ref. 24, clearly shows that one of them is more appropriate for oxygen determination at medium and high concentrations, while the other is better for low and ultra-low concentrations.

The multi-frequency measurements can be also used to fit a global multi-site phosphorescence model of the sensing phase, according to eqn (6.9). In ref. 2, multi-frequency modulation-factor and phase-shift were measured at different oxygen concentrations and a global model was fitted, estimating the constants $M_{0i}$, $\tau_{0i}$ and $k_i$ ($i$ index for each site) by minimizing the mean squared error between the frequency response model and the measurements. Figure 6.2 illustrates the frequency response fitting for a well-known oxygen-sensing film coated at the end of an optical fiber (a Pt(II) porphyrin immobilized in

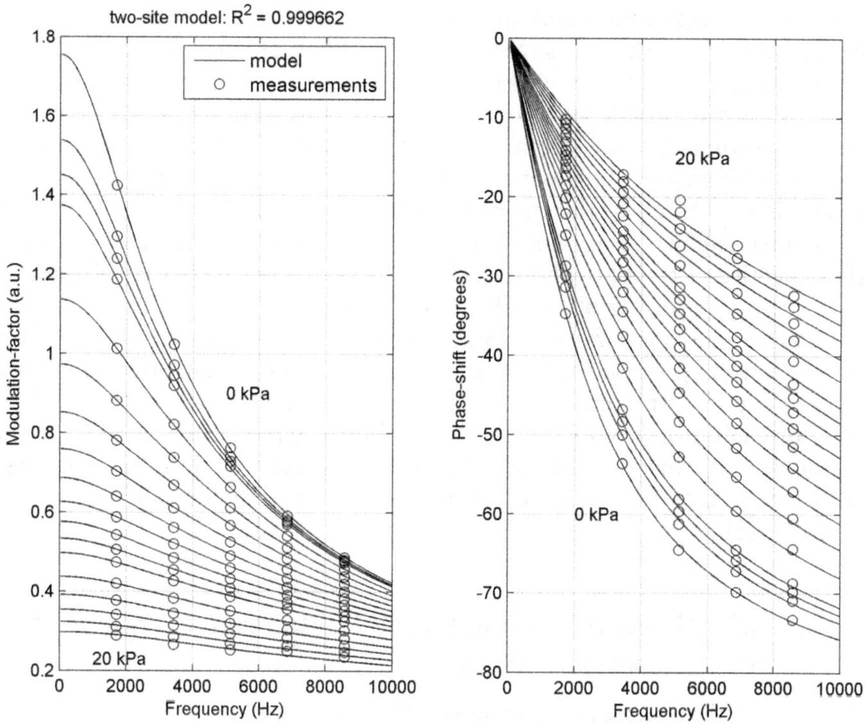


**Figure 6.2** Frequency response of the two-site model (solid lines) fitting the measured modulation-factor and phase-shift (circles). Measurements recorded at concentrations 0, 0.5, 0.75, 1, 2, 3, 4, 5, 6, 7, 8, 9, 10, 12, 14, 16, 18, and 20 kPa $O_2$. Model parameters: $M_{01} = 1.4413$; $k_1 = 0.32278$ $kPa^{-1}$; $\tau_{01} = 60.31$ μs; $M_{02} = 0.3134$; $k_2 = 0.10108$ $kPa^{-1}$; $\tau_{02} = 89.60$ μs. Adapted with permission from *Anal. Chem.*, 2014, **86**, 5245.[2] Copyright 2014 American Chemical Society.

polystyrene). This figure represents the frequency response measurements (circles) and the frequency response of the model (solid lines) at different concentrations ranging between 0 and 20 kPa $O_2$. The measurements were recorded using a 10% duty-cycle rectangular-wave as excitation signal. This fitting corresponds to a two-site model, and the resulting model parameters are indicated in the figure legend. The determination coefficient of the fitting was $R^2 = 0.999662$, while it was $R^2 = 0.995437$ for a one-site (or monoexponential) model, and $R^2 = 0.999695$ for a three-site model. These determination coefficients reveal that a one-site model is obviously insufficient for describing the sensing phase ($R^2$ significantly increases when a two-site model is fitted), and that a three-site model is inappropriate (the number of free parameters increases but it just provides a marginal increase of $R^2$). Therefore, the multi-frequency measurements together with the fitting of the global frequency response provide an appropriate modelling of the sensing phase that is reasonably described, in this case, with a two-site model.

In this example, the residual of the two- and three-site models is associated to noise in the measurements or to the mismatch between the real phosphorescent system and the multi-site model (for this reason no substantial increase of $R^2$ is observed when modelling with the three-site model). Fitting the measurements at only a single frequency (*i.e.* the third harmonic) increases the determination coefficient to $R^2 = 0.999905$, but this model is over-fitted to the measurements at this specific frequency and it is inconsistent when applied to other frequencies. For this reason, when the objective of the phosphorescence instrument is the characterization of the sensing phase, a global multifrequency fitting is preferable (in order to estimate consistent parameters $M_{0i}$, $\tau_{0i}$ and $k_i$ valid for most frequencies and concentrations). However, when the phosphorescence instrument is aimed to measure oxygen concentrations at a given range, an individual calibration for each individual frequency is preferable, since it better reduces the residual and provides more accurate oxygen determination, (even though consistency across frequencies of the model parameters is not guaranteed in this case).

### 6.4.2 Selecting the Optimal Single Modulation Frequency for Analyte Determination

The modulation-factor and the phase-shift are functions of the oxygen concentration and the modulation frequency ($m = m(C,f)$; $\phi = \phi(C,f)$). Conventionally the modulation frequency is considered to be optimal when a small change of the oxygen concentration provides the largest change in the frequency response.[1,19] Mathematically, according to this criterion, the optimal modulation frequency is the one maximizing the partials $\partial m(C,f)/\partial C$ and $\partial \phi(C,f)/\partial C$ for modulation-factor and phase-shift based oxygen determinations, respectively. Since the modulation-factor and the phase-shift are functions of both $C$ and $f$, the optimal modulation frequency is expected to depend on the oxygen concentration and is different, in general, for modulation-factor and for phase-shift based oxygen determinations. The multi-frequency characterization of the sensing phase at different oxygen concentrations provides a global model describing $m(C,f)$ and $\phi(C,f)$ that can be applied to directly determine the optimal modulation frequency at a given oxygen concentration or for a given range of concentrations. In general, with this criterion, the modulation-factor based determinations are optimal at low modulation frequencies (lower than the cut-off frequency), while the phase-shift based determinations are optimal at a modulation frequency around the cut-off frequency (that strongly depends on the oxygen concentration).[1,4]

However, the previously described procedure does not take into account the effect of the noise in the oxygen determinations. The noise usually modifies the optimal modulation frequency because the phosphorescence intensity decreases with the modulation frequency, and the modulation-factor or phase-shift estimations are more affected by noise at higher modulation frequencies. In order to take into account the noise, the optimal modulation frequency should be defined as the one minimizing the standard error in the oxygen determination:[2]

$$\mathrm{SE}(C_{\mathrm{m}}) = \left|\frac{\partial C}{\partial m}\right| \mathrm{SE}(m) = \left|\frac{\partial m(C,f)}{\partial C}\right|^{-1} \mathrm{SE}(m) \tag{6.19}$$

$$\mathrm{SE}(C_{\phi}) = \left|\frac{\partial C}{\partial \phi}\right| \mathrm{SE}(\phi) = \left|\frac{\partial \phi(C,f)}{\partial C}\right|^{-1} \mathrm{SE}(\phi) \tag{6.20}$$

where the standard errors of the modulation-factor and the phase-shift can be estimated from the experimental data (by repeating the measurements), and the partials can directly be obtained from the multifrequency model describing the phosphorescent system.

Medina *et al.*[2] have combined the multi-frequency characterization of the sensing film and the standard errors SE($m$) and SE($\phi$) for the estimation of the standard error of the oxygen concentration. This standard error is represented in Figure 6.3 as a function of the modulation frequency, for different oxygen concentrations, and for modulation-factor (left panel) and phase-shift (right panel) based oxygen determinations. In the case of modulation-factor based determinations, the lowest modulation frequency provides the most accurate oxygen determinations. In the case of phase-shift based determinations, there is a well-defined optimal modulation frequency which depends on the oxygen concentration. This dependence with the frequency substantially differs when the effect of the noise is ignored.[2] Usually, under laboratory conditions, the noise can be minimized and appropriately characterized, and for this reason, the determination of the optimal modulation frequency usually ignores the noise when a sensing phase is characterized in laboratory.[24,25] However, when measurements are acquired with portable instruments, the noise conditions could change significantly and therefore the noise should be considered for selecting the optimal modulation frequency.

According to the results in Figure 6.3, taking into account that the oxygen determinations based on the modulation-factor are more accurate than those based on the phase-shift, the former should be preferred for oxygen determination purposes. However, because of the recalibration requirements of modulation-factor based determinations, phase-shift measurements are often preferred.[1,4]

### 6.4.3 Using Multi-frequency Information for Oxygen Determination

Even though the characterization of the phosphorescent system is based on the estimation of the frequency response (either $H(C,f)$ or equivalently $m(C,f)$ and $\phi(C,f)$) at different oxygen concentrations, oxygen determination is usually performed with calibration curves describing the changes of

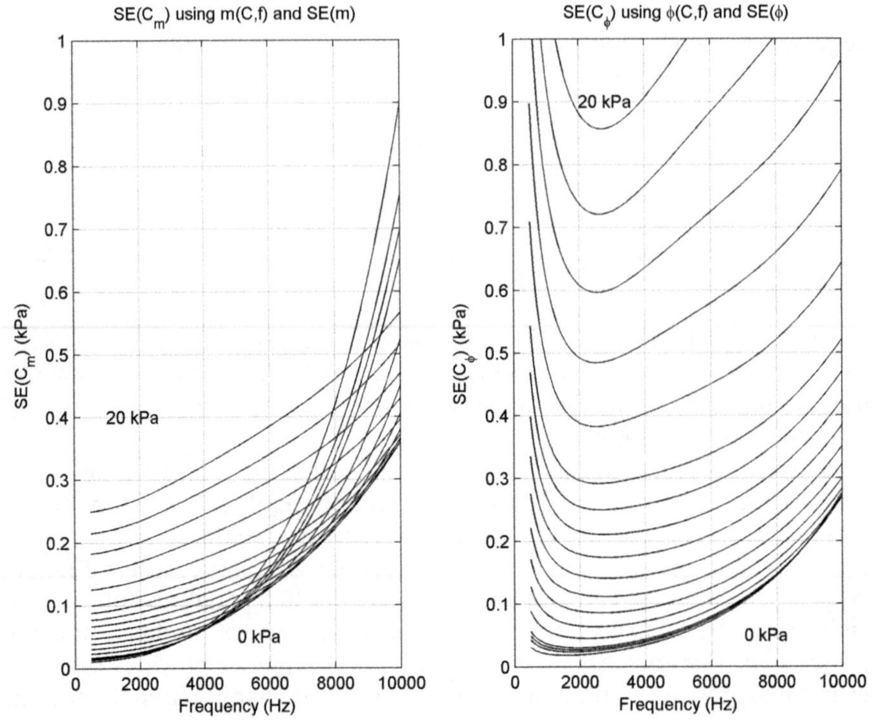


**Figure 6.3** Standard error of oxygen determination, SE($pO_2$), as a function of modulation frequency when the determination is based on modulation-factor (left panel) or phase-shift (right panel). Curves correspond to the concentrations 0, 0.5, 0.75, 1, 2, 3, 4, 5, 6, 7, 8, 9, 10, 12, 14, 16, 18, and 20 kPa $O_2$. The estimation of SE($pO_2$) includes the standard error associated with the measured parameters. Adapted with permission from *Anal. Chem.*, 2014, **86**, 5245.[2] Copyright 2014 American Chemical Society.

the apparent lifetime with the oxygen concentration.[18,26,32–35] The apparent lifetime can be estimated from the phase-shift or from the modulation-factor (eqn (6.10) and (6.11)), at one or at several modulation frequencies. A multi-frequency measuring system allows the simultaneous estimation of the apparent lifetimes ($\tau_m(f_i)$ and $\tau_\phi(f_i)$) at several modulation frequencies ($f_i$) from a single recording. Due to the inconsistence inherent to the definition of the apparent lifetimes,[2] the limitations of a multi-site model for describing the actual phosphorescent system[35] and in order to optimally fit the experimental data and provide accurate determinations, usually an independent curve is calibrated for each individual apparent lifetime (for each modulation frequency and for both modulation-factor and phase-shift).

Therefore, a multi-frequency phosphorescence instrument appropriately calibrated and using $I$ modulation frequencies, provides up to $2I$ independent determinations of the oxygen concentration ($I$ derived from $\tau_m(f_i)$ and $I$ from $\tau_\phi(f_i)$ obtained from a single signal recording). Medina *et al.*[21] have compared the accuracy in the oxygen determination for different strategies for combining them: average, weighted average, selecting the optimal determination for each condition (*i.e.* the one closer to the optimal modulation frequency at each measuring condition) or using the estimation corresponding to a globally optimal modulation frequency. They found that a substantial improvement of the instrument accuracy can be achieved, but optimal combination of the independent determinations is not immediate.

A more detailed analysis of the problem showed that the optimal combination of several statistically independent oxygen determinations is a weighted average of the independent determinations with weights inversely proportional to the variance of each one:[2]

$$\hat{C} = \frac{\sum w_i C(\tau_i)}{\sum w_i} \qquad w_i = \frac{1}{\sigma_i^2} \tag{6.21}$$

where $\tau_i$ is an apparent lifetime, $C(\tau_i)$ is the corresponding calibration curve and $\sigma_i$ is the standard error associated to the $i$th individual determination $C(\tau_i)$, which can be estimated according to the error propagation theory using the corresponding calibration curve and the standard error of the lifetime:[2]

$$\sigma_i = \mathrm{SE}\left(C(\tau_i)\right) = \left|\frac{\partial C(\tau_i)}{\partial \tau_i}\right| \mathrm{SE}(\tau_i) \tag{6.22}$$

The standard error of the lifetimes can be obtained from several measurements, and can be easily estimated during the instrument calibration. Even though the standard errors should be estimated under the specific noise conditions in which the oxygen is being determined, according to eqn (6.21) the optimal combination remains invariant if the standard errors are all scaled with the same constant. Therefore, if the standard errors are estimated under calibration conditions and these standard errors are applied to measurements at different noise conditions, the main effect of the change in the SNR is a global scaling of the standard errors and the combination proposed in eqn (6.21) would be still valid.

The combination of oxygen determinations based on eqn (6.21) was demonstrated by Medina *et al.*[2] using short duty-cycle rectangular-wave signals as excitation. Figure 6.4 shows root-mean-square error (RMSE) in the oxygen determination at different concentrations, when a conventional sinusoidal signal is used as excitation and when the rectangular signal is used. Results for the rectangular signal including information from 1 to 5 harmonics are shown. Since both excitation signals were configured for the same average intensity, and since the rectangular signal provides more amplitude in the harmonics at the same average intensity, there is a significant improvement for rectangular signals with respect to sinusoidal modulation. This figure also shows that the combination of information from different harmonics decreases the RMSE of the measurements, and the error is smaller as more harmonics are included. The first harmonic of

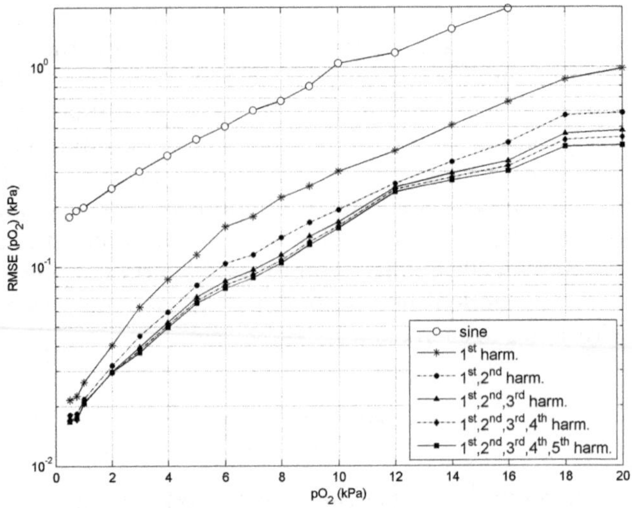


**Figure 6.4** RMSE in determination of $pO_2$ using single-frequency modulation (sinusoidal) and multi-frequency modulation (rectangular signals). Oxygen determinations based on phase-shift estimations. Determinations with rectangular signals are obtained by combining different harmonics. Adapted with permission from *Anal. Chem.*, 2014, **86**, 5245.[2] Copyright 2014 American Chemical Society.

the rectangular signal achieved an average RMSE 4.4 times smaller than the sinusoidal signal, and the RMSE was 7.4 times smaller when the 5 harmonics were combined. These results correspond to oxygen determinations based on phase-shift lifetimes. Similar results are obtained for modulation-factor lifetimes (lower RMSE as more harmonics are combined) or for the combination of both (lower RMSE when modulation-factor and phase-shift lifetimes are combined).

These methods can be easily implemented in multi-frequency photoluminescence instruments since the algorithms just involve digital signal processing of the excitation and emission signals and some mathematical operations that can be computed with software.

## 6.5 Noise Analysis and Applications

As previously described, the in-phase and quadrature components of a signal at a given modulation frequency can be estimated with the I/Q method (by multiplying the signal with the corresponding cosine and sine waveforms)[2,21] or with the FFT.[13] Both methods are identical if the duration of the signal to be analyzed contains an integer number of cycles of the corresponding harmonic (otherwise the harmonic spectral power is spread around several FFT components). One advantage of the FFT analysis is that the I/Q information is obtained not only at the harmonic of interest, but also at all the

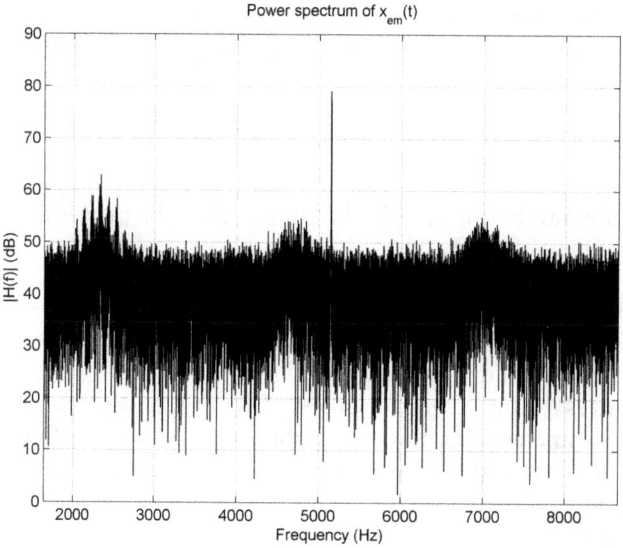


**Figure 6.5** FFT of an emission signal for sinusoidal excitation with modulation frequency $f_0$ = 5145 Hz.

frequency components, and therefore the FFT can be applied to analyze the noise affecting the measurements. Additionally, the FFT algorithm is very efficient and computational cost is low even for relatively long signals. Figure 6.5 shows the FFT spectrum of an emission signal obtained with a sinusoidally modulated excitation. The spectral peak corresponding to the emission signal can be identified at the modulation frequency, and the FFT components in the vicinity of the modulation frequency provide information about the noise affecting this harmonic.

Using the FFT, the power spectral density of the noise (*i.e.* a statistical characterization of the noise) can be estimated around each harmonic of interest. The real and imaginary parts of the FFT components corresponding to background noise follow a Gaussian probability distribution with null mean and variance equal to the power spectral density of the noise.[36] Medina *et al.* proposed and evaluated an accurate method for estimating the variance (or the standard deviation) of the noise around the harmonic of interest.[13] This method involves an FFT computation, the selection of the frequency range used for characterizing the noise, removing outliers (corresponding to other signals or interferences but not to background noise) and estimating the standard deviation of the distribution. The method, therefore provides a characterization of the noise around the modulation frequency valid for white noise (*i.e.* if the spectral distribution of the noise is flat) or even for colored noise (for noise spectral distribution changing with the frequency).

Using the statistical characterization of the noise (*i.e.* the standard deviation of the real or imaginary parts in the FFT domain $\sigma_N$), and assuming that the modulation frequency component is affected by a noise with the same statistics, the standard error of the I/Q estimations can be easily computed:[13]

$$\sigma_{nIQ} = \frac{2}{L}\sigma_N \tag{6.23}$$

where $L$ is the number of samples in the signals to be processed $x_e[l]$.

### 6.5.1 Application to Uncertainty Estimation

The standard error affecting the I/Q components can be used to estimate the uncertainty of all the parameters related to the oxygen determination. It simply requires to apply the error propagation theory with the equations providing the parameters (from I/Q components to the oxygen determinations). Medina *et al.*[13] have applied the FFT-based noise analysis of the excitation and emission signals to estimate the uncertainty of the modulation-factor, the phase-shift, the lifetimes and the oxygen determinations. A comparison of the standard error prediction (derived from the FFT-based noise analysis) with the standard errors estimated from a collection of measurements demonstrated the utility of this method.

Even though the uncertainty estimation based on the FFT noise analysis involves some mathematical complexity, the fundamentals are simple (the uncertainty of the primary parameters can be estimated from the FFT and is then propagated to the oxygen determination) and the implementation in a computational system (a microcontroller or a computer) does not involve any difficulty.

This method allows the phosphorescence instrument to provide not only the oxygen determinations but also the corresponding standard errors (or the confidence intervals). Since the standard errors are derived from the recorded signals involved in the oxygen determination, the uncertainty estimation is specifically describing the effect of the particular measuring conditions (taking into account the amplitudes of the excitation and emission signals and the noise affecting them at the modulation frequency). This uncertainty describes the effect of the electronic noise in the excitation or the emission signal, the SNR reduction of the emission signal due to the quenching at a given oxygen concentration or due to photodegradation of the sensing phase. The uncertainty also describes the effect of the modulation frequency taking into account both, the emission intensity (that decreases with the frequency) and the spectral distribution of the noise.

The uncertainty estimation is very useful for the phosphorescence instrument. Under low illumination conditions (for example, when the phosphorescence instrument is configured to measure small samples with an optical fiber) or when the emission efficiency decreases (if the sensing dye is changed)[4,12,15,19,27] the operator can immediately evaluate the impact in the instrument accuracy at the current measuring conditions. Additionally, from a single measurement, the operator can determine how many times the measurement should be repeated in order to achieve a predefined accuracy (the uncertainty decreases with the square root of the number of independent measurements). The uncertainty can be also applied to evaluate the degree of photodegradation of the sensing phase and decide when it should be replaced. It can be also used to decide when the instrument should be recalibrated (when the measurement and the confidence interval using a calibrating pattern are not statistically compatible with the calibrating value).[13]

### 6.5.2 Application to Optimal Combination of Harmonics

The FFT-based noise analysis provides an estimation of the standard error affecting the oxygen determination. This method can be applied to modulation-factor or to phase-shift based oxygen determinations, and can be also applied to single-frequency signals or to multi-frequency signals.

Additionally, the FFT analysis of a pair of multi-frequency signals $x_{exc}[l]$ and $x_{em}[l]$ provides several independent oxygen determinations (for both, modulation-factor and phase-shift based lifetimes and for different harmonics). Since a set of independent oxygen determinations, each one with its corresponding standard error, is available, different determinations can be optimally combined using eqn (6.21) in order to obtain a more accurate oxygen determination. Compared with the combination proposed in ref. 2 (where the standard errors are estimated during calibration), this combination is better adapted to the specific measuring conditions (since both, the independent oxygen determinations and the corresponding standard errors are directly estimated from the signals $x_{exc}[l]$ and $x_{em}[l]$).

The real-time combination of multi-frequency oxygen determinations using the FFT-based noise analysis has been applied with short duty-cycle rectangular signals.[31] Figure 6.6 shows the RMSE achieved by the phosphorescence instrument when different independent determinations are combined according to their respective standard errors. As in the case of pre-calibrated standard errors (Figure 6.4), the combination of determinations using the estimated uncertainties reduces the RMSE (lower as more harmonics are combined and lower when modulation-factor and phase-shift estimations are combined).

## 6.6 Instrument Development

Since the beginning of phase modulation phosphorescence detection, many instruments, based on different technologies and principles, have been developed.[1,8,37] Conventionally, most of the measurement schemes were based on sinusoidally modulated excitation sources and phase detection was performed at a single frequency with analog lock-in amplifiers[12,16,23] or specialized analog circuits.[17,20,38] Nowadays, digital technology is displacing the analog instruments, with the consequent reduction of costs and improvement of the flexibility of the modern phosphorescence instruments.

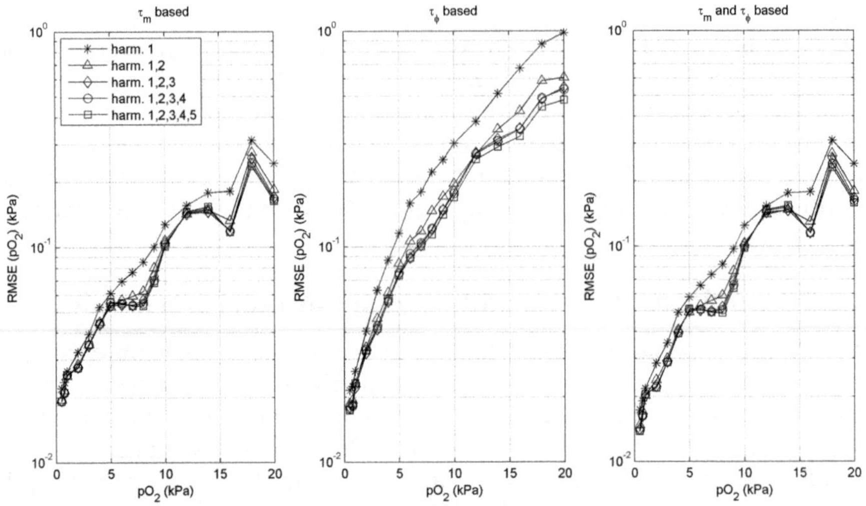


**Figure 6.6** RMSE in determination of $pO_2$ using multi-frequency modulation (rectangular signals) and combining different harmonics. Oxygen determinations based on modulation-factor (left panel), phase-shift (center panel) and combining both (right panel). The procedure for combining the independent determinations was based on the FFT noise analysis.

The technological advances in the design and manufacturing processes of micro-computers, microcontrollers, electronic/optoelectronic components, *etc.* (with lower costs, better capabilities and smaller sizes) allow the design of modular low-cost phosphorescence instruments according to the architecture previously proposed in this chapter and incorporating the methods described in the previous sections. The proposed modular architecture allows the configuration of the excitation signal, the modulation of the illumination source, the acquisition of both the excitation and the emission signals, and digital signal processing of both signals as well as the associated mathematical computations in order to obtain the instrument response (that could be a characterization of the phosphorescent system or an oxygen determination).

In our laboratories, the experimental set-up prepared for different research purposes has been modified according to the specific requirements of each experiment. In this way, the excitation signal has been generated either with a DA board or with a microcontroller;[21,27] different LEDs, optical filters and sensing phases have been mounted;[23–25] sometimes the photodiode[16] has been substituted by a photomultiplier tube;[2,9,21] DA conversion has been performed with a digital oscilloscope or with a microcontroller,[2,9,13,31] *etc.* However, even though one or several elements have been changed for different experiments, the modular architecture has always been maintained, and in all the cases, the digital signal processing and the mathematical computations are applied to the digital excitation and emission signals. The advantages of all the algorithms described in this chapter just involve numerical computation with the digital signals that can be easily programmed in a computer or a microcontroller, and

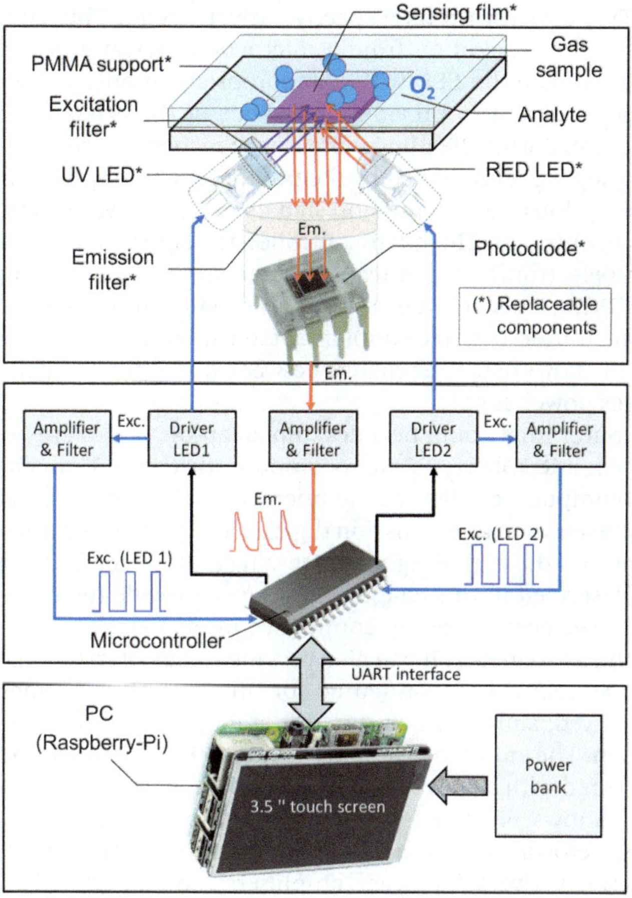


**Figure 6.7** Main components of the modular and portable phosphorescence instrument.

therefore the implementation of these methods does not involve significant changes in the instrument architecture or in the instrument costs.

Figure 6.7 shows a schematic diagram of the elements in a portable phosphorescence instrument for oxygen determination developed at our laboratories. The optical subsystem includes two LEDs, the photodetector, the optical filters, a support for the sensing film and the sensing film. One of the LEDs is used for exciting the sensing film (UV LED) while the other is a red LED (wavelength in the same range of the phosphorescence emission) used for calibrating the response of the photodetector (*i.e.* for measuring delay of the photodetector and the analog electronics for signal conditioning),[19] which is relevant for accurate characterization of the sensing film. The photodetector is a preamplifier photodiode. All these optical and optoelectronic elements are replaceable (in order to substitute degraded components or for changing the sensing dye and accommodate the new excitation and emission wavelengths).

Each LED is driven with its respective driver circuit. This circuit receives the signal used as excitation from a microcontroller and provides enough current to modulate the LED with an appropriate intensity. The LED driver also includes a circuit to provide a reference of the excitation signal. The photodiode transforms the phosphorescence response into an electrical signal. The analog signal provided by the photodiode is amplified and filtered. Both the excitation and the emission signals are AD converted and registered by the microcontroller. The microcontroller, the signal conditioning circuits and the optoelectronic and optical elements are integrated in a printed circuit board (PCB). This board also includes a UART (universal asynchronous receiver/transmitter) interface implemented with a standard micro-USB connector which allows data interchange between a computer and the board and also provides power supply.

The computer (that could be a desktop, a laptop or a micro-computer) is, in our design, a Raspberry-Pi microcomputer (with a 3.5″ touch screen display). This computer configures the microcontroller for the measurements, receives the excitation and emission digital signals from the microcontroller and performs all the digital signal processing and the mathematical computations as described in this chapter in order to provide the oxygen determination (with the corresponding confidence intervals) when it is configured for oxygen measurements. It can also provide a characterization of the phosphorescent system if it is configured for this purpose. The measurements can be displayed, submitted or stored in a log file according to the system configuration. The entire instrument (including the micro-computer and the PCB) is powered with a conventional power-bank.

Figure 6.8 shows, on the left side, a picture of the printed circuit board containing the microcontroller, the UART communication module, all the specific electronics (LED drivers, filters and amplifiers), and the optical and optoelectronic elements (LEDs, optical filters, photodiode). The picture on the right

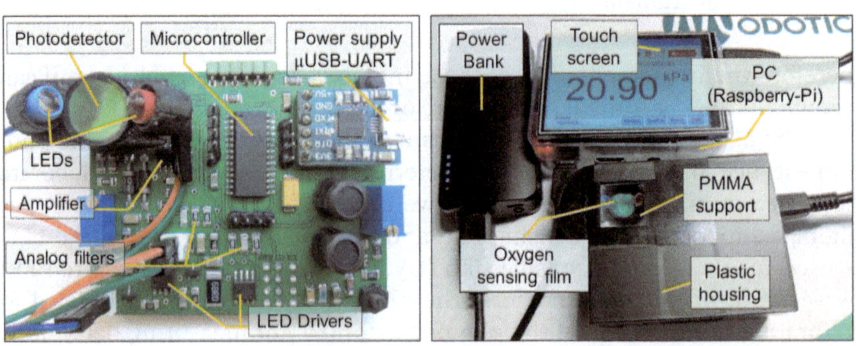


**Figure 6.8** Left panel: picture of the PCB containing the optical and electronic components of the phosphorescence instrument. Right panel: picture of the complete phosphorescence instrument including the central unit (a Raspberry- Pi PC with a touch screen display), the PCB and the power bank.

side shows the complete phosphorescence instrument including the circuit (in a plastic housing), the micro-computer Raspberry-Pi and the power bank. A compact and portable instrument has been designed (60 × 85 × 30 mm; 180 gr). This example illustrates how the technological advances (in optoelectronics, electronics, microcontrollers and micro-computers) allow the design of flexible, modular and low-cost phosphorescence instruments incorporating advanced signal processing to optimize the acquisition of analytical information.

## 6.7 Conclusions

In this chapter we have presented the phosphorescent system from a new perspective: it can be considered as a linear system with an input (excitation) and an output (response) for which all the signal theory related to linear systems can be applied and exploited in order to optimize the performance of phosphorescence instruments. Based on this concept and bearing in mind the technological advances (better and cheaper components in the field of electronics, optics, computation, *etc.*), we have described an instrument architecture where the phosphorescent system is excited with a configurable excitation signal and where the analytical information is obtained by processing the excitation and response signals. Since both signals are digitized, all the operations required to obtain the analytical information involve digital signal processing and mathematical computation that can be implemented in a microcontroller or a computer without significant increase in the instrument cost.

In this chapter we have described some applications of signal processing, easily implementable under this architecture, with demonstrated utility for phosphorescence lifetime based oxygen sensing. Some applications are associated to the multi-frequency phosphorimetry (using multi-frequency modulated signals as excitation and applying a multifrequency processing to the excitation and emission recorded signals). This paradigm provides an easier and more complete characterization of the phosphorescent systems, it can be applied for selecting the optimal modulation frequency for oxygen determination, and it provides mechanisms for combining independent oxygen determinations (estimated for different modulation frequencies from just one multi-frequency recording) into an improved and more accurate oxygen determination. Other applications of signal processing are associated with the analysis of the background noise affecting the recorded signals. The noise characterization provides an estimation of the uncertainty affecting the oxygen determinations, which is a valuable add-on for the phosphorescence instrument. This uncertainty estimation also provides the optimal weights for combining the independent oxygen determinations obtained in multi-frequency measurements. The described procedures are easily implementable in a phosphorescence instrument designed according to the proposed architecture, as illustrated with a prototype developed at our laboratories.

The proposed architecture and the linear system perspective can be also useful for future challenges of luminescence instrumentation, such as (a) a more detailed modelling of the phosphorescent/fluorescent systems (fitting higher order differential equations, including non-linear effects in the differential equations, or including non-linear dependence of the quenching with the quencher concentration); (b) a more detailed modelling of the instrument, including a description of both the luminescent system and also the transducers; (c) an analysis of interfering quenchers and the design of multi-parametric phosphorescence instruments; or (d) adaptation of the luminescent instruments to ultra-low illumination conditions (which requires a very detailed managing of the noise in the signal processing) in order to apply them for microfluidics devices in control processes, fiber optic based sensors for small samples or micro-biosensors with optical oxygen transduction.

## Acknowledgement

This work has been partly supported by the Spanish Ministry of Economy, Industry and Competitiveness (CTQ2014-53442-P, Grant BES-2009-026919 and Torres Quevedo Grants PTQ-15-07922 and PTQ-15-07912), the CEI BioTic Granada Campus (Project CEIbioTIC14-2015) and the CEMIX UGR-MADOC-Santander Bank (Project MSOADKISS-10-2016).

## References

1. J. R. Lakowicz, *Principles of Fluorescence Spectroscopy*, Kluwer Academic, New York, 2nd edn, 1999.
2. S. Medina-Rodríguez, A. de la Torre-Vega, F. J. Sainz-Gonzalo, M. Marín-Suárez, C. Elosúa, F. J. Arregui, I. R. Matías, J. F. Fernández-Sánchez and A. Fernández-Gutiérrez, *Anal. Chem.*, 2014, **86**, 5245.
3. D. B. Papkovsky and T. C. O'Riordan, *J. Fluoresc.*, 2005, **15**, 569.
4. H. Szmacinski and J. Lakowicz, in *Topics in Fluorescence Spectroscopy*, ed. C. D. Geddes and J. R. Lakowicz, Springer, USA, 2002, vol. 4, Lifetime-based sensing, p. 295.
5. J. G. Proakis and D. G. Manolakis, *Digital Signal Processing*, Pearson Education Inc., Prentice Hall, 4th edn, 2007.
6. D. Sundararajan, *The Discrete Fourier Transform: Theory, Algorithms and Applications*, World Scientific Publishing Co. Pte. Ltd, USA, 2001.
7. R. D. Hippenstiel, *Detection Theory: Applications and Digital Signal Processing*, CRC Press, 2002.
8. O. S. Wolfbeis, *Fiber Optic Chemical Sensors and Biosensors*, CRC Press, Boca Raton, MA, 1991.
9. S. Medina-Rodríguez, A. de la Torre-Vega, C. Medina-Rodríguez, J. F. Fernández-Sánchez and A. Fernández-Gutiérrez, *Sens. Actuators, B*, 2015, **212**, 278.

10. C. M. McGraw, G. Khalil and J. B. Callis, *J. Phys. Chem. C*, 2008, **112**, 8079.
11. H. M. Rowe, S. P. Chan, J. N. Demas and B. A. DeGraff, *Anal. Chem.*, 2002, **74**, 4821.
12. M. Valledor, J. C. Campo, I. Sánchez-Barragán, J. M. Costa-Fernández, J. C. Álvarez and A. Sanz-Medel, *Sens. Actuators, B*, 2006, **113**, 249.
13. C. Medina-Rodríguez, S. Medina-Rodríguez, A. de la Torre-Vega, A. Fernández-Gutiérrez and J. F. Fernández-Sánchez, *Sens. Actuators, B*, 2016, **224**, 521.
14. C. McDonagh, C. S. Burke and B. D. MacCraith, *Chem. Rev.*, 2008, **108**, 400.
15. M. Schäferling, *Angew. Chem., Int. Ed.*, 2012, **51**, 3532.
16. O. Ergeneman, G. Chatzipirpiridis, J. Pokki, M. Marín-Suárez, G. A. Sotiriou, S. Medina-Rodríguez, J. F. Fernández-Sánchez, A. Fernández-Gutiérrez, S. Pané and B. J. Nelson, *IEEE Trans. Biomed. Eng.*, 2012, **59**, 3104.
17. G. A. Holst, T. Köster, E. Voges and D. W. Lübbers, *Sens. Actuators, B*, 1995, **29**, 231.
18. W. Trettnak, C. Kolle, F. Reininger, C. Dolezal and P. O'Leary, *Sens. Actuators, B*, 1996, **36**, 506.
19. M. Cajlakovic, A. Bizzarri, C. Konrad and H. Voraberger, *Optochemical Sensors Based on Luminescence Encyclopedia of Sensors*, American Scientific Publishers, 1st edn, 2006, vol. 7.
20. C. McDonagh, C. Kolle, A. K. McEvoy, D. L. Dowling, A. A. Cafolla, S. J. Cullen and B. D. MacCraith, *Sens. Actuators, B*, 2001, **74**, 124.
21. S. Medina-Rodríguez, A. de la Torre-Vega, J. F. Fernández-Sánchez and A. Fernández-Gutiérrez, *Sens. Actuators, B*, 2013, **176**, 1110.
22. P. Langer, R. Müller, S. Drost and T. Werner, *Sens. Actuators, B*, 2002, **82**, 1.
23. S. Medina-Rodríguez, M. Marín-Suárez, J. F. Fernández-Sánchez, A. de la Torre-Vega, E. Baranoff and A. Fernández-Gutiérrez, *Analyst*, 2013, **138**, 4607.
24. S. Medina-Rodríguez, F. J. Orriach-Fernández, C. Poole, P. Kumar, A. de la Torre-Vega, J. F. Fernández-Sánchez, E. Baranoff and A. Fernández-Gutiérrez, *Chem. Commun.*, 2015, **51**, 11401.
25. S. Medina-Rodríguez, S. A. Denisov, Y. Cudré, L. Male, M. Marín-Suárez, A. Fernández-Gutiérrez, J. F. Fernández-Sánchez, A. Tron, G. Jonusauskas, N. D. McClenaghan and E. Baranoff, *Analyst*, 2016, **141**, 3090.
26. J. N. Demas, B. A. DeGraff and W. Xu, *Anal. Chem.*, 1995, **67**, 1377.
27. S. Medina-Rodríguez, A. de la Torre-Vega, J. F. Fernández-Sánchez and A. Fernández-Gutiérrez, *Sens. Actuators, B*, 2014, **192**, 334.
28. M. Marín-Suárez, S. Medina-Rodríguez, O. Ergeneman, S. Pané, J. F. Fernández-Sánchez, B. J. Nelson and A. Fernández-Gutiérrez, *Nanoscale*, 2014, **6**, 263.
29. M. L. Meade, *Principle and Applications in Lock-in Amplifiers*, Peter Pegrins Ltd, London, 1983.
30. M. L. Mandelis, *Rev. Sci. Instrum.*, 1994, **65**, 3309.

31. S. Medina-Rodríguez, C. Medina-Rodríguez, A. de la Torre-Vega, J. C. Segura-Luna, S. Mota-Fernández and J. F. Fernández-Sánchez, *Sens. Actuators, B*, 2017, **238**, 221.
32. V. I. Ogurtsov and D. B. Papkovsky, *Sens. Actuators, B*, 1998, **51**, 377.
33. S. Lehrer, *Biochemistry*, 1971, **10**, 3254.
34. J. N. Demas and B. A. DeGraff, *Sens. Actuators, B*, 1993, **11**, 35.
35. E. R. Carraway, J. N. Demas, B. A. DeGraff and J. R. Bacon, *Anal. Chem.*, 1991, **63**, 337.
36. A. León-García, *Probability, Statistics and Random Processes for Electrical Engineering*, Pearson Education Inc., Prentice Hall, 3rd edn, 2008.
37. C. Kolle, Ph.D. Thesis, *Development and Evaluation of a Phase Fluorometric Instrumentation for Luminescence Based Optical Oxygen Sensors*, University of Leoben, July 1999.
38. G. O'Keeffe, B. D. MacCraith, A. K. McEvoy, C. M. McDonagh and J. F. McGilp, *Sens. Actuators, B*, 1995, **29**, 226.

CHAPTER 7

# *Optical $O_2$ Sensing in Aquatic Systems and Organisms*

KLAUS KOREN[a] AND MICHAEL KÜHL*[a,b]

[a]Marine Biological Section, Department of Biology, University of Copenhagen, Helsingør, DK-3000, Denmark; [b]Plant Functional Biology and Climate Change Cluster, University of Technology Sydney, Ultimo, New South Wales 2007, Australia
*E-mail: mkuhl@bio.ku.dk

## 7.1 Introduction

Molecular oxygen ($O_2$) is a key analyte in aquatic systems, where the balance between: (i) production *via* oxygenic photosynthesis, (ii) consumption *via* aerobic respiration and re-oxidation of reduced chemical species, and (iii) transport *via* diffusion and advection controls the $O_2$ concentration. Measurements of net $O_2$ consumption/production are thus important proxies for primary production and carbon mineralization in aquatic habitats.[1] The solubility and diffusion of $O_2$ in water is relatively low, where the content of $O_2$ in 1 L air-saturated water is about 1/30 of the $O_2$ content in 1 L atmospheric air, and where the molecular diffusion coefficient of $O_2$ in water is about 10 000 times lower than in air.[2] As aquatic organisms, microbial cell aggregates, biofilms and impermeable sediments exchange solutes with the surrounding/overlaying water phase *via* a diffusive boundary layer (DBL),[3,4] the $O_2$ supply often becomes limiting in systems with high availability of organic matter leading to the formation of pronounced

Detection Science Series No. 11
Quenched-phosphorescence Detection of Molecular Oxygen: Applications in Life Sciences
Edited by Dmitri B. Papkovsky and Ruslan I. Dmitriev

Published by the Royal Society of Chemistry, www.rsc.org

$O_2$ concentration gradients. The quantification of $O_2$ concentration and dynamics is thus a key component of many ecophysiological and biogeochemical studies in freshwater and marine systems, and many measuring schemes have been applied. Historically, one of the most applied (and precise) methods for measuring $O_2$ is the well-established Winkler titration method.[5] While this method is sufficient to quantify bulk water $O_2$ levels, it requires sampling of relatively large volumes and has a limited time resolution, and the preferred method for measuring $O_2$ in aquatic samples now relies on the use of chemical $O_2$ sensors with high specificity, reversibility, and sensitivity.

Electrochemical $O_2$ sensors, and especially polarographic cathode-type and Clark-type $O_2$ electrodes,[6,7] have been the predominant $O_2$ sensors in aquatic science for many years, including the development of $O_2$ microelectrodes[8] and trace $O_2$ sensors,[9] which have changed our understanding of $O_2$ dynamics in biofilms, sediments, and the water column.[3,10–14] However, electrochemical $O_2$ sensors are now increasingly complemented or exchanged with luminescence quenching-based optical $O_2$ sensors, *i.e.* $O_2$ opt(r)odes, for aquatic analyses. Although the principle of luminescence quenching by $O_2$ has been known for a long time,[15] optical $O_2$ sensing was mostly used in blood gas analysis before it was introduced to aquatic science in 1995.[16] In the past 20 years, a variety of $O_2$ sensors have been developed and commercialized, and optodes have also enabled new measuring schemes and applications in terms of $O_2$ imaging, distributed sensing with sensor particles, and trace analysis.[16–22]

In this chapter, we review different optical $O_2$ sensor platforms and their application in aquatic science following a chronological as well as application-oriented order. Additionally, we highlight the demands and challenges related to $O_2$ sensing in aquatic systems.

## 7.2 Optical $O_2$ Sensing Platforms

All $O_2$ opt(r)ode sensor platforms rely on the same basic sensing mechanism, *i.e.*, collision-based quenching of an excited indicator dye embedded in a matrix (the so-called receptor). Collision-based dynamic quenching lowers both the luminescence intensity and decay time of the indicator. This quenching mechanism does not consume $O_2$ as the produced singlet oxygen can return to the ground state. In an ideal case, such quenching can be described by the Stern–Volmer relationship (eqn (7.1)) that relates the change in luminescence intensity ($I$) or decay time ($\tau$) to the $O_2$ concentration:

$$\frac{I_0}{I} = \frac{\tau_0}{\tau} = 1 + K_{SV}\left[O_2\right], \tag{7.1}$$

where $I_0$ and $\tau_0$ represent the luminescence intensity and decay time, respectively, in the absence of the quencher ($O_2$).

However, in practice an ideal Stern–Volmer relationship is not achieved with most sensors, where the immobilization of the indicator dye in a polymer matrix leads to the establishment of more quenchable and a less-quenchable fraction, which is often described by the so-called two-site model[23,24] (eqn (7.2)).

$$\frac{I}{I_0} = \frac{\tau}{\tau_0} = \frac{f_1}{1 + K_{SV}{}^1[O_2]} + \frac{f_2}{1 + K_{SV}{}^2[O_2]} \tag{7.2}$$

The factors $f_1$ and $f_2$ describe the percentage of indicator within the two fractions, and $K_{SV}{}^1$ and $K_{SV}{}^2$ denote the corresponding quenching coefficients. If one of the $K_{SV}$ values is much higher than the other it is possible to set the lower value equal to zero, *i.e.*, introduce a non-quenchable fraction of the indicator. This simplifies the mathematical expression. Generally, it is important to describe the calibration function in detail and to determine the mentioned parameters in order to enable simple two-point calibration of the sensor. For two-point calibration, measurements in air-saturated and anoxic medium are typically employed. It is important to notice that temperature compensation is essential in order to obtain reliable $O_2$ measurements,[25] as temperature not only effects the $O_2$ solubility, but also the calibration of the sensor. Salinity has an additional effect on $O_2$ solubility and is thus relevant in marine applications.

The sensor sensitivity can be tuned by changing either the decay time of the indicator[26] or the $O_2$ permeability of the matrix.[27] Signal intensities can also be enhanced by adding scattering particles to the sensor layer, which enhances excitation efficiency and homogenizes the dye exposure and emission in the polymer matrix.[28,29] Further details on the basics of $O_2$ opt(r)ode measuring schemes and characteristics can be found in recent reviews[25,26] and in other chapters of this book.

In order to enable optical $O_2$ measurements, the receptor needs to be excited and the obtained luminescence needs to be recorded. Excitation light as well as the obtained luminescence signal can be guided to the receptor and towards the detector unit (photodiode, photomultiplier or camera) by several means. Depending on the optical design of the light paths and the construction of the sensor, several sensing platforms can be distinguished. The most important sensing platforms for aquatic applications are briefly described in the following.

### 7.2.1 Fiber-optic $O_2$ Opt(r)odes

The probably most widely used sensor platform for optical $O_2$ sensing uses optical fibers to transport light towards and from the receptor. Total internal reflection within the optical fiber(s) transports excitation light from the light source to the receptor (indicator within polymer matrix) immobilized at the tip of the fiber, and the $O_2$-dependent luminescence is guided back to a detector. This setup is highly flexible and enables several measuring geometries.

While initially bulky instrumentation like spectrometers, lasers or large lamps[17] were used for excitation and readout purposes, rapid advancements in optoelectronic components have enabled significant miniaturization of the detection systems used with $O_2$ opt(r)odes. Nowadays, optoelectronic measuring systems for opt(r)odes are based on relatively cheap and low-power consuming components like light emitting diodes (LEDs) and photodiodes. In combination with advancements in electronics and battery technology, standalone instruments capable of long term *in situ* monitoring are now increasingly employed in aquatic science.

Fiber-optic opt(r)odes have the receptor immobilized on the analyte-exposed end of an optical fiber. Different types of immobilization are possible and chosen depending on the application. Measurements in "bulk" water are done with a large receptor area, which is often glued or screwed on to the tip of glass or plastic fibers (see Figure 7.1). This enables straightforward exchange of the receptor and uncouples the optics from the receptor. Such sensors normally have a sensor tip of 1–10 mm and find application in water monitoring or gas exchange measurements in respirometers or fermenters. They are mechanically robust and easy to handle.

In applications involving small sample volumes, measurements of $O_2$ concentration gradients or requiring fast response times, the glass fiber tip is miniaturized or tapered *e.g.* by chemical etching[30] or local heating.[31,32] Afterwards, the receptor (indicator and matrix dissolved in suitable solvent) is typically applied by dip coating. Sometimes, it is necessary to modify the glass surface prior to the coating process in order to increase adhesion.[33] Micro-opt(r)odes[16,18,34,35] are characterized by a small sensor tip of <0.1 mm (Figure 7.1). Very fast response times (<0.5–1 s) and high spatial resolution (<50 μm) can be obtained due to the small receptor volume at the measuring tip. Micro-opt(r)odes find application in for example profiling or eddy-correlation experiments.

Fiber-optic $O_2$ opt(r)odes can be regarded as the optical counterpart to the classical amperometric Clark-type electrodes that also can be manufactured for bulk water sensing or for microscale measurements.[8] In contrast to Clark-type sensors, optical sensors generally do not consume the analyte (see above) and do not display stirring sensitivity of the measuring signal. Additionally, fiber-optical opt(r)odes need less maintenance than electrolyte-containing electrodes and generally exhibit long-term stability in their calibration and other measuring characteristics. This makes opt(r)odes very attractive for long-term monitoring, and almost all commercial electrode/sensor providers have nowadays added $O_2$ opt(r)odes to their portfolio.

### 7.2.2 Planar Opt(r)odes for $O_2$ Imaging and Remote Read-out

A particular interesting sensing scheme for optical sensors is to use them for imaging chemical parameters using cameras as a detector. This enables 2D (or even 3D) measurements of $O_2$. Shortly after the first fiber-based

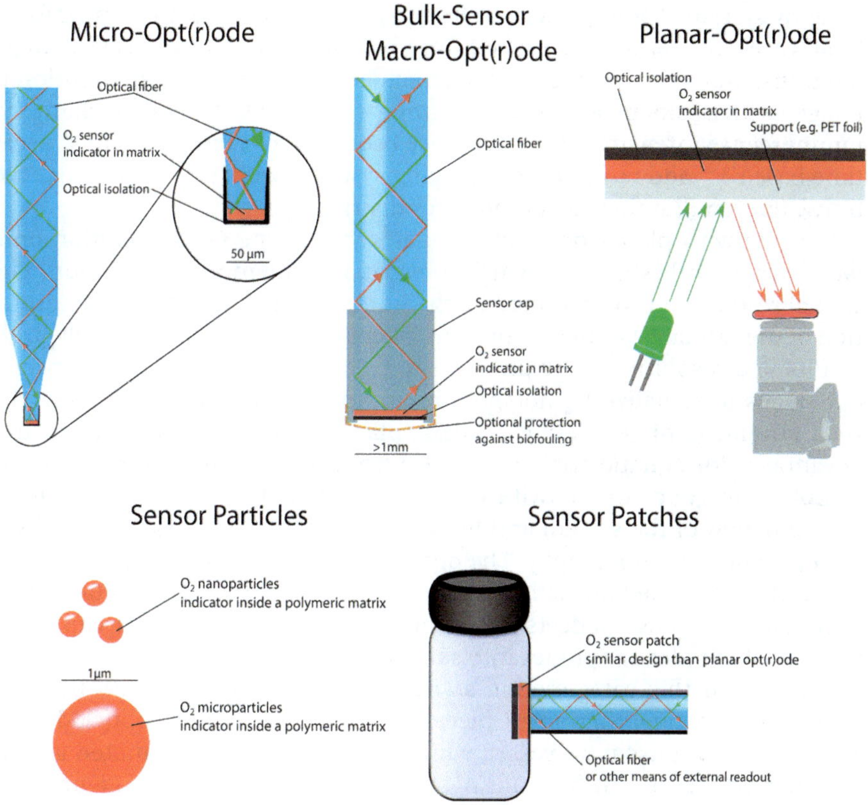


**Figure 7.1** Commonly used sensor platforms for optical $O_2$ measurements in aquatic systems. Fiber-optic sensors are constructed either as micro-opt(r)odes or as macro-opt(r)odes. Micro-opt(r)odes enable high resolution $O_2$ measurements, while macro-opt(r)odes are more mechanically robust and typically used for longer term bulk measurements. Planar opt(r)odes enable chemical imaging using camera based read-out. Sensor nano- or microparticles can be dispersed within the sample or used to coat tissues, biofilms or other (a)biotic structures. By integrating a sensor patch (small piece of a planar opt(r)ode) within a vessel, the internal $O_2$ levels can be read out non-invasively through the transparent wall.

$O_2$ sensors were introduced,[16] such 2D imaging setups were also developed for use in aquatic systems.[28,36] For 2D imaging, the receptor is immobilized on a transparent $O_2$ impermeable planar support (*e.g.* a glass cover slip or plastic polymer foil) (Figure 7.1). Large areas of such solid support can be covered with a homogeneous receptor layer by simple knife-coating[28,37] or spin-coating processes.[33] The planar support for the receptor can also be replaced by a fiber-optic faceplate in order to increase the spatial resolution of the sensor.[38] Such planar opt(r)odes can be put in contact with the sample *e.g.* by pressing a sample against the sensor foil,[28] inserting it into the sample *via* a periscopic imaging system,[39] or by growing cells and biofilms on top of the sensor foil.[33] In all cases, the receptor is excited from behind (typically by LED's) through the transparent carrier, and the $O_2$-dependend luminescence distribution is imaged with a camera. First systems simply used the $O_2$-dependent change in luminescence intensity,[28] but have since been replaced by more robust decay time based[36,40] and ratiometric[41] imaging systems, which are less influenced by *e.g.* uneven illumination or indicator distribution.[42]

Imaging with planar opt(r)odes enables visualizing $O_2$ distribution on a larger scale, while still preserving high spatial and temporal resolution. The spatial resolution is determined by the imaged size compared to the resolution of the camera; *e.g.* an area of 1 cm × 1 cm imaged with 1000 × 1000 pixels results in a resolution of 10 μm per pixel. Several studies have used planar opt(r)odes to visualize $O_2$ hotspots in marine environments.[39,43,44] Despite their advantage of generating 2D data, planar optodes also have a few disadvantages for aquatic applications. Up front it has to be ensured that the sample is in good contact with the planar opt(r)ode, which often requires manipulation of the system and limits their application to sample surfaces with pronounced topography. The opt(r)ode itself acts as a diffusion barrier and will change the flow pattern at the surface, which can lead to unwanted "smearing" or "edge" effects.[28] Another type of smearing effect can occur when pushing planar optodes into samples.[39,45]

Also within this category, the so-called sensor patches should be mentioned. This means that small pieces of sensor foil (planar opt(r)ode) are fixed inside a transparent vessel, while sensor readout is facilitated *via* the transparent vessel wall by pointing a fiber at the sensor spot (Figure 7.1). Such patches can be integrated in all sorts of vessels such as fermenters or respirometry chambers, and readout can be realized with different optical emitter/detector units.[46]

### 7.2.3 Particle-based Optical $O_2$ Sensors

An imaging-based sensor platform that has been introduced to aquatic systems relatively recently[22] employs distributed sensing with micro- or nano-meter sized $O_2$ sensor particles. Rather than coating the receptor (indicator within a matrix) onto a fiber tip or a planar carrier, solid support particles can be made out of the matrix containing the indicator. Several straightforward methods for fabrication and use of $O_2$ sensitive particles are available.[47–50] Such sensor particles may also contain additional functionalities, like surface functionalization and magnetic properties,[47] making them attractive *e.g.* in applications mapping $O_2$ dynamics on flow-exposed surfaces that are coated with $O_2$ sensitive particles.[22,51] Sensor particles can also be immobilized into (semi)transparent gels such as agar and agarose, which can be used for embedding biological samples. This has *e.g.* been used for mapping the distribution and dynamics of $O_2$ in the rhizosphere of aquatic plants, which were planted in artificial sediments containing $O_2$ sensor particles.[20]

## 7.3 Challenges Related to Optical $O_2$ Measurements in Aquatic Systems

Aquatic habitats exhibit an extreme variety in environmental conditions, which present challenges to environmental $O_2$ monitoring. In polar or alpine regions, sensors have to deal with low temperatures, in the deep-sea extreme pressures prevail, and in light-exposed habitats strong solar radiation (including UV radiation) can lead to sensor bleaching and deterioration. Furthermore, several environmental parameters might change in parallel *e.g.* in intertidal zones, where temperature, salinity and light co-vary. In order to meet these and many other challenges, special sensor designs have been developed. Below, we discuss some of the common challenges and solutions for optical $O_2$ measurements in various aquatic systems.

### 7.3.1 Large $O_2$ Dynamics

While $O_2$ levels within certain environments and organisms are highly regulated (*e.g.* blood oxygenation), aquatic systems can exhibit extremely varying $O_2$ levels. Photosynthetic systems like microbial mats or dense algal cultures can easily reach $O_2$ levels exceeding air-saturation several times and approaching 100% $O_2$ when exposed to light.[16] At the other extreme, the so-called oxygen minimum zones (OMZ) in the oceans are characterized by extremely low $O_2$ availability in the nM range.[52–54] An advantage of optical $O_2$ sensors is that they can relatively easily be optimized for specific measuring characteristics needed in a particular habitat without changing the basic optical readout methods.

The dynamic range of an optical $O_2$ sensor can be tuned by altering the decay time of the indicator or the permeability of the immobilization polymer matrix. Due to intense research on indicator synthesis, many different classes and variants of indicators are described[25,26] and can be used to customize the measuring characteristics of $O_2$ opt(r)odes. Additionally, the permeability of the polymer can be altered by either changing the type of polymer[25] or by altering the monomer structure.[27] The latter may be the preferred method when only slight (2–5 times) changes in sensitivity are desirable. Changing to an entirely different polymer (*e.g.* between less permeable polystyrene and highly permeable silicone) can result in more dramatic changes in the sensitivity, but one has to be careful that the indicator and the polymer are compatible. If the indicator does not dissolve well in the polymer or even aggregates within the polymer, this leads to signal drift and changing sensor calibration.

Many commercial providers of $O_2$ opt(r)odes (*e.g.* Pyro-Science GmbH, Presens GmbH, Ocean Optics) offer sensors optimized for different dynamic ranges, *i.e.*, normal range $O_2$ sensors (typically 0–50% $O_2$ with a limit of detection (LOD) of 0.02% $O_2$) and trace $O_2$ sensors (typically 0–10% $O_2$ with LOD of 0.005% $O_2$). Commonly used oceanographic sensors have an LOD of ~1 μM, which is generally acceptable. However, in oxygen minimum zones (OMZs)

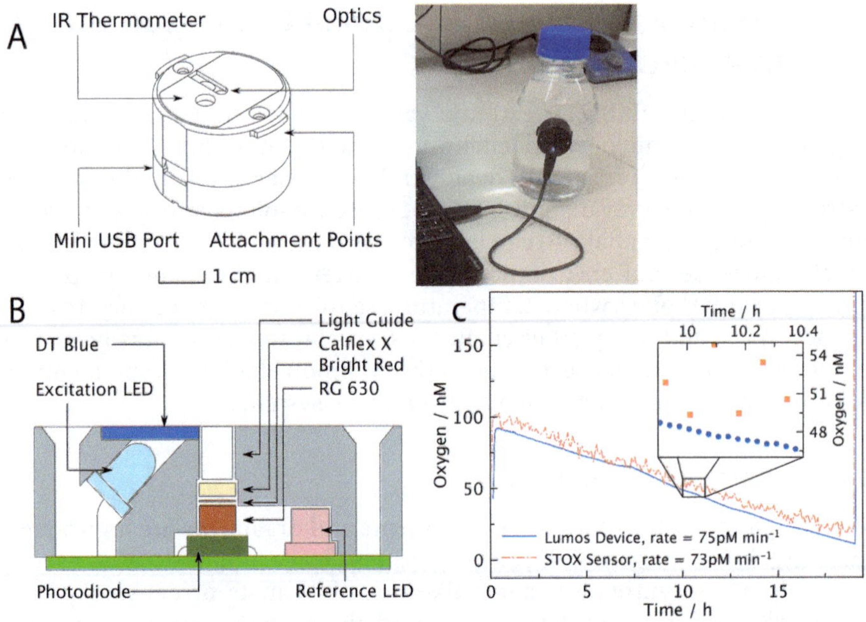


**Figure 7.2** Design and application of the LUMOS[46] sensor device for trace $O_2$ measurements (A). The entire optics (panel B), an IR thermometer and the electronic components are packed into a small device that can be attached to a vessel containing the trace $O_2$ sensor patch. The optical trace sensor provides similar respiration rates than an electrochemical trace $O_2$ sensor (STOX), but the LUMOS data is less noisy (C). Reprinted from *PLoS One*, 2015, **10**, e0128125,[46] https://doi.org/10.1371/journal.pone.0128125. © 2015 Lehner *et al.* Published under a CC BY 4.0 licence, https://creativecommons.org/licenses/by/4.0/.

a significantly lower LOD is needed to describe the true $O_2$ availability. Therefore, highly sensitive trace $O_2$ opt(r)odes were developed.[46,55] Larsen *et al.* (2016) used a combination of a standard (LOD 200 nM), a mid-range (LOD 50 nM) and a trace sensor (LOD 5 nM) in order to fully describe the $O_2$ profile within an OMZ.[55] Also, $O_2$ consumption measurements at nM concentrations were performed using a specially developed sensor material and optical setup (LUMOS)[46,56] (see Figure 7.2). Optical sensors for ultra-trace $O_2$ measurements enable $O_2$ measurements down to a LOD of 7 pM.[57] Optical $O_2$ sensors thus outcompete electrochemical sensors in terms of LOD; although also highly sensitive trace electrochemical $O_2$ sensors were developed and successfully applied.[9,58]

### 7.3.2 Extreme Environments Require Special Designs

As mentioned, extreme conditions can be found within aquatic habitats. Probably some of the harshest environments on Earth are found in the deep marine trenches under extreme hydrostatic pressure and low temperature.

Nevertheless, optical $O_2$ sensors have also been applied successfully in such extreme deep-sea habitats. First, micro-opt(r)odes were adopted and tested for use at high pressures,[59] showing that the performance of $O_2$ micro-opt(r)odes is similar to pressure-compensated $O_2$ microelectrodes. Also 2D imaging setups have been developed and tested for imaging $O_2$ in the seafloor[39,44] both in coastal regions[39] and the deep sea.[44] This enabled *in situ* visualization of the spatio-temporal $O_2$ dynamics in the seafloor as affected by worm burrows and other animal activities (bioturbation and bio-irrigation),[60] sedimentation of marine snow particles, as well as photosynthetic $O_2$ production in shallow waters.[39,61]

Besides extreme hydrostatic pressure, also extreme temperatures can be challenging for optical $O_2$ sensors. Besides the need to apply the appropriate corrections for temperature and salinity effects, opt(r)odes are well suited for measurements below 0 °C as the sensor does not contain any electrolyte or other liquid phases that could freeze; this can be a problem for electrode-based sensors. Microopt(r)odes have for example been applied within sea ice, where sensors have either been frozen into forming sea ice[62] or used to measure gradients at the sea ice water interface.[63] Also planar opt(r)odes have been used at temperatures below 0 °C to investigate the $O_2$ dynamics in sea ice during melting.[64]

$O_2$ opt(r)odes are also well suited for measurements in hot water. Many commercial sensors are specified for measurements in water up to 55 °C, but as long as the mechanical properties of the sensor matrix are not affected, measurements at much higher temperatures are possible *e.g.* in geothermal springs where microopt(r)odes have been applied up to 85 °C (M. Kühl, unpublished data).

A further challenge for optical $O_2$ sensors arises from background irradiance. When measuring $O_2$ production within photosynthetic systems such as microbial mats, biofilms or a coral reef, the sensor may be exposed to high levels of solar radiation including high levels of UV light. This can lead to bleaching of the indicator and saturation of the detector by background light (despite strong spectral filtering). Solutions for this include the use of an $O_2$ permeable optical isolation layer on microopt(r)odes[16] or planar sensors,[65] which may however result in a longer response time. Nevertheless, it remains important to check that the background light is not affecting the measurement. Background light can be better controlled in the laboratory, and it might also be possible and useful to switch the light source (*e.g.* used to stimulate photosynthesis) off, *e.g.* at the onset of imaging the $O_2$ distribution.[51,65] Besides background light, also background fluorescence of chlorophyll and other fluorescent pigments in the sample can interfere with the luminescence signal used to sense $O_2$. Especially in dense photosynthetic systems such as biofilms and coral tissue, this can be a significant problem and needs special sensor optimization in terms of optical isolation, tuning of the sensor composition enabling excitation outside the absorption maxima of chlorophyll, or by using a time gated readout where the fast decaying chlorophyll fluorescence is filtered out.

In conclusion, optical $O_2$ sensors have proven useful within extreme aquatic habitats, although special developments are often needed in order to meet the challenges presented by extreme environmental conditions.

### 7.3.3 Sensor Stability – from Mechanical Stress to Biofouling

A key challenge for sensors in aquatic environments is mechanical and optical stability. Although often overlooked, this issue is far from solved and presents a key challenge for future sensor development and optimization. Several aspects affect the stability of optical sensors. Obviously, photostability (especially of the $O_2$ indicator) is a key issue. However, several $O_2$ indicators exhibit very good photostability and withstand extended illumination without significantly affecting the measurement.[26] The mechanical stability of optical $O_2$ sensors is generally high, especially when robust sensor designs are used rather than microsensors.

The major factor limiting stability and applicability of $O_2$ sensors in aquatic systems is biofouling. Biofouling describes the generally unwanted formation of a biological cover on any surface submerged in water. A sensor surface will first be colonized by bacterial or algal biofilms, which primes the ground for settlement of larvae and propagules of other organisms that further colonize the sensor, this includes also larger algae or even barnacles and mussels on the submerged sensor. Obviously, the formation of any sort of biological cover of the sensor (in particular the receptor) will affect the readings. Biofouling results in changed $O_2$ diffusion towards the sensor as well as incorrect readings due to respiration or $O_2$ production in the biofilm coating the sensor. Biofouling can thus strongly compromise the long term application of optical sensors for monitoring purposes in aquatic habitats.

Solutions to the biofouling problem are manifold and include mechanical as well as chemical solutions.[66] Automated wipers or brushes are commonly used for mechanical cleaning, which will prevent large organism settling on the sensor, while often not preventing the formation of thin biofilms on the sensor. Critical factors are robustness of the mechanical cleaning system as well as power consumption. Alternative mechanical systems include shutters that hide the sensor surface and only open during the measurement period. Here, copper-based shutters have proven useful, as copper also inhibits microbial growth[66] by releasing the biocide $Cu^{2+}$. In general, the use of biocide-releasing paints was found to be an effective anti-fouling strategy, in line with experiences from ship paints. Nevertheless, the release of a biocide might also affect the conditions around the sensor and therefore the validity of the sensor reading. Chemical modification of the sensor surface with anti-fouling or repelling layers was also investigated and showed promising results. For example a coating with a phosphorylcholine-containing polymer resulted in a sensor with reduced bio-fouling, and showed only minor effects on the sensor characteristics.[67]

Biofouling is a key limitation, when it comes to using optical $O_2$ sensors for long-term monitoring in aquatic habitats. Several attempts have been made to reduce biofouling, but no "golden solution" has been identified.

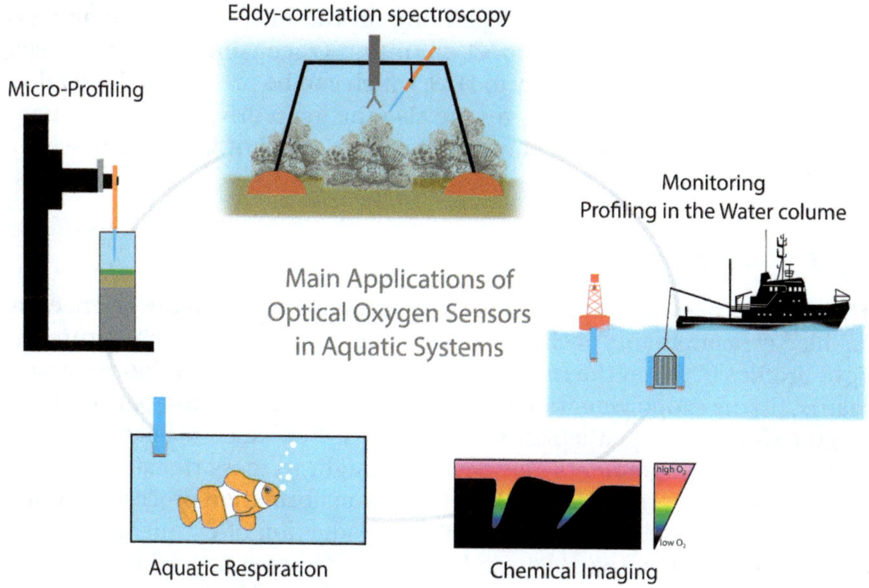


**Figure 7.3** Optical $O_2$ sensors are applied in multiple areas of aquatic science ranging from high-resolution profiling or imaging measurements to long-term monitoring.

Most likely, such a universal solution will not be found, as the rate and species composition of biofouling on sensors is highly dependent on the environment. Nutrient availability, temperature, light exposure, flow and depth are just a few of the many environmental factors that influence biofouling. Detailed studies on both sensor materials as well as environmental factors are thus needed to get a better understanding of how to prevent biofouling and optimize long-term deployment of optical $O_2$ sensors in aquatic environments (Figure 7.4).

## 7.4 Applications of Optical $O_2$ Sensors

Optical $O_2$ sensors have become widely applied in aquatic science (Figure 7.3) and are now available (also commercially) in a large variety of designs and measuring characteristics. Currently, commercial optical and electrochemical $O_2$ sensors share the market more or less equally.

### 7.4.1 Application of Micro-opt(r)odes

While electrochemical microsensors have been used since the mid/end 1960s[8,11,12,68] to measure $O_2$ in aquatic systems, optical $O_2$ microsensors were first introduced to aquatic science in 1995.[16,18,35] While at first the luminescence intensity was used to quantify $O_2$,[16] lifetime-based measurements were introduced soon thereafter as a prerequisite for more robust sensor readout.[34] The main advantages of $O_2$ micro-opt(r)odes are: (i) high spatial resolution (around 30–50 μm depending on the measuring tip diameter), (ii) fast response time (down to <1 s), (iii) no $O_2$ consumption by the sensor, and (iv) no cross-sensitivity to H2S, which can be present in the seafloor and in biofilms. These advantages are also the main drivers of using micro-opt(r)odes. In the following some of the main areas of application will be described.

#### 7.4.1.1 *Profiling*

The development and application of microsensors in aquatic science has enabled measurements of chemical concentration profiles, temperature and light profiles between the mixed water and (i) the porous seafloor sediment matrix, (ii) the exopolymer matrix of biofilms and marine snow particles, and (iii) the tissues of aquatic plants and animals.[12,14,32] Such measurements of physico-chemical gradients give detailed insight to mass transfer, radiative exposure, optical properties, and chemical transformation processes in such surface-associated aquatic systems. The introduction of $O_2$ microelectrodes to aquatic science in the 1960s and 1970s[8,11,12,68] has thus lead to a detailed understanding of important mechanisms such as the diffusive boundary layer,[3] the $O_2$ penetration depth into sediments,[11] and the $O_2$ dynamics in corals.[69] The solubility of $O_2$ in water is much lower than in air (about 30 times) and $O_2$ diffuses 10 000 times slower in water as in air. At the same time, $O_2$ is the most energetically favorable electron acceptor for respiratory metabolism and is produced by oxygenic photosynthesis in aquatic plants, algae and cyanobacteria. This leads to the formation of $O_2$ concentration gradients in aquatic systems that require fine scale measurements for accurate quantification, typically at <0.1 mm spatial resolution.

Optical $O_2$ sensing was actually first introduced to aquatic biology by the development of $O_2$ microopt(r)odes.[16] Such optical microsensors have similar measuring characteristics in terms of response time and temperature dependence as electrochemical microsensors, but they are much easier to fabricate, have no intrinsic $O_2$ consumption and are not susceptible to *e.g.* $H_2S$ interference. They can be used in the same types of applications as electrochemical $O_2$ microsensors, but also in settings where microelectrodes are challenged *e.g.* by low temperatures[62] or strong electromagnetic fields. In terms of measuring systems, very similar setups for motorized microprofiling in 1D, 2D, or even 3D configurations are commercially available for both $O_2$ microelectrodes and micro-opt(r)odes, which can be regarded as largely complementary techniques in most applications.

#### 7.4.1.2 *Eddy Correlation Spectroscopy*

The flux of $O_2$ across the sediment–water interface is a good proxy for the integrated benthic carbon mineralization as well as primary production of biofilms and sediments.[1] There are several established methods to determine the $O_2$ uptake of sediments in the field including measurements of

$O_2$ concentration microprofiles[70] or by using chamber incubations of known sediment areas/volumes.[71] Both established methods use either optical or electrochemical $O_2$ sensors. In 2003, a new method based on eddy correlation spectroscopy was introduced to aquatic science enabling non-invasive quantification of the net $O_2$ uptake/production of larger areas of sediments and other surfaces.[71] This so-called eddy-covariance or eddy-correlation method measures the covariance between two parameters simultaneously, namely the fluctuating vertical velocity component using an acoustic Doppler velocimeter and the fluctuating $O_2$ concentration using a microsensor within the same sampling volume. Measurements are typically performed in the turbulent boundary layer 10–50 cm above the sediment at a frequency of 14–25 Hz. As the method is non-invasive and does not need any sensor to penetrate the surface, it can be used to measure integral $O_2$ fluxes over large areas of complex habitats (not accessible by other means) like coral reefs,[72] hard bottom substrates,[73] seagrass meadows[74] or sea ice.[75,76]

In recent years, this technique has seen a significant boost and commercial systems are now on the market further stimulating the use of eddy correlation systems in aquatic systems. Many of these systems employ fast-responding electrochemical $O_2$ microsensors. As such sensors consume $O_2$ as part of the measurement process (contrary to optical sensors), $O_2$ microelectrodes have an inherent, albeit small, stirring sensitivity. It was shown that this stirring sensitivity, even if very low, can result in an artificial flux.[77] Optical $O_2$ microsensors were also tested in the same study and did not exhibit such artefacts. Consequently, due to the lack of stirring sensitivity and the fast response time of optical $O_2$ microsensors, they are now recommended as the preferred $O_2$ microsensor for eddy covariance measuring systems.[77] Recently a dual sensor measuring $O_2$ as well as temperature at the same time was developed and successfully applied for eddy covariance measurements.[78] In terms of sensor design, this sensor is rather a bulk sensor than a microsensor. Nevertheless, the achieved response time is remarkably short ($t_{90}$ of ~0.5 s for $O_2$), while the sensor at the same time is more sturdy and does not show any stirring sensitivity. In the field of eddy correlation measurements, the sensor performance is essential for the data quality and the conclusions that can be drawn from the measurements. Therefore, optical sensors have a great potential in this field and have already provided excellent data.

### 7.4.2 Applications of Bulk Sensors and Sensor Patches

Bulk sensors or $O_2$ macro-opt(r)odes are characterized by a sensor tip size of ~1 mm or more. Spatial resolution is thus low and generally the response time is longer than for $O_2$ micro-opt(r)odes, *i.e.*, several seconds. The main benefit of these macro sensors is their improved mechanical stability. Bulk sensors come in a variety of designs and are often made to fit a certain application. In contrast to micro-opt(r)odes, bulk sensors are often constructed using cheaper plastic fibers to guide the light between the optics and the receptor.[79] Alternatively, entire sensor modules or heads can also be constructed,[80] where the sensor material (receptor) is directly placed on top of the optical part. Sensor patches may also be seen as bulk sensors integrated into a vessel enabling external readout through a transparent wall *via* a fiber, camera or other types of detector modules (see *e.g.* Figure 7.2).

### *7.4.2.1 Respirometry*

The $O_2$ consumption or production of aquatic organisms is an important physiological and environmental parameter, and such respirometric measurements involve species (or system samples) of many different sizes ranging from microorganisms to fish. Three different experimental approaches are used for respirometry: (i) closed systems, (ii) flow-through systems, and (iii) intermittent flow systems.[81] In closed systems, the sample of interest is placed inside a closed airtight vessel and a change in $O_2$ concentration in the water within the vessel is followed over time. Closed system respirometry is the easiest to use and the only option for small species like bacteria or algae that would be washed out in a flow-through system. Flow through-systems provide a constant in and out flow of water and measure the difference in $O_2$ concentration between inlet and outlet to determine the respiration rate. Due to the flow regime and constant replenishment of aerated water, this system is well suited for respirometry on larger organisms during different modes of their activity (*e.g.* during swimming).[82] Intermittent flow systems combine the two other methods by repeatedly closing and opening the system. Respiration is measured in the closed system state, where after the system is flushed with aerated water to ensure removal of waste products and to reestablish ambient $O_2$ levels; then the cycle can start again.[83] The advantages and potential errors of the different types of respirometry are discussed elsewhere.[84,85]

In terms of sensors, a variety of $O_2$ sensor designs are used in respirometry. As mentioned, optical $O_2$ sensors do not consume $O_2$ during the measurement. This becomes particularly important when very low respiration rates are measured or in measurements on very small volumes. Micro-opt(r)odes can be inserted into specially designed low volume respiration chambers and are used to measure *e.g.* respiration of invertebrates or sponges.[86] This early study[86] compared micro-opt(r)odes with other commonly used methods and concluded that opt(r)ode technology is advantageous when compared to other methods due to negligible drift, ease of use and stability. Micro-opt(r)odes are used for small volume applications and are commercially available. Alternatively, also respiration vessels with integrated sensor spots (sensor patches) can be used and read out through a transparent wall using optical fibers. A variety of bulk sensors can also be inserted into respirometers *e.g.* for *in situ* sensing of $O_2$ in incubation chambers.[87]

An advantage of optical sensors is that they can be used at different $O_2$ regimes and that their sensitivity can be tuned to fit the problem. For example, trace sensors were used to measure $O_2$ consumption and production at very low $O_2$ levels.[56,88] Photosynthetic $O_2$ net production can also be determined using optical sensors within closed systems.[89] When studying photosynthetic $O_2$ production it is, however, important to protect the sensor with an optical isolation (especially at high light levels).

#### 7.4.2.2 *Environmental Monitoring*

Oxygen monitoring in the environment typically involves long-term measurements, which are important means to better understand *e.g.* biogeochemical dynamics within the environment or to monitor water quality or processes involved in water treatment. Low power consumption makes opt(r)ode systems very interesting for long term monitoring, but biofouling has to be addressed and can be a limiting factor for the actual deployment time. There are several companies on the market providing optical sensors for such $O_2$ monitoring. Sensors can *e.g.* be mounted on conductivity-temperature-depth (CTD) profilers used for oceanographic or lake measurements[55] or be implemented on buoys for continuous measurement.[90] Electrochemical sensors still dominate this field of application, mainly for historical reasons and because this technology is well established within the field. However, opt(r)ode based devices are now increasingly used in many monitoring applications due to their better long-term stability and mechanical robustness.

Besides water monitoring, also monitoring within wetlands or riverbeds is highly relevant. For such purposes, systems based on multiple robust fiber-based $O_2$ sensors have been developed for long-term deployment in the environment (Figure 7.4). For example a multi-fiber $O_2$ opt(r)ode consisting of 100 individual sensors was constructed enabling simultaneous readout of all 100 sensors[91] by fixing the ends of all the fibers in a matrix in front of a camera system imaging all sensors at once. The fiber measuring tips can then be placed at any position within the environment,[92] and the detector system records sensor readout autonomously. In other systems, sensor fibers were arranged (along with thermocouples) along a spear that was inserted into a wetland, which enabled combined long-term (several months) *in situ* monitoring of $O_2$ and temperature in peat soil at 10 position over a distance of 1 m.[79] In a similar spear concept, the fibers are not fixed at one position but can be moved within the spear that is entirely coated on one side with a planar $O_2$ opt(r)ode.[93,94]

### 7.4.3 Chemical Imaging

Contrary to fiber- or electrode-based sensors, planar opt(r)ode-based sensors can also be used to image the distribution and dynamics of $O_2$. Imaging is achieved by using a camera as detector unit in combination with appropriate filters and excitation sources. The great advantage of this method is that 2D information on the $O_2$ distribution and dynamics over larger areas can be obtained, thus alleviating the inherent limitation of microsensors

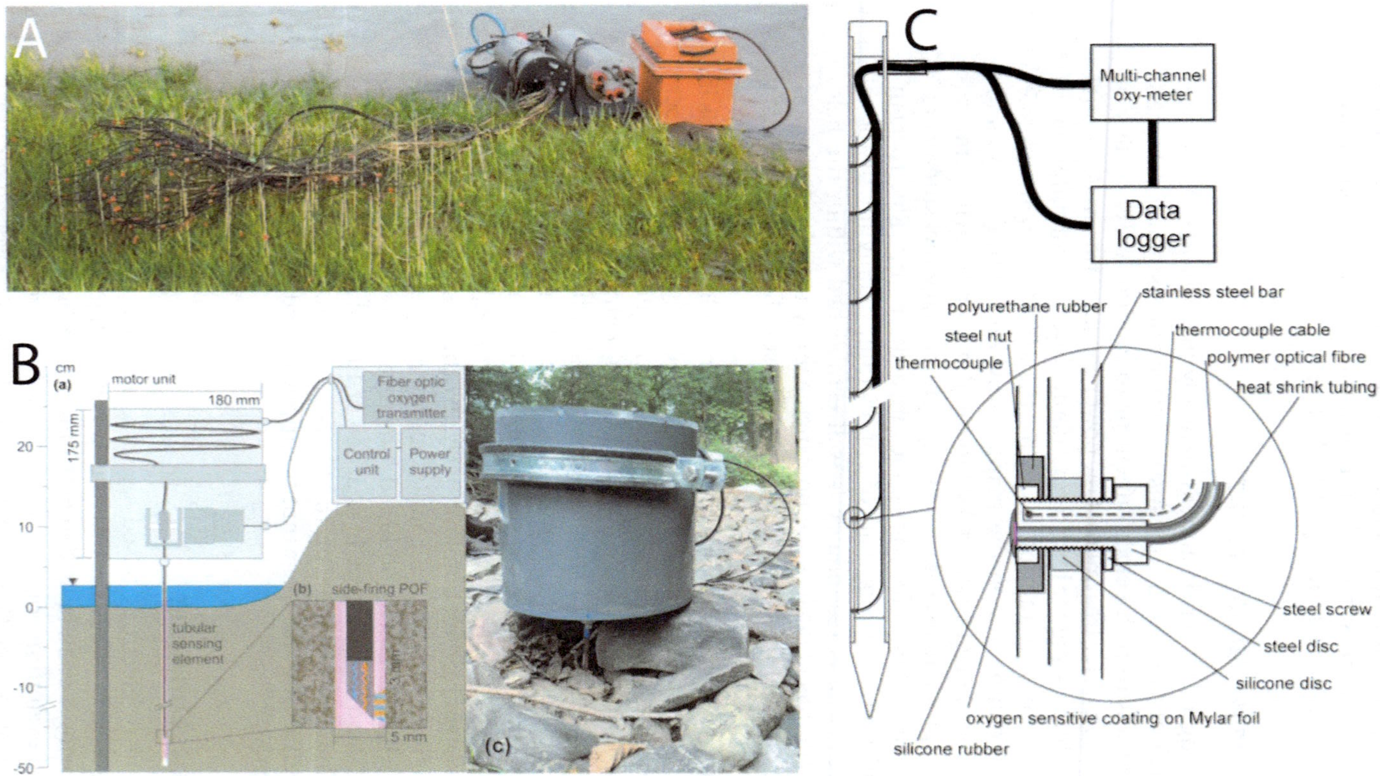


**Figure 7.4** Examples of bulk optical $O_2$ sensors for environmental monitoring. (A) The multi-fiber $O_2$ opt(r)ode consisting of 100 individual sensors ready to be used in the field.[92] Reprinted from *Science of The Total Environment*, **589**, K. Koop-Jakobsen, J. Fischer and F. Wenzhöfer, Survey of sediment oxygenation in rhizospheres of the saltmarsh grass – *Spartina anglica*, 191–199,[92] Copyright © 2017 The Authors, with permission from Elsevier. (B) Visualization of the coated spear sensor that enables $O_2$ profiling in the environment by moving a side-firing optical fiber within the hollow spear.[93] Reprinted with permission from ref. 93. Copyright © 2017 American Chemical Society. (C) Ten $O_2$ sensor spots with individual fiber-optic readout and thermocouple temperature sensing elements are distributed within a 1 m long spear enabling simultaneous monitoring of $O_2$ and temperature.[79] Reprinted with permission from ref. 79. Copyright © 2013, the American Society of Agronomy, Crop Science Society of America, and Soil Science Society of America, Inc.

to a few point measurements. This enables visualization of the spatio-temporal heterogeneity in $O_2$ concentration at different scales. Depending on the chosen setup, $O_2$ images with high spatial resolution (<50 μm) can be obtained or larger areas (up to entire root systems) can be imaged simultaneously.[95] The possibility to visualize heterogeneities and dynamics at once as well as the possibility of combining $O_2$ imaging with structural or other imaging techniques (*e.g.* chlorophyll fluorescence) makes this technique highly interesting for different applications in aquatic systems.[45,96] Such measurements have mainly employed planar-opt(r)odes but recently also particle based methods have been applied.[19,20,51] While such chemical imaging was initially purely intensity-based, more robust referenced measurements now use either lifetime-based[40,97] or ratiometric[41] imaging systems. Such $O_2$ imaging has now been implemented in a variety of applications ranging from microscopic imaging (Figure 7.5(1)) of $O_2$ dynamics in biofilms,[33,98] organisms[51,99] and in sediments,[28,41] including the development of special imaging systems for *in situ* $O_2$ imaging[39] (Figure 7.5(2)). In the following, we present just a few of these applications representative of $O_2$ imaging at different spatial scales.

### 7.4.3.1 Microscopic $O_2$ Imaging

By mounting a lifetime-based or ratiometric luminescence imaging system on a microscope[22,33,98,100] or *e.g.* by using a simple USB-based microscope equipped with excitation LED's and a long-pass filter,[101–103] the $O_2$ distribution and dynamics can be imaged at very high (μm) spatial resolution at the level of a few cell clusters or biofilms. For microscopic imaging with planar sensors, it is important to apply very thin planar opt(r)odes to avoid the receptor layer acting as an $O_2$ reservoir and/or $O_2$ diffusion matrix, which would lead to smearing effects and low spatio-temporal resolution. Thin sub-μm receptor layers can *e.g.* be immobilized *via* spin-coating onto glass cover slips, which are silanized to facilitate better binding between the glass and receptor layer.[33] Such coverslip sensors can be constructed to exhibit near-ideal Stern–Volmer quenching by $O_2$, and they have a well-defined temperature response. By mounting coverslip sensors as lids in flow-chambers or membrane filtration reactors (Figure 7.5), it is possible to monitor the $O_2$ concentration gradients and dynamics relative to the distribution of the cellular biomass in biofilms using simple ratiometric imaging approaches.[98,100] Higher resolution studies combining $O_2$ imaging with laser confocal scanning microscopy on aquatic biofilms have also been realized.[33]

While the above mentioned examples of application mainly have involved cell/biofilm growth on planar opt(r)odes, microscopic $O_2$ imaging can also employ sensor micro- or nano-particles,[104–106] albeit few applications have so far been realized in aquatic science, where most particle-based imaging has been done at a more macroscopic level.[19,20,51,107]

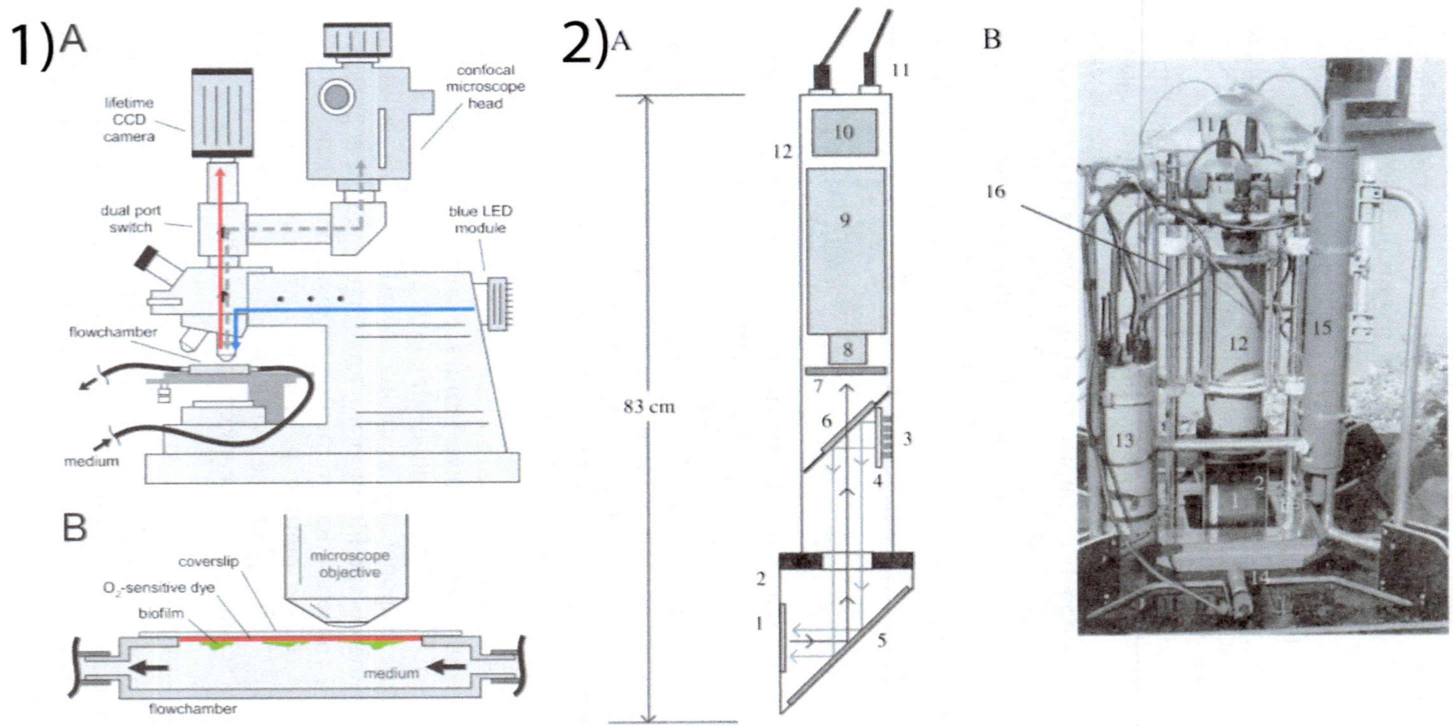


**Figure 7.5** (1) Setup for microscopic $O_2$ imaging[33] consisting of a luminescence lifetime-based camera system mounted on a confocal microscope (A) and a flow chamber with an integrated $O_2$ opt(r)ode. (2) *In situ* $O_2$ imaging system[39] consisting of multiple components: (1) Planar opt(r)ode, (2) sensor window, (3) LEDs, (4) diffuser, (5) mirror, (6) dichroic mirror, (7) excitation filter, (8) lens, (9) CCD camera, (10) electronics controlling the excitation light, (11) connectors, (12) central housing, (13) battery for video and light, (14) video camera, (15) Niskin bottle, and (16) positioning elevator (further details in ref. 39). Reproduced with permission from ref. 39, © 2001, by the Association for the Sciences of Limnology and Oceanography, Inc. and ref. 33, Copyright 2007, American Society for Microbiology.

### *7.4.3.2 $O_2$ Imaging in Sediments*

Knowledge about the $O_2$ distribution and dynamics in sediments is crucial for the understanding of overall carbon mineralization and the zonation of different aerobic and anaerobic processes involved in overall carbon and nutrient cycling. Dependent on the composition of the sediment and its associated fauna and flora, a distinct mosaic of oxic and anoxic micro-niches can establish within the sediment.[108] When first introduced, planar opt(r)odes already showed heterogeneities in the diffusive boundary layer due to the topography of the sediment and also hotspots of $O_2$ consumption due to variations in available labile carbon.[28] In the presence of macrofauna such as burrowing animals, the $O_2$ availability within the sediment can exhibit extreme spatio-temporal heterogeneity.[64] Planar opt(r)odes are ideal tools to visualize such effects of burrowing animals on the $O_2$ availability within sediments.[108] By confining burrowing animals close to the planar opt(r)ode, it is *e.g.* possible to study the sediment $O_2$ dynamics associated with the bio-irrigation of the burrow and consequently the changes in the surrounding sediment layers (Figure 7.6).[41,109] Such studies can be done at various scales depending on the sediment characteristics and animal sizes, as planar opt(r)odes and the measuring systems can easily by adapted for measurements even over several 10s of $cm^2$. Besides lab-based studies, such effects of bio-irrigation and bioturbation can also be observed *in situ* using special autonomous instruments platforms for planar opt(r)ode-based chemical imaging.[39,110]

Recently, also sensor particles were used to study the dynamics of bioturbation at the water sediment interface (Figure 7.6). For this purpose a special lifetime-based method was developed that uses a laser sheet to excite $O_2$ sensor nanoparticles dispersed within the overlaying water.[107] This enabled the imaging of the outburst of anoxic water from a ventilating burrow, which could be followed without disturbing the natural flow.

Besides the effects of fauna, aquatic plants can also strongly impact the $O_2$ dynamics within sediments. As an example, seagrasses transport $O_2$ from the leaves to their roots, where $O_2$ is released at their root tips[111] protecting the most sensitive parts of below-ground plant tissue against intrusion of toxic $H_2S$ from the surrounding sediment. Chemical imaging enables visualization of such hotspots and can be used to study dynamic $O_2$ changes in the plant rhizosphere *e.g.* as a function of light exposure of the leaves or the $O_2$ content of the surrounding water (Figure 7.7).[43] Similar $O_2$ transport towards the roots is found in other aquatic[110] or water locked plants like rice.[112] For example, a recent study monitored the effect of $O_2$ release from the roots of the salt marsh grass *Spartina anglica* on the bulk sediment. By using both planar-opt(r)odes and bulk sensors, the authors could show that the effect is very localized and is not sufficient to significantly aerate the bulk sediment.[97]

Planar opt(r)odes were first applied to map the $O_2$ dynamics in the seagrass rhizosphere,[43,113] while sensor nanoparticles were applied more recently enabling similar resolution but larger coverage of the complete rhizosphere. For particle-based $O_2$ imaging, seagrass specimens were grown in an artificial

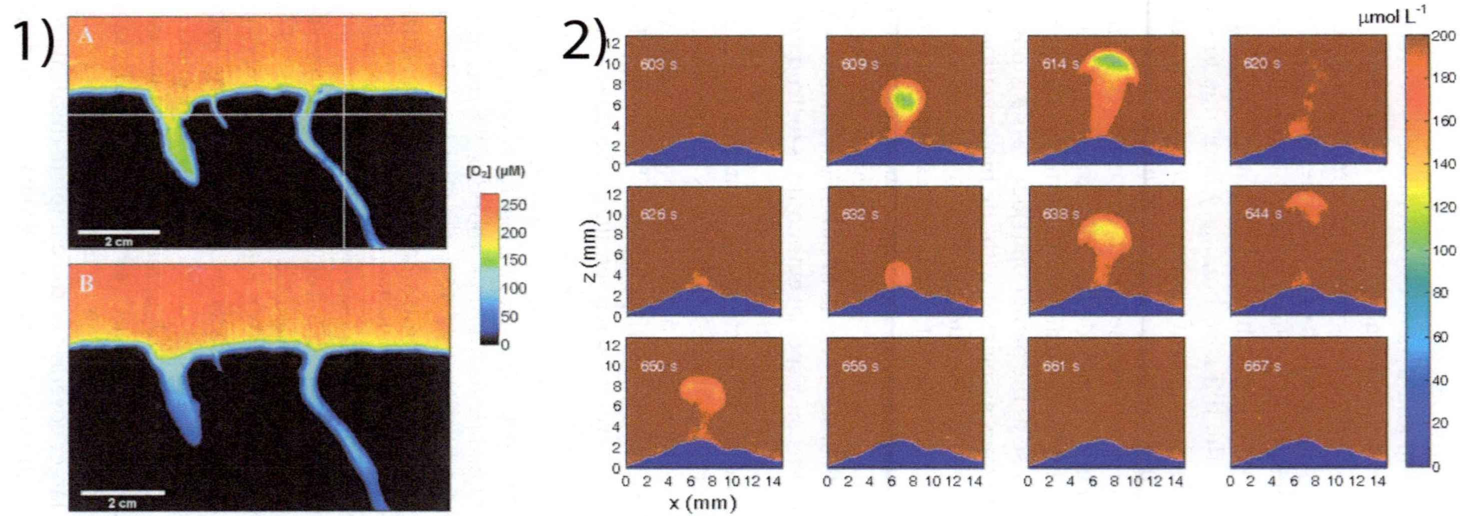


**Figure 7.6** Examples of $O_2$ imaging in and above aquatic sediments. (1) $O_2$ imaging using ratiometric readout of a planar opt(r)ode inserted in a sediment with burrowing animals.[41] (2) Ejection of anoxic water from a worm burrow visualized with sensor nanoparticles in the water above the water–sediment interface.[107] Reproduced with permission from ref. 41, © 2011, by the Association for the Sciences of Limnology and Oceanography, Inc. and ref. 107, © 2016 Association for the Sciences of Limnology and Oceanography.

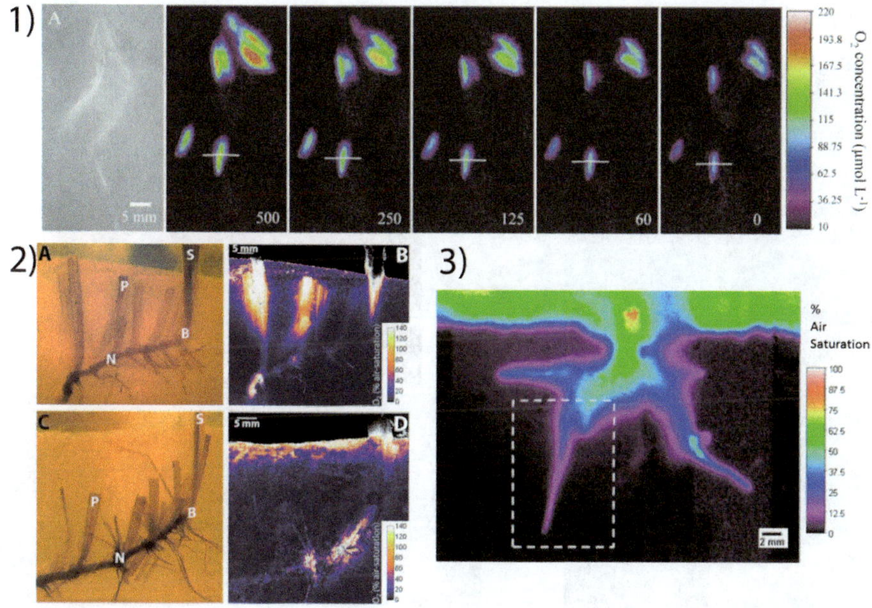


**Figure 7.7** Examples of planar opt(r)ode- and nanoparticle-based imaging of $O_2$ in the rhizosphere of aquatic plants. Oxygenation around the roots of aquatic plants can be visualized by using planar opt(r)odes like shown in (1)[43] and (3)[112] or by embedding the plant in a sensor nanoparticle containing agar (2).[20] Reproduced with permission from ref. 43, © 2006, by the Association for the Sciences of Limnology and Oceanography, Inc.; ref. 20. Copyright 2015 American Chemical Society, and P. N. Williams, *et al.*, *Environ. Sci. Technol.*, 2014, **48**, 8498–8506,[112] under a CC BY 4.0 license, https://creativecommons.org/licenses/by/4.0/. Copyright © 2014 American Chemical Society.

agar-based sediment resembling the natural sediment in terms of $H_2S$ content and pH. Addition of $O_2$ sensor nanoparticles to the artificial sediment matrix enabled visualization of $O_2$ microenvironments over the entire root system.[19,20] Such measurements were used to study the effects of rising water temperature and hypoxia on the $O_2$ transport to the rhizosphere.

In summary, planar opt(r)ode-based as well as more recently developed particle-based $O_2$ imaging has yielded numerous novel insights to factors regulating the $O_2$ microenvironment in sediments and rhizospheres.[108]

### 7.4.3.3 $O_2$ Imaging on Structurally Complex Samples

Soft samples like sediments are easy to manipulate, and it is relatively easy to bring the opt(r)ode into contact with the sample or to insert microsensors for profiling, while measurements on hard and/or structurally complex samples are more challenging. For such systems, both planar opt(r)ode- and particle-based $O_2$ imaging has great advantages (see examples in Figure 7.8).

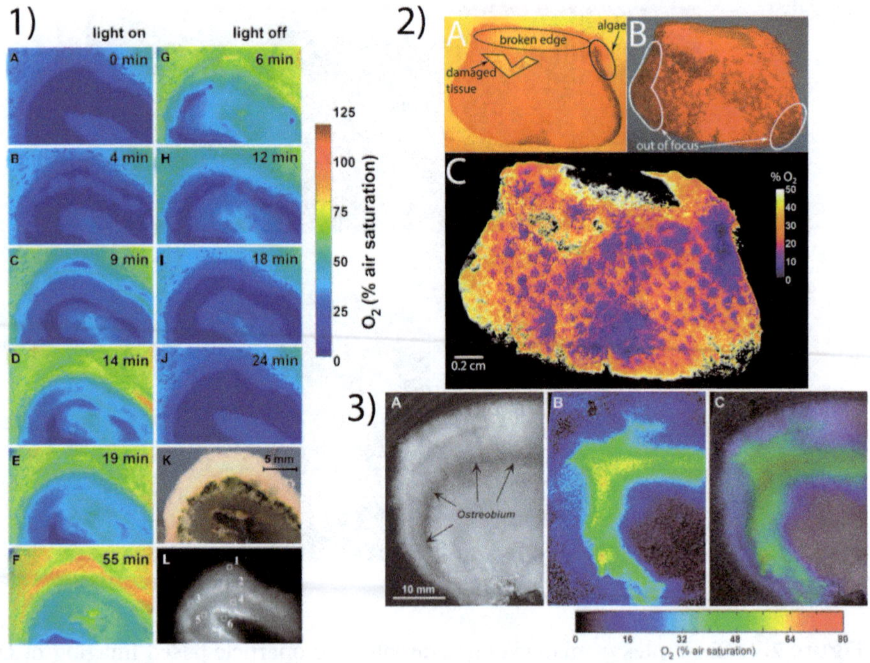


**Figure 7.8** Examples of planar opt(r)ode- and particle-based $O_2$ imaging of structurally challenging samples: (1) light dependent $O_2$ dynamics of a ascidian sample visualized by pressing it against a planar opt(r)ode.[99] Reprinted from Kühl *et al.*, *Front. Microbiol.*, 2012, **3**, 402,[99] https://doi.org/10.3389/fmicb.2012.00402. © 2012 Kühl, Behrendt, Trampe, Qvortrup, Schreiber, Borisov, Klimant and Larkum. Published under the CC BY 3.0 licence, https://creativecommons.org/licenses/by/3.0/. (2) Surface $O_2$ imaging of a coral fragment enabled by spray-painting the sample with sensor nanoparticles.[51] Reprinted from *Sensors and Actuators B: Chemical*, **237**, K. Koren, S. L. Jakobsen and M. Kühl, *In vivo* imaging of $O_2$ dynamics on coral surfaces spray-painted with sensor nanoparticles, 1095–1101,[51] Copyright 2016, with permission from Elsevier; (3) $O_2$ dynamics in the endolithic subsurface algal communities of a stony coral visualized by pressing the sample against a planar opt(r)ode.[21] Reproduced from ref. 21 with permission from John Wiley and Sons Inc. © 2008 Phycological Society of America.

Imaging of the $O_2$ dynamics in endolithic subsurface algal communities of stony corals have *e.g.* been done with planar opt(r)odes.[21] For this, a smooth vertical cross-section through the hard coral skeleton was put in contact with a planar-opt(r)ode.[21] Another structurally complex sample is the symbiont-containing ascidian *Lissoclinum patella*, which forms a cartilage-like structure with internal populations of microbial symbionts.[99] Cross-sections through the ascidians were pressed against a planar opt(r)ode[99] enabling mapping of strong $O_2$ dynamics due to photosynthesis and respiration at a scale of several $cm^2$.

While cutting is a possible way to enable $O_2$ imaging for certain samples, it is certainly not a universal solution. A way to study the $O_2$ dynamics on complex surfaces is to coat the surface with sensor material. Potentially a planar opt(r)ode could be pressed onto the surface, but in this case the sensor foil will act as a diffusion barrier imposing strong artefacts. The application of optical $O_2$ sensor particles can overcome or minimize this problem. It was demonstrated that micro-[22] as well as nano-particle[51] based $O_2$ sensors can be used to visualize the dynamics on the surface of corals. The magnetic microparticles were held in place on the coral with the help of a strong magnet placed underneath the coral. For the nanoparticles a spray-painting approach was chosen, which yielded homogeneous coverage of the studied species without any detectable short-term harm of the investigated coral. Such distributed sensing with $O_2$ sensor particles has just been introduced to aquatic science and holds a strong potential for novel applications in natural samples with complex surface topography.

## 7.5 Future Challenges

Optical $O_2$ sensors are the most advanced optical sensors on the market. Many applications have been made possible by this technology especially in aquatic environments. So the question arises – what is coming next for optical $O_2$ sensors?

### 7.5.1 Multi-analyte Measurements

There is now an increasing interest in further combining and developing optical sensors for multi-analyte sensing and imaging. As $O_2$ sensors are so well developed and many different indicators are available, they are ideal for muli-analyte chemical imaging systems. Dual optical sensors for imaging $O_2$ and pH;[65] $O_2$ and $CO_2$,[114] and $O_2$ and temperature[29] have already been described and partly tested on aquatic samples.[65] Besides combining two optical sensors within a planar setup, the combination of $O_2$ sensors with other analytical techniques is of interest. For example, diffusive gradients in thin films (DGT) is an exciting technology enabling 2D ion sensing. Similar to planar-opt(r)odes, this method also uses thin films and can be put on top of an opt(r)ode[45,115] and such combined sensing bears a strong potential in various aquatic systems.[116]

Multiplexing of different sensor materials for multi-analyte measurements is also of interest for the development of novel fiber-based sensors. This can be facilitated either by combining several fibers or sensing positions for different analytes[117,118] or by combining several sensing functionalities within one fiber.[119] Either way, cross-sensitivities need to be minimized and easy calibration protocols need to be established in order to allow multi-analyte sensing to find its way into real world applications. Once the usability is there, the possible applications are endless.

### 7.5.2 Special $O_2$ Microniches and Their Impact on Other Processes

Several processes in aquatic systems are dependent on certain $O_2$ concentrations. For example, dinitrogen ($N_2$) fixation and denitrification are favored at hypoxic or anoxic conditions. Dinitrogen fixing filamentous cyanobacteria localize this process in specialized cell, the so-called heterocysts, that exhibit hypoxia or anoxia. Also within marine snow, low $O_2$ concentrations are expected to favor $N_2$ fixation and other anaerobic processes. However, only few microsensor based studies have investigated the $O_2$ microenvironment in these systems. Nanoparticle-based $O_2$ sensors or dissolved $O_2$ indicators that find use in cell culture experiments,[104,120–122] could be used to stain such cyanobacteria or marine snow particles enabling mapping of $O_2$ concentrations using *e.g.* lifetime-based $O_2$ microscopy. First experiments in this direction were made when looking into the $O_2$ levels within copepod carcasses.[123] These measurements showed that the inside of carcasses can easily be anoxic, even in fully air saturated waters. Knowledge of such anoxic microniches in aerated aquatic habitats is still scarce, but the application of high-resolution optical $O_2$ sensing and imaging methodology for studying $O_2$ microenvironments seems very promising.

### 7.5.3 Low Cost Instrumentation for Long-term Monitoring

A better understanding of the $O_2$ dynamics at larger spatial and temporal scales is still needed in many, if not all, aquatic systems. This calls for relatively low cost robust $O_2$ sensors that can be widely deployed for long-term environmental monitoring, and here optical $O_2$ sensor technology is the way to go as such sensor systems consume less energy and can be produced relatively cheaply as compared to electrochemical sensors. This is a rapidly developing field, which will supplement existing monitoring sensor networks. As mentioned, a major challenge for such monitoring is sensor stability and in particular biofouling.

## 7.6 Conclusions

Since its introduction to aquatic science in 1995,[16] the technology for optical $O_2$ sensing in aquatic systems has undergone astonishing development now encompassing fiber-optic micro- and macro-sensors, sensor patches for remote readout, planar opt(r)odes, as well as nano- and micro-particle based sensors for mapping $O_2$ concentration and dynamics at high spatial resolution. Furthermore, optical sensors for multi-parameter chemical sensing are now also available for application in aquatic science. While this development has largely been driven by interdisciplinary efforts of analytical chemists, biologist and engineers, the market for optical $O_2$ sensors has also increased enormously. A large diversity of optical $O_2$ sensors is commercially available, and optical $O_2$ sensing is now considered a standard technique in many areas of aquatic science.

## References

1. R. N. Glud, *Mar. Biol. Res.*, 2008, **4**, 243–289.
2. T. R. Marrero and E. A. Mason, *J. Phys. Chem. Ref. Data*, 1972, **1**, 3–118.
3. B. B. Jørgensen and N. P. Revsbech, *Limnol. Oceanogr.*, 1985, **30**, 111–122.
4. *The Benthic Boundary Layer: Transport Processes and Biogeochemistry: Transport Processes and Biogeochemistry*, ed. B. P. Boudreau and B. B. Jørgensen, Oxford University Press, New York, USA, 2001.
5. L. W. Winkler, *Ber. Dtsch. Chem. Ges.*, 1888, **21**, 2843–2854.
6. L. C. Clark, R. Wolf, D. Granger and Z. Taylor, *J. Appl. Physiol.*, 1953, **6**, 189–193.
7. E. Gnaiger and H. Forstner, *Polarographic Oxygen Sensors: Aquatic and Physiological Applications*, Springer, Berlin Heidelberg, 1983.
8. N. P. Revsbech, *Limnol. Oceanogr.*, 1989, **34**, 474–478.
9. N. P. Revsbech, L. H. Larsen, J. Gundersen, T. Dalsgaard, O. Ulloa and B. Thamdrup, *Limnol. Oceanogr.: Methods*, 2009, **7**, 371–381.
10. N. P. Revsbech, J. Sorensen, T. H. Blackburn and J. P. Lomholt, *Limnol. Oceanogr.*, 1980, **25**, 403–411.
11. N. P. Revsbech, B. B. Jørgensen and T. H. Blackburn, *Science*, 1980, **207**, 1355–1356.
12. N. P. N. Revsbech and B. B. B. Jørgensen, *Adv. Microb. Ecol.*, 1986, **9**, 293–352.
13. M. Fazli, T. Bjarnsholt, K. Kirketerp-Møller, A. Jørgensen, C. B. Andersen, M. Givskov and T. Tolker-Nielsen, *Wound Repair Regen.*, 2011, **19**, 387–391.
14. M. Kühl and N. P. Revsbech, in *The Benthic Boundary Layer: Transport Processes and Biogeochemistry*, ed. B. B. Boudreau and B.P. Jørgensen, Oxford University Press, New York, 2001, pp. 180–210.
15. H. Kautsky and A. Hirsch, *Ber. Dtsch. Chem. Ges. (A B Ser.)*, 1931, **64**, 2677–2683.
16. I. Klimant, V. Meyer and M. Kuhl, *Limnol. Oceanogr.*, 1995, **40**, 1159–1165.
17. R. McFarlane and M. C. Hamilton, *International Society for Optics and Photonics*, ed. A. M. Verga Scheggi, 1987, pp. 324–330.
18. I. Klimant, G. Hoist and M. Kühi, *Proc. SPIE Int. Soc. Opt. Eng.*, 1995, **2508**, 375–386.
19. K. E. Brodersen, K. Koren, M. Lichtenberg and M. Kühl, *Plant, Cell Environ.*, 2016, **39**, 1619–1630.
20. K. Koren, K. E. Brodersen, S. L. Jakobsen and M. Kühl, *Environ. Sci. Technol.*, 2015, **49**, 2286–2292.
21. M. Kühl, G. Holst, A. W. D. Larkum and P. J. Ralph, *J. Phycol.*, 2008, **44**, 541–550.
22. J. Fabricius-Dyg, G. Mistlberger, M. Staal, S. M. Borisov, I. Klimant and M. Kühl, *Mar. Biol.*, 2012, **159**, 1621–1631.
23. J. N. Demas, B. A. DeGraff and W. Xu, *Anal. Chem.*, 1995, **67**, 1377–1380.
24. E. R. Carraway, J. N. Demas, B. A. DeGraff and J. R. Bacon, *Anal. Chem.*, 1991, **63**, 337–342.
25. X.-D. Wang and O. S. Wolfbeis, *Chem. Soc. Rev.*, 2014, **43**, 3666–3761.
26. M. Quaranta, S. M. Borisov and I. Klimant, *Bioanal. Rev.*, 2012, **4**, 115–157.

27. K. Koren, L. Hutter, B. Enko, A. Pein, S. M. Borisov and I. Klimant, *Sens. Actuators, B*, 2013, **176**, 344–350.
28. R. Glud, N. Ramsing, J. K. Gundersen and I. Klimant, *Mar. Ecol.: Prog. Ser.*, 1996, **140**, 217–226.
29. S. M. Borisov, A. S. Vasylevska, C. Krause and O. S. Wolfbeis, *Adv. Funct. Mater.*, 2006, **16**, 1536–1542.
30. L. F. Rickelt, L. D. M. Ottosen and M. Kühl, *Sens. Actuators, B*, 2015, **211**, 462–468.
31. G. Holst, I. Klimant, O. Kohls and M. Kühl, *Chemical Sensors in Oceanography*, 2000, pp. 143–188.
32. M. Kühl, *Methods Enzymol.*, 2005, **397**, 166–199.
33. M. Kühl, L. F. Rickelt and R. Thar, *Appl. Environ. Microbiol.*, 2007, **73**, 6289–6295.
34. I. Klimant, M. Kühl, R. N. Glud and G. Holst, *Sens. Actuators, B*, 1997, **38**, 29–37.
35. G. Holst, M. Kühl and I. Klimant, *Proc. SPIE*, 1995, **2508**, 387–398.
36. G. Holst, O. Kohls, I. Klimant, B. König, M. Kühl and T. Richter, *Sens. Actuators, B*, 1998, **51**, 163–170.
37. S. Schreml, R. J. Meier, M. Kirschbaum, S. C. Kong, S. Gehmert, O. Felthaus, S. Küchler, J. R. Sharpe, K. Wöltje, K. T. Weiß, M. Albert, U. Seidl, J. Schröder, C. Morsczeck, L. Prantl, C. Duschl, S. F. Pedersen, M. Gosau, M. Berneburg, O. S. Wolfbeis, M. Landthaler and P. Babilas, *Theranostics*, 2014, **4**, 721–735.
38. J. P. Fischer and F. Wenzhöfer, *Limnol. Oceanogr. Methods*, 2010, **8**, 254–268.
39. R. N. Glud, A. Tengberg, M. Kühl, P. O. J. Hall and I. Klimant, *Limnol. Oceanogr.*, 2001, **46**, 2073–2080.
40. G. Holst and B. Grunwald, *Sens. Actuators, B*, 2001, **74**, 78–90.
41. M. Larsen, S. M. Borisov, B. Grunwald, I. Klimant and R. N. Glud, *Limnol. Oceanogr. Methods*, 2011, **9**, 348–360.
42. R. J. Meier, L. H. Fischer, O. S. Wolfbeis and M. Schäferling, *Sens. Actuators, B*, 2013, **177**, 500–506.
43. M. S. Frederiksen and R. N. Glud, *Limnol. Oceanogr.*, 2006, **51**, 1072–1083.
44. R. N. Glud, F. Wenzhöfer, A. Tengberg, M. Middelboe, K. Oguri and H. Kitazato, *Deep Sea Res., Part I*, 2005, **52**, 1974–1987.
45. J. Santner, M. Larsen, A. Kreuzeder and R. N. Glud, *Anal. Chim. Acta*, 2015, **878**, 9–42.
46. P. Lehner, C. Larndorfer, E. Garcia-Robledo, M. Larsen, S. M. Borisov, N.-P. Revsbech, R. N. Glud, D. E. Canfield and I. Klimant, *PLoS One*, 2015, **10**, e0128125.
47. G. Mistlberger, K. Koren, E. Scheucher, D. Aigner, S. M. Borisov, A. Zankel, P. Pölt and I. Klimant, *Adv. Funct. Mater.*, 2010, **20**, 1842–1851.
48. S. M. Borisov, T. Mayr, G. Mistlberger, K. Waich, K. Koren, P. Chojnacki and I. Klimant, *Talanta*, 2009, **79**, 1322–1330.
49. K. Koren, G. Mistlberger, D. Aigner, S. M. Borisov, A. Zankel, P. Pölt and I. Klimant, *Monatsh. Chem.*, 2010, **141**, 691–697.

50. S. M. Borisov, T. Mayr and I. Klimant, *Anal. Chem.*, 2008, **80**(3), 573–582.
51. K. Koren, S. L. Jakobsen and M. Kühl, *Sens. Actuators, B*, 2016, **237**, 1095–1101.
52. A. Paulmier and D. Ruiz-Pino, *Prog. Oceanogr.*, 2009, **80**(3–4), 113–128.
53. D. E. Canfield, F. J. Stewart, B. Thamdrup, L. De Brabandere, T. Dalsgaard, E. F. Delong, N. P. Revsbech and O. Ulloa, *Science*, 2010, **330**(6009), 1375–1378.
54. J. J. Childress and B. A. Seibel, *J. Exp. Biol.*, 1998, **201**(8), 1223–1232.
55. M. Larsen, P. Lehner, S. M. Borisov, I. Klimant, J. P. Fischer, F. J. Stewart, D. E. Canfield and R. N. Glud, *Limnol. Oceanogr.: Methods*, 2016, **14**, 784–800.
56. E. Garcia-Robledo, C. C. Padilla, M. Aldunate, F. J. Stewart, O. Ulloa, A. Paulmier, G. Gregori and N. P. Revsbech, *Proc. Natl. Acad. Sci. U. S. A.*, 2017, **114**, 8319–8324.
57. P. Lehner, C. Staudinger, S. M. Borisov and I. Klimant, *Nat. Commun.*, 2014, **5**, 4460.
58. N. P. Revsbech, B. Thamdrup, T. Dalsgaard and D. E. Canfield, *Methods Enzymol.*, 2010, **486**, 325–341.
59. R. N. Glud, I. Klimant, G. Holst, O. Kohls, V. Meyer, M. Kühl and J. K. Gundersen, *Deep Sea Res., Part I*, 1999, **46**, 171–183.
60. T. Fenchel and B. Finlay, *Biol. Rev. Cambridge Philos. Soc.*, 2008, **83**, 553–569.
61. K. Hancke, B. K. Sorrell, L. Chresten Lund-Hansen, M. Larsen, T. Hancke and R. N. Glud, *Limnol. Oceanogr.*, 2014, **59**, 1599–1611.
62. T. Mock, G. S. Dieckmann, C. Haas, A. Krell, J.-L. Tison, A. L. Belem, S. Papadimitriou and D. N. Thomas, *Aquat. Microb. Ecol.*, 2002, **29**, 297–306.
63. T. Mock, M. Kruse and G. S. Dieckmann, *Aquat. Microb. Ecol.*, 2003, **30**, 197–205.
64. H. Satoh and S. Okabe, *Microbes Environ.*, 2013, **28**, 166–179.
65. M. Moßhammer, M. Strobl, M. Kühl, I. Klimant, S. M. Borisov and K. Koren, *ACS Sens.*, 2016, **1**, 681–687.
66. L. Delauney, C. Compare and M. Lehaitre, *Ocean Sci.*, 2010, **6**, 503–511.
67. F. Navarro-Villoslada, G. Orellana, M. C. Moreno-Bondi, T. Vick, M. Driver, G. Hildebrand and K. Liefeith, *Anal. Chem.*, 2001, **73**, 5150–5156.
68. H. R. Bungay, W. J. Whalen and W. M. Sanders, *Biotechnol. Bioeng.*, 1969, **11**, 765–772.
69. M. Kühl, Y. Cohen, T. Dalsgaard, B. B. Jørgensen and N. P. Revsbech, *Mar. Ecol. Prog. Ser.*, 1995, **117**, 159–172.
70. C. E. Reimers, K. M. Fischer, R. Merewether, K. L. Smith and R. A. Jahnke, *Nature*, 1986, **320**, 741–744.
71. R. Nøhr Glud, J. K. Gundersen, N. P. Revsbech, B. B. Jorgensen and M. Hüttel, *Deep Sea Res., Part I*, 1995, **42**, 1029–1042.
72. M. H. Long, P. Berg, D. de Beer and J. C. Zieman, *PLoS One*, 2013, **8**, e58581.

73. R. Glud, P. Berg, A. Hume, P. Batty, M. Blicher, K. Lennert and S. Rysgaard, *Mar. Ecol. Prog. Ser.*, 2010, **417**, 1–12.
74. A. C. Hume, P. Berg and K. J. McGlathery, *Limnol. Oceanogr.*, 2011, **56**, 86–96.
75. M. H. Long, D. Koopmans, P. Berg, S. Rysgaard, R. N. Glud and D. H. Søgaard, *Biogeosciences*, 2012, **9**, 1957–1967.
76. B. G. T. Else, S. Rysgaard, K. Attard, K. Campbell, O. Crabeck, R. J. Galley, N. X. Geilfus, M. Lemes, R. Lueck, T. Papakyriakou and F. Wang, *Cold Reg. Sci. Technol.*, 2015, **119**, 158–169.
77. M. Holtappels, C. Noss, K. Hancke, C. Cathalot, D. F. McGinnis, A. Lorke and R. N. Glud, *PLoS One*, 2015, **10**, e0116564.
78. P. Berg, D. J. Koopmans, M. Huettel, H. Li, K. Mori and A. Wüest, *Limnol. Oceanogr. Methods*, 2016, **14**, 151–167.
79. L. F. Rickelt, L. Askaer, E. Walpersdorf, B. Elberling, R. N. N. Glud and M. Kühl, *J. Environ. Qual.*, 2013, **42**, 1267.
80. A. Tengberg, J. Hovdenes, H. J. Andersson, O. Brocandel, R. Diaz, D. Hebert, T. Arnerich, C. Huber, A. Körtzinger, A. Khripounoff, F. Rey, C. Rönning, J. Schimanski, S. Sommer and A. Stangelmayer, *Limnol. Oceanogr. Methods*, 2006, **4**, 7–17.
81. M. B. S. Svendsen, P. G. Bushnell and J. F. Steffensen, *J. Fish Biol.*, 2016, **88**, 26–50.
82. S. Gerile and J. Pirhonen, *Aquaculture*, 2017, **479**, 616–618.
83. J. W. Behrens, M. Van Deurs and E. A. F. Christensen, *PLoS One*, 2017, **12**, e0176038.
84. J. F. Steffensen, *Fish Physiol. Biochem.*, 1989, **6**, 49–59.
85. D. Chabot, J. F. Steffensen and A. P. Farrell, *J. Fish Biol.*, 2016, **88**, 81–121.
86. S. Gatti, T. Brey, W. Müller, O. Heilmayer and G. Holst, *Mar. Biol.*, 2002, **140**, 1075–1085.
87. J. Woelfel, K. Sørensen, M. Warkentin, S. Forster, A. Oren and R. Schumann, *Aquat. Microb. Ecol.*, 2009, **56**, 263–273.
88. X. Gong, E. Garcia-Robledo, A. Schramm and N. P. Revsbech, *Appl. Environ. Microbiol.*, 2015, **82**, 1412–1422.
89. E. Trampe, P. J. Hansen and M. Kuhl, *Algae*, 2015, **30**, 103–119.
90. M. Martini, B. Butman and M. J. Mickelson, *J. Atmos. Ocean Tech.*, 2007, **24**, 1924–1935.
91. J. P. Fischer and K. Koop-Jakobsen, *Sens. Actuators, B*, 2012, **168**, 354–359.
92. K. Koop-Jakobsen, J. Fischer and F. Wenzhöfer, *Sci. Total Environ.*, 2017, **589**, 191–199.
93. T. Brandt, M. Vieweg, G. Laube, R. Schima, T. Goblirsch, J. H. Fleckenstein and C. Schmidt, *Environ. Sci. Technol.*, 2017, **51**, 9970–9978.
94. M. Vieweg, N. Trauth, J. H. Fleckenstein and C. Schmidt, *Environ. Sci. Technol.*, 2013, **47**, 9858–9865.
95. N. Rudolph-Mohr, P. Vontobel and S. E. Oswald, *Ann. Bot.*, 2014, **114**, 1779–1787.
96. M. Kühl and L. Polerecky, *Aquat. Microb. Ecol.*, 2008, **53**, 99–118.
97. K. Oguri, H. Kitazato and R. N. Glud, *Mar. Chem.*, 2006, **100**, 95–107.

98. M. Staal, S. M. M. Borisov, L. F. F. Rickelt, I. Klimant and M. Kühl, *J. Microbiol. Methods*, 2011, **85**, 67–74.
99. M. Kühl, L. Behrendt, E. Trampe, K. Qvortrup, U. Schreiber, S. M. Borisov, I. Klimant and A. W. D. Larkum, *Front Microbiol.*, 2012, **3**, 402.
100. E. I. Prest, M. Staal, M. Kühl, M. C. M. van Loosdrecht and J. S. Vrouwenvelder, *J. Membr. Sci.*, 2012, **392–393**, 66–75.
101. F. Zhu, D. Baker, J. Skommer, M. Sewell and D. Wlodkowic, *Cytometry, Part A*, 2015, **87**, 446–450.
102. H. Tschiersch, G. Liebsch and A. Stangelmayer, *Microsensors*, InTech, 2011, pp. 281–285.
103. K. E. Brodersen, D. A. Nielsen, P. J. Ralph and M. Kühl, *Mar. Biol.*, 2014, **161**, 2921–2930.
104. A. V. Kondrashina, R. I. Dmitriev, S. M. Borisov, I. Klimant, I. O'Brien, Y. M. Nolan, A. V. Zhdanov and D. B. Papkovsky, *Adv. Funct. Mater.*, 2012, **22**, 4931–4939.
105. A. Fercher, S. M. Borisov, A. V. Zhdanov, I. Klimant and D. B. Papkovsky, *ACS Nano*, 2011, **5**, 5499–5508.
106. R. I. Dmitriev, S. M. Borisov, H. Düssmann, S. Sun, B. J. Müller, J. Prehn, V. P. Baklaushev, I. Klimant and D. B. Papkovsky, *ACS Nano*, 2015, **9**, 5275–5288.
107. E. Murniati, D. Gross, H. Herlina, K. Hancke, R. N. Glud and A. Lorke, *Limnol. Oceanogr. Methods*, 2016, **14**(8), 506–517.
108. A. Stockdale, W. Davison and H. Zhang, *Earth-Sci. Rev.*, 2009, **92**, 81–97.
109. L. Pischedda, P. Cuny, J. L. Esteves, J.-C. Poggiale and F. Gilbert, *Hydrobiologia*, 2012, **680**, 109–124.
110. F. Wenzhöfer and R. N. Glud, *Limnol. Oceanogr.*, 2004, **49**, 1471–1481.
111. K. E. Brodersen, D. A. Nielsen, P. J. Ralph and M. Kühl, *New Phytol.*, 2015, **205**, 1264–1276.
112. P. N. Williams, J. Santner, M. Larsen, N. J. Lehto, E. Oburger, W. Wenzel, R. N. Glud, W. Davison and H. Zhang, *Environ. Sci. Technol.*, 2014, **48**, 8498–8506.
113. S. Jensen, M. Kühl, R. Glud, L. Jørgensen and A. Priemé, *Mar. Ecol.: Prog. Ser.*, 2005, **293**, 49–58.
114. C. R. Schroeder, G. Neurauter and I. Klimant, *Microchim. Acta*, 2007, **158**, 205–218.
115. C. Han, J. Ren, Z. Wang, H. Tang and D. Xu, *Water Res.*, 2017, **108**, 179–188.
116. N. J. Lehto, M. Larsen, H. Zhang, R. N. Glud and W. Davison, *Sci. Rep.*, 2017, **7**, 11369.
117. K. L. Michael, L. C. Taylor, S. L. Schultz and D. R. Walt, *Anal. Chem.*, 1998, **70**, 1242–1248.
118. J. A. Ferguson, B. G. Healey, K. S. Bronk, S. M. Barnard and D. R. Walt, *Anal. Chim. Acta*, 1997, **340**, 123–131.
119. S. M. Borisov, R. Seifner and I. Klimant, *Anal. Bioanal. Chem.*, 2011, **400**, 2463–2474.
120. D. B. Papkovsky and R. I. Dmitriev, *Chem. Soc. Rev.*, 2013, **42**, 8700–8732.

121. R. I. Dmitriev, A. V. Kondrashina, K. Koren, I. Klimant, A. V. Zhdanov, J. M. P. Pakan, K. W. McDermott and D. B. Papkovsky, *Biomater. Sci.*, 2014, **2**, 853.
122. R. I. Dmitriev, A. V. Zhdanov, G. Jasionek and D. B. Papkovsky, *Anal. Chem.*, 2012, **84**, 2930–2938.
123. R. N. Glud, H.-P. Grossart, M. Larsen, K. W. Tang, K. E. Arendt, S. Rysgaard, B. Thamdrup and T. Gissel Nielsen, *Limnol. Oceanogr.*, 2015, **60**, 2026–2036.

CHAPTER 8

# *Monitoring of Extracellular and Intracellular $O_2$ on a Time-resolved Fluorescence Plate Reader*

MICHELLE POTTER[†], TIFFANY A. LODGE[†] AND KARL J. MORTEN*

Nuffield Department of Obstetrics and Gynaecology, University of Oxford, John Radcliffe Hospital, Oxford, OX39DU, UK
*E-mail: karl.morten@obs-gyn.ox.ac.uk

## 8.1 Introduction

Oxygen is a life-sustaining molecule with a crucial role in cell homeostasis and metabolism, integral to both health and disease in eukaryotic organisms. Approximately 90% of the oxygen we consume contributes to the process of mitochondrial oxidative phosphorylation (OxPhos).[1] Oxygen is required by normal differentiated cells for the efficient oxidative metabolism of substrates including glucose and fatty acids. Mitochondrial oxidation converts substrates into carbon dioxide and water yielding the high energy substrate ATP. The remaining 10% of oxygen is used for enzymatic reactions *e.g.* oxidation and hydroxylation. A residual proportion of oxygen contributes to

[†]Authors contributed equally to this chapter.

Detection Science Series No. 11
Quenched-phosphorescence Detection of Molecular Oxygen: Applications in Life Sciences
Edited by Dmitri B. Papkovsky and Ruslan I. Dmitriev

Published by the Royal Society of Chemistry, www.rsc.org

the production of reactive oxygen species (ROS) and this can vary in disease states and when the respiratory chain compromised.[1] Therefore, oxygen has an important role to play in maintaining mammalian cell function providing the necessary ATP for cellular work, assisting in enzymatic reactions and generating ROS, which plays a role in a number signaling pathways.[2] The capability to study and quantify the effects of oxygen utilisation and the level of oxygenation in relation to mitochondrial function is extremely important for understanding the dynamic relationship between cellular bioenergetics and the effect it can have on cell phenotype.

## 8.2 Measuring Oxygen

### 8.2.1 Clark Electrodes

Today, there are numerous methods available that allow us to measure oxygen consumption and respiration *in vitro*. Clark type electrodes were, for many years, the workhorses in oxygen assessment and measuring mitochondrial oxidative metabolism. Clark type electrodes use amperometric oxygen sensors based on the principle that when oxygen is dissolved in a solution and placed in the sample chamber oxygen can be detected and monitored by polarography.[3] The instrumental setup consists of a positively charged anode (silver) and a negatively charged cathode (platinum) immersed in a saturated potassium chloride solution.[4] An oxygen permeable Teflon membrane surrounds the probe (Figure 8.1A). Upon the application of the electric current oxygen contained in the test solution of interest will diffuse through the Teflon membrane towards the cathode and get reduced, resulting in the production of hydrogen peroxide. The $H_2O_2$ subsequently oxidises the silver on the anode producing an electrical current. Oxygen tension is proportional to the measured current generated by the electrode (Figure 8.1C). There are advantages of using such a system: (1) it can be used on intact cells as well as permeabilised cells and isolated mitochondria; (2) ease of addition of drug/inhibitor/substrate combinations for monitoring electron transport chain (ETC) complex activity.[5] However, some of the shortcomings of this technique are: (1) it can be time consuming to assemble the apparatus, thus making it unsuitable for high throughput applications; (2) the addition of inhibitors/drugs will require the sample chamber to be thoroughly cleaned to remove any residual compounds from interfering with subsequent measurements; (3) large volumes are required which would need a higher number of cells; (4) the electrode is known to introduce interferences as it consumes oxygen.[6]

### 8.2.2 Phosphorescent Porphyrin-based Probes

Porphyrin-based sensors, first developed in the 1980s, provide the basis of current plate based oxygen sensing systems.[7] These phosphorescence-based sensors are oxygen-sensitive and subject to quenching by high oxygen

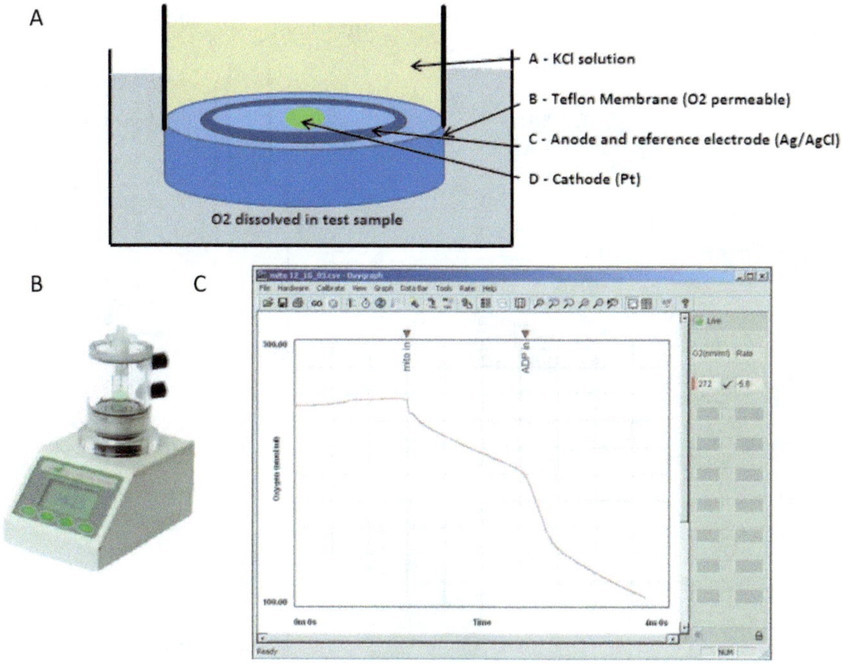


**Figure 8.1** A typical Clark electrode setup showing the design of the measuring probe (A), a reading chamber (B) and a representative Oxygraph data output (C).

concentrations. In this chapter we will describe two plate-based approaches, the Seahorse Bioscience XF Analyser (Agilent) and Luxcel MitoXpress assay which uses porphyrin-based oxygen-sensing probes.[3] The key difference between the two systems is that the Seahorse XF Analyser requires a specialised instrument to read the specialised plates while the Luxcel MitoXpress® system is compatible with many standard fluorescence plate readers and microplates.

### 8.2.2.1 *Seahorse XF Analyser*

The Seahorse Bioscience XF24 Extracellular Flux Analyser uses oxygen-sensitive fluorophores to determine the oxygen consumption rate (OCR) of intact cells (Figure 8.2). It couples this with a measurement of media acidification giving information on extracellular acidification rate (ECAR), linked to glycolytic flux. The solid-state sensors (oxygen and proton) sit on the ends of plastic pistons that make up the XF sensor cartridge. This cartridge then fits over the assay plate (24- and 96-well plate formats are available for higher throughput analyses). When the sensor cartridge is in the reading position thus measuring, the oxygen and proton sensors are approximately 5 millimetres above the cells. Such a distance would

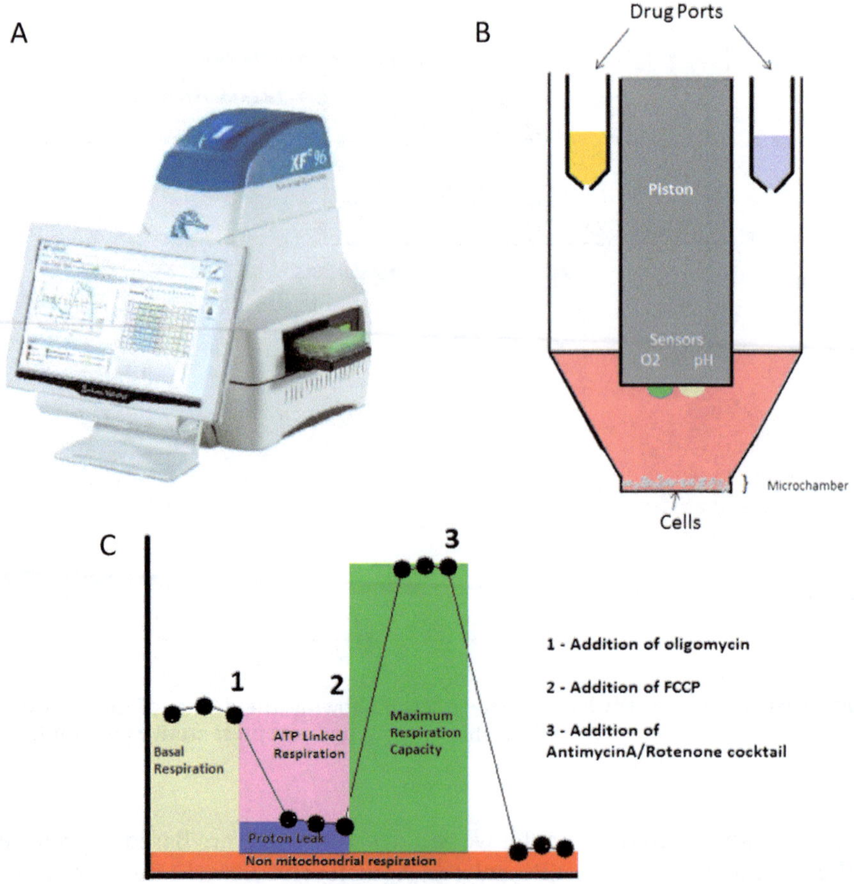


**Figure 8.2** The Seahorse XF Flux Analyser (A) is a dedicated stand-alone instrument used for the measuring and assessment of mitochondrial oxidative metabolism and extracellular acidification due to glycolysis. The oxygen and pH sensors are mounted on a moving probe (B) adjacent to injection ports for substrate/drug additions. The resulting data output resembles the graph shown in (C) giving information about various aspects of cellular respiration upon the addition of drugs at time point 1 (oligomycin), 2 (FCCP) and 3 (antimycin A/rotenone).

require several hours to detect a significant change in oxygen or proton concentrations. In addition, the well would need to be sealed to prevent back diffusion of atmospheric oxygen from dissolving into the media as is the case with the MitoXpress®-Xtra assay, discussed later. To avoid the need for a sealed system and speed up measurement times, the sensors are lowered to within 200 µm of the cells. This close proximity creates a better sealed micro-chamber of either 3 or 7 µL, in the 96- and 24-well plate formats respectively, where back diffusion interference is minimal.[3] After reading, the cartridge is raised to its original position and the medium re-equilibrates. Subsequent measurements can be made every few minutes for the duration of the assay.[8] Some of the major advantages of this system include:

(1) Much smaller volumes of sample and lower numbers of cells are required compared with Clark type electrode measurements.
(2) OCR as a standalone measurement will give an indication of mitochondrial function in intact cells allowing the impact of cellular metabolism (*i.e.* substrate flux and metabolic pathways) on mitochondrial function to be determined. This information cannot be obtained from studies of isolated mitochondria. OCR can also be paired with acidification rate (ECAR) to provide parallel information on glycolysis within the same wells.
(3) Has a higher throughput multi-well plate format allowing for either 24 or 96 wells to be measured per run. This enables a variety of treatments with multiple replicates to be measured simultaneously.
(4) Enhanced sensitivity due to the creation of a microchamber.
(5) The presence of four injection ports allows for compound/substrate addition.
(6) The assays are considered non-destructive so the cells can be used for subsequent downstream analyses such as ATP or viability tests, which can serve to normalise the Seahorse XF data.
(7) With the addition of mitochondrial respiratory modulators such as oligomycin, FCCP and antimycin A, an insight into the maximal rate of mitochondrial respiration and spare reserve capacity is obtained, Figure 8.2C.

Notwithstanding the many benefits of this system over older methods there are still a number of drawbacks including the expense of the instrument and the associated consumables (sensor cartridges and plates). Furthermore, the addition of mitochondrial inhibitors and uncouplers to the cells may affect cellular function and impede downstream assays thus limiting the potential for a multiplexed analytical approach. Overall, the Seahorse XF assays offer users an easy to measure platform. The real-time kinetic information generated is far superior to more single read endpoint type assays. The facility to inject compounds that perturb the mitochondria gives a more rounded evaluation of the cellular bioenergetics of the cell being measured. However, the cost of the instrument and its very specialised applicability means it may not be achievable for many end users.

### 8.2.3 Oxygen Sensing Plate Based Assays Using Standard Tissue Culture Plates

A complementary system to the Seahorse Bioscience XF and one that also uses an oxygen sensitive fluorophore is MitoXpress developed by Luxcel Biosciences. Similar to the Seahorse XF, MitoXpress assays are based on a phosphorescent probe being quenched by high oxygen concentrations in the surrounding cellular environment.[9] Both systems facilitate high throughput research using whole cells or isolated mitochondria making them preferable in a drug discovery setting. However, there are some significant differences between the two approaches. As mentioned previously, the Seahorse XF is a standalone instrument, using bespoke plates and pre-prepared probe assay kits. In contrast, the MitoXpress kit can be used on standard black tissue culture plates with clear bottom and the assay can be performed on a standard fluorescence plate reader. Readers with time-resolved fluorescence mode (TRF) provide enhanced performance and ones that incorporate an atmospheric control unit (ACU) have a large advantage over the Seahorse XF allowing for tighter regulation of oxygen and carbon dioxide. The data outputs for the two systems are quite different. The Seahorse XF approach does not present the raw data but provides an output showing oxygen consumption rates following the sequential modulators of mitochondrial function (Figure 8.2C). These are then used as an output of mitochondrial fitness providing information on different aspects of mitochondrial function. The MitoXpress assays provide an output linked to the software available on the plate reader. Raw data is available and can be used to calculate rates of oxygen consumption with the benefit of observing mitochondrial respiration over several hours. Plate reader manufacturers *i.e.* BMG Labtech now provide specific software modifications which allow for a rapid analysis of raw data. The MitoXpress assay can be used in the same way as the Seahorse XF system to identify aspects of mitochondrial fitness with the modulator compounds added during the assay set up.

#### *8.2.3.1 MitoXpress® Oxygen Probes*

The Luxcel Biosciences MitoXpress probe system uses a phosphorescent Pt-porphyrin based oxygen-sensitive probe to measure oxygen consumption rates. Oxygen reversibly quenches the probe phosphorescence so depletion of oxygen in the media due to the activity of respiring cells will result in an increase in measured phosphorescence intensity and lifetime signals.[10] The probe can be measured on any standard fluorescence capable plate reader although ones with TRF capability are preferred. The instrumental set up consists of dual delay time resolved measurements (Exc. 340–400 nm/Em. 640–660 nm, with delay times of 30 and 70 µs and the subsequent conversion of RFU values to phosphorescence lifetime values using the following equation: Lifetime = $(t_2 - t_1)/\mathrm{Ln}(D_1/D_2)$ where $t$ is delay time and $D$ is measured TRF intensity value). From this equation, a lifetime curve is generated where time ($x$-axis) and lifetime ($y$-axis) are plotted (Figure 8.3). The OCR is then determined from the slope of the linear portion of the lifetime curve. Lifetime is defined as the time in which a phosphorescent $O_2$-sensitive molecule exists in an excited state before returning to the ground state. Lifetime values are an important means of normalising the data from these probes, minimising and often

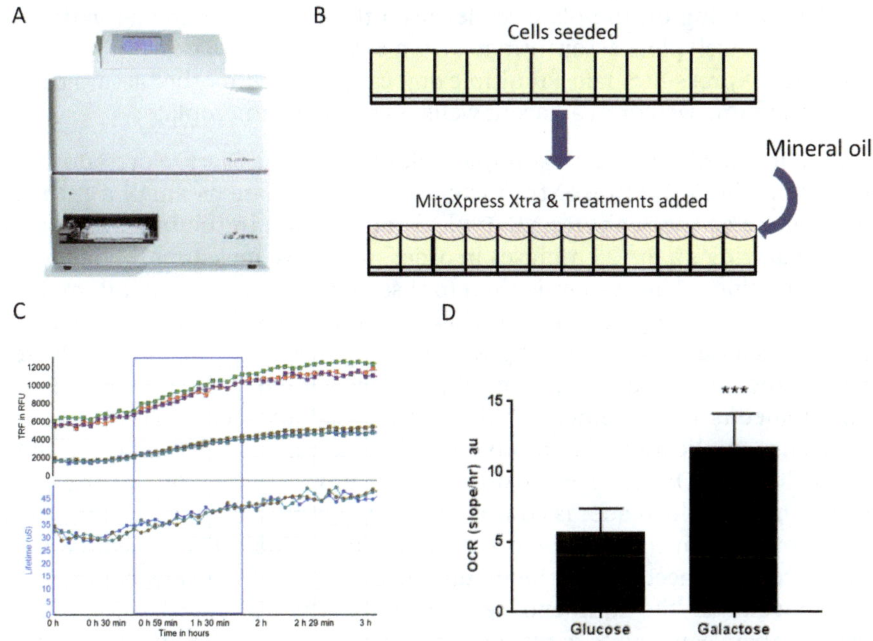


**Figure 8.3** The MitoXpress oxygen sensing probe can be used on standard plate readers, pictured (A) is a BMG Labtech CLARIOstar reader with an ACU. Assays are set up in standard black clear bottom culture plates (B). The conversion of TRF to lifetime (C) is required for data utilisation where the blue box is placed on the linear portion of the curve to determine slope (μs/h), which corresponds to oxygen consumption rate (OCR). (D) Example data of cells incubated for 72 hours in glucose/glucose free (galactose) medium.

eliminating variables like concentration of fluorophore added or variations in the intensity of excitation and allowing simple calculation of oxygen concentration.

### *8.2.3.2 MitoXpress-Xtra*

Due to some shared features with the Seahorse XF, MitoXpress-Xtra comes with many of the same advantages including ease of use, small sample volumes and having a high sample throughput. Both systems measure oxygen levels in the culture media. Some additional advantages of the MitoXpress-Xtra system include:

(1) Relatively inexpensive consumables allow for smaller experiments to be set up where only part of the plate needs to be used.
(2) MitoXpress will work on any fluorescence plate reader ruling out the need for expensive dedicated equipment;

(3) Depending on the plate reader used there can be a greater variety of multi-well plate assay formats, from 6- to 384-well plates;
(4) MitoXpress-Xtra has a multiplexing capability with other assays run in the same well or in adjacent wells on the same microplate;

Despite needing smaller sample volumes than a Clark electrode measurement the MitoXpress-Xtra system does not create as small a measuring volume as the Seahorse XF. Typical volumes are in the range of 150 μL thus requiring up to 120 minutes in order to achieve measurable changes in lifetime values. The system is therefore sealed using a mineral oil, making it more difficult to perform some downstream assays, in addition to MitoXpress-Xtra assay. If multiplexing with other assays they must be conducted before the oil is added or simultaneously with oxygen measurements. Simultaneous measurement of time resolved fluorescence with other fluorescence applications is possible with many readers, including the BMG Labtech CLARIOstar system using MARS software. Injections to the wells can be made if the reader is equipped with injectors and any additions to be made must be at the start before mineral oil is added. The measured OCR rates are influenced by (A) the metabolic activities of the cells being examined but also (B) the number of cells present in the well. Any deviation in the cell numbers can lead to compromised reproducibility. For MitoXpress-Xtra assays it is useful to run parallel standard reader based quantification techniques including Hoescht 33342 and Crystal violet (DNA) or BCA (protein), to normalise the rate of oxygen consumption to cell number/protein content. Cell number cannot be determined on the cells used in the OCR measurement, as the oil used to seal these samples does not allow for the recovery of the cells for further analysis. Due to the larger well volume cells with slower rates of oxygen consumption, which often need to be assayed in suspension, are easily measured using the MitoXpress-Xtra assay. These include human peripheral blood mononuclear cells (PBMC). A typical experiment is shown in Figure 8.4: for human control PBMCs 500 000 cells are required to be added to each well. As can be observed oxygen consumption rates in these cells are low in non-activated cells with a considerable amount of oxygen consumption due to mitochondrial respiration (total respiration rate assessed by antimycin A treatment). When stimulated with the phorbal ester PMA, PMBCs demonstrate a large increase in oxygen consumption which is unaffected by co-incubation with mitochondrial inhibitors (Figure 8.4). This suggests that the burst of oxygen consumption is not derived from mitochondrial respiration and is most likely due to the NADPH oxidase, a known response to a pathogen with cells generating high levels of superoxide. As the MitoXpress-Xtra assay can be followed for several hours, an interesting difference is observed between PMA stimulation with and without mitochondrial inhibitors. The cells treated with PMA alone maintain a higher rate of oxygen consumption than those treated with inhibitors suggesting that mitochondrial function is important post PMA stimulation in maintaining cellular function.

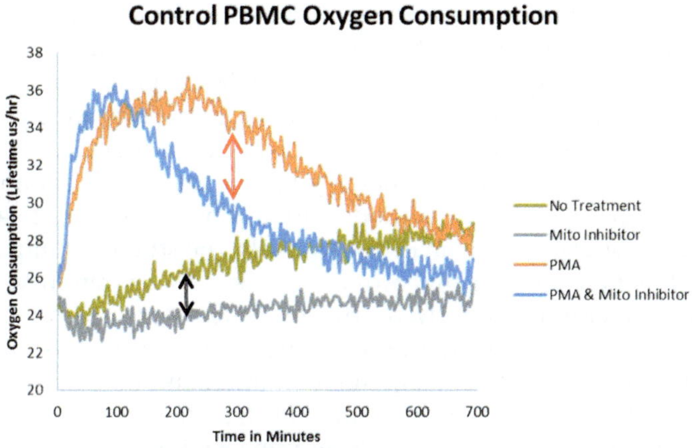


**Figure 8.4** Lifetime profiles of PBMC cells untreated (yellow), treated with mitochondrial inhibitors antimycinA and rotenone (grey), PMA (orange) or PMA + mitochondrial inhibitors (blue). The black and red arrows indicate the mitochondrial contribution of the PBMC to the depicted profiles.

In the majority of eukaryotic cells, ATP is produced by oxidative phosphorylation in the mitochondrial respiratory chain. The integrity of the mitochondrial membrane potential is critical for maintaining the physiological function of the respiratory chain to generate ATP. Therefore, the ability to investigate dual parameters of cellular health such as mitochondrial function and cell viability is of great value as it saves time, material and cost. Parallel assays in other wells on the plate could be used in conjunction with MitoXpress-Xtra giving further information about cell death, *e.g.* CellTox Green (Promega) and mitochondrial membrane potential (TMRM or JC-1, Cayman Chemicals). For instance, using the oxygen sensing MitoXpress Xtra probe in conjunction with the mitochondrial membrane dye JC-1 enable to measure the bioenergetic state of the cells by monitoring both the mitochondrial membrane potential and the oxygen consumption in response to stimuli.

## 8.3 Measurement of Intracellular Oxygen

The importance of oxygen availability cannot be overstated. There are many physiological and indeed pathophysiological requirements for oxygen especially in cancer and the adaptive response to hypoxia. Cells and tissues are exposed to different levels of oxygen. Most tissues have a partial pressure of oxygen ($pO_2$) of 20–40 mm Hg but a large tumor can be severely hypoxic or anoxic with $pO_2$ levels <0.1 mm Hg.[11,12] Hypoxia is observed in most solid tumors with up 60% exposed to <1% oxygen.[13] Hypoxia and HIF signaling are important in the metabolic reprogramming of cancer cells so measuring *in situ* hypoxia and oxygenation is essential in studies looking to target hypoxic cancer cells which tend to be resistant to current drug therapies. Cells vary in their energy needs and depending on their location within the body can be exposed to varying oxygen tensions. Epithelial cells lining the bronchi are exposed to atmospheric oxygen levels while chondrocytes are exposed to low oxygen.[14] Most standard tissue culture experiments are carried out in an atmosphere of 20% oxygen, which is higher than found in well oxygenated parts of the body and considerably higher than levels found in lower oxygenated regions. Because methods have not been available to accurately measure and control intracellular oxygen levels, the studies of impact of oxygen on cell metabolism *in vitro* beyond hypoxia (<1% oxygen) have not been carried out properly. It is conceivable that oxygen levels between 1–20% could have an impact on cell function. Oxygen levels have been shown to have a significant impact on metabolism, differentiation and development[15–17] and senescence.[18,19] With ROS heavily involved in signalling[20,21] levels of oxygen in the environment could have a significant impact on a variety of cellular processes. As oxygen levels fluctuate *in vivo* having a different impact on cells in an organ depending upon their proximity to blood vessels, the ability to control and determine intracellular levels during *in vitro* experiments is of vital importance. Oxygen concentrations and consumption rates can be difficult to monitor as there is constant diffusion between the cells and the extracellular environment and between the atmosphere and cell culture media solution.[22] As described earlier techniques to measure oxygen based on phosphorescence quenching are well established[23] but mainly as an extracellular means of assessing oxygen levels and consumption rates.[10] However, monitoring this analyte within a cell has remained problematic for years. Previous attempts to quantify intracellular oxygen have included using probes that needed to be microinjected,[24] electroporation or induce endocytosis.[25] Complex, time consuming and invasive such methods have hindered advances in this area.

### 8.3.1 MitoXpress-Intra Probe

In an effort to address some of these issues a new product recently added to the Luxcel Biosciences portfolio is the MitoXpress®-Intra probe. MitoXpress-Intra is potentially an extremely valuable *in vitro* tool to investigate the level of intracellular oxygenation. The MitoXpress®-Intra probe consists of cationic hydrogel nanoparticles loaded with an oxygen-sensitive PtTFPP (Pt(II)-5,10,15,20-tetrakis-(2,3,4,5,6-pentafluorophenyl)-porphyrin) dye (Figure 8.5). It works on the same principle as the MitoXpress-Xtra, discussed earlier, whereby molecular oxygen present in the cells is able to quench the emission of the probe reversibly. However, the difference lies in its cell-penetrating properties that enable the real-time quantification of oxygen within the cell monolayer using plate reader and microscopy-based techniques.[26] Cells are cultured to 80–90% confluence in appropriate culture media and left to adhere. Media is then replaced with 150 µL fresh culture media containing 10 µg mL$^{-1}$ of probe. The particles are internalised *via* endocytosis[26] and the probe remains in the cells and does not leak back over time. However, upon cell division it will become less concentrated. Maximum probe uptake generally occurs in 6–16 hours. The small particle size (<50 nm) allows high penetration with minimum damage/toxicity to the cells.[22]

Unlike previous extracellular oxygen sensing probes, which acted in sealed environments, the Seahorse XF (plunging probe) and MitoXpress-Xtra (oil), the wells are left unsealed so steady-state is reached between oxygen consumption by the cells and back diffusion into the well from the environment. Cells will reach a "resting steady-state" characterised by a flat lifetime signal, which correlates with the intracellular oxygen level. When cells are stimulated to consume more oxygen *e.g.* with uncouplers like FCCP or inhibited from using their mitochondria *e.g.* with antimycin A treatment, changes in phosphorescence lifetime signal will be observed. Reduced ETC activity results in a decrease in probe signal (*i.e.* high cellular $O_2$) as a result of reduced oxygen consumption by the cells. Conversely, an increase in probe signal will be seen where there is a reduction in cellular oxygen due to stimulated respiration (*i.e.* increased ETC activity). This is depicted in Figure 8.5.

To facilitate quantification of intracellular oxygen levels a calibration curve for MitoXpress-Intra, which converts measured lifetime values into the percentage oxygen within the cell monolayer, is provided by the vendor. To verify this calibration, cells loaded with MitoXpress-Intra probe are treated with a cocktail of 1 μM rotenone and 10 μM antimycin A. These compounds inhibit cellular respiration thereby increasing intracellular oxygen levels within cells to environmental oxygen levels in the plate reader chamber. The BMG Labtech CLARIOstar plate reader comes equipped with an atmospheric control unit that allows the user to control the temperature, oxygen (range

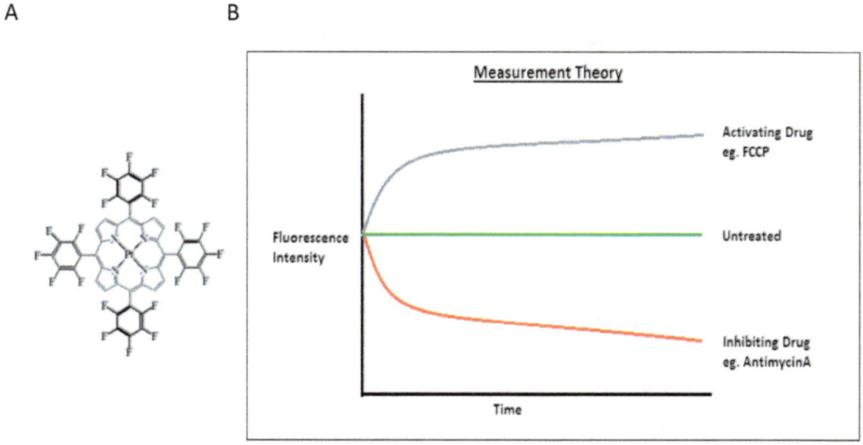


**Figure 8.5** The structure of the MitoXpress-Intra probe (A) and the theory behind the drug response observed when using it (B). If a treatment increases oxidative phosphorylation it will cause an increase in fluorescent signal and conversely a decrease in signal will be seen if a treatment results in inhibition. (A) reprinted with permission from A. Fercher, S. M. Borisov, A. V. Zhdanov, I. Klimant and D. B. Papkovsky, *ACS Nano*, 2011, **5**, 5499–5508. Copyright (2011) American Chemical Society.

from 19 to 0.1%) and carbon dioxide levels in the microplate chamber. To construct a calibration curve the antimycin A and rotenone- treated cells are placed in the reader at 37 °C with 5% $CO_2$. When the assay is started the oxygen level should be set to atmospheric ~19% and allowed to run until the fluorescence equilibrates to a flat line. Once this equilibrium is reached the oxygen levels can be dropped in a stepwise manner to 12, 8, 4, 2, 1 and 0.1% allowing for a steady state to be reached at each concentration before reducing the oxygen levels. At this point, glucose oxidase can be injected into the wells thereby chemically deoxygenating them and allowing a 0% oxygen concentration to be achieved. Once the lifetime kinetic trace has been produced (Figure 8.6A), a calibration curve of lifetime *versus* % inspired oxygen

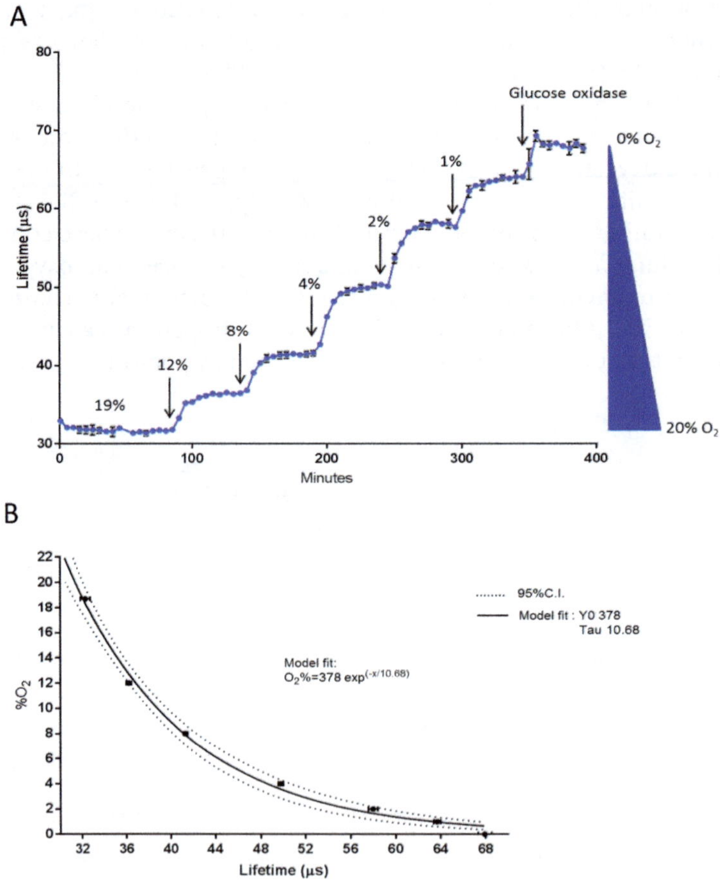


**Figure 8.6** Kinetic trace of the intracellular probe showing the response to decreasing atmospheric oxygen concentrations (A). Starting atmospheric oxygen concentration was 19%, once the signal had equilibrated the oxygen was dropped to 12%, 8%, 4%, 2%, 1%. Complete de-oxygenation was achieved by the addition of glucose oxidase. (B) Intracellular probe calibration curve of Lifetime *versus* % inspired Oxygen with a monoexponential fit curve.

is plotted (Figure 8.6B). To this plot a mono-exponential decay fit is applied giving the model fit equation (Figure 8.6B).

Once the model fit has been calculated, this equation allows future determinations of the intracellular oxygen levels within a cell monolayer by inputting the lifetime value into the equation, where % $O_2 = 378\ \exp^{(-x/10.68)}$ and $x$ is the measured lifetime value.

The calibration curve is only generated once and can be used to determine the oxygen concentration for that cell line in all subsequent assays. In practice, the calibration curve should work for all cell lines as long as the inhibition of respiration by the respiratory chain inhibitors is complete. The utilisation of MitoXpress-Intra will be of great benefit in future studies by allowing more accurate determination of oxygen concentrations within a cell population. The impact of metabolic stimulation and drug effects can now be easily examined and compared under controlled oxygen concentration, more closely simulating their physiological environment. This probe enables us to further our understanding of intracellular oxygen gradients, which will be vital in studies investigating metabolic disorders and mitochondrial function.

### 8.3.2 Application of the MitoXpress-Intra Probe

MitoXpress-Intra has proven to be a useful tool in the area of cancer metabolism and the Warburg effect. Potter *et al.* employed the method to examine the metabolism of two established cancer cell lines, RD (rhabdomyosarcoma) and U87MG (glioblastoma multiforme) under different applied oxygen concentrations with the results highlighting some interesting findings.[27] RD cancer cells cultured on standard DMEM with high glucose (25 mM) do not rely solely on glycolysis for ATP production. In fact, they displayed high rates of oxygen consumption on high glucose media consistent with a cell with an aerobic phenotype. When ambient oxygen is set to 18% oxygen one might assume that the cells are at 18% as it has been reported that the rate of oxygen diffusion into the cells is faster than the oxygen consumption rate.[28] However, when the intracellular oxygen levels were assessed the aerobically proficient RD cells, at ambient oxygen concentrations of 18, 10 and 5%, had intracellular levels of 13.67, 6.15 and 2.5% respectively. In contrast, U87MG cells, which are well known for their glycolytic metabolism, do not deplete the intracellular oxygen and as such they reflect the applied oxygen concentration. In the study, U87MG cells displayed oxygen levels of 17.9, 9.65 and 4.6% when incubated at 18, 10 and 5% oxygen. Overall this study highlighted that in aerobically competent cells intracellular oxygen levels can be significantly lower than what you might anticipate and this may have far reaching implications especially in the field of hypoxia with the applied oxygen not necessarily reflecting intracellular levels. Further studies have now expanded this work demonstrating that cells with high rates of oxygen consumption including INS-1 and HEPG2 cells have intracellular oxygen levels below 1% when incubated in an atmosphere of 2% $O_2$.[29] Not all cell types respond to hypoxia in the same way. Some cells can survive in anoxia while others will die. Chi *et al.* found significant heterogeneity in the hypoxic responses between a variety of cell types.[14] Three mechanisms underpinning the difference in hypoxia response in tumours have been proposed including (i) variations in oxygen tension in the tumours; (ii) cell type-specific differences in the extent/threshold of the response to hypoxia; or (iii) genetic changes that result in incorrect activation of the hypoxia response (pseudohypoxia).[14] While the hypoxic response in some tumours is caused by mechanisms (ii) and (iii) such as loss of von Hippel Lindau (VHL) gene in renal clear cell carcinomas the hypoxia response in most tumours is thought to lie in the variations in oxygen tension. Accurately measuring the transient changes in oxygen tension is therefore crucial for hypoxia studies and MitoXpress-Intra could prove to be a vital tool.

Standard 2D cell culture conditions are ubiquitous in cell biology and it is well documented that these models do not accurately portray the complexity of *in vivo* environments. 3D spheroids provide heterogeneous microenvironments that more accurately reflect tissue, including cell-to-cell interactions, cell–matrix interactions as well as gradients of oxygen, nutrients, waste products and drugs. MitoXpress®-Intra in 3D models has the potential to further our understanding of individual oxygen concentration gradients in 3D systems and is an important area to develop. Oxygen gradients will exist in the structure that create heterogeneous sub-populations of quiescent, hypoxic and anoxic cells each of which will have distinct metabolic profiles.[30] Measuring oxygen consumption externally has proven to be useful but has its limitations, which the MitoXpress-Intra can overcome. It can assess drug effects on 3D structures in a real time and non-invasive manner. Traditional methods of assessing spheroid "health" are time-consuming and involve spheroid dissociation or fixing and staining prior to sectioning and immunohistochemistry. Using the intracellular probe potentially allows the assessment of spheroid health in minutes with the ability to assess the effects of a drug screen in real time.

## 8.4 Future Applications of Plate-based Oxygen Monitoring Systems

Molecular oxygen has an important role in the ETC and is therefore a valuable marker of mitochondrial function and cellular metabolism crucial to many disease research areas. Standard cell culture protocols use hyperoxic conditions of 21% (160 mm Hg). However, apart from alveolar cells, which experience 110 mm Hg (14.5%) not many cells in the body are actually ever exposed to such high levels of oxygen under normal physiological conditions. In the circulation dissolved oxygen can be anywhere between 5–13% (40–100 mm Hg). Cells and tissues that are not in close proximity to blood vessels display considerably lower oxygen concentrations depending on: (1) the distance from the nearest blood vessel, and (2) the metabolism (oxygen consumption) of a particular cell type. Cellular levels of oxygen have been reported to lie between 1.3–2.5% while in the mitochondria oxygen is often less than 1%.[31] Using more appropriate, biologically relevant conditions is important as cells are likely to behave differently according to changes in their environment which may also impact on the efficacy of potential drug treatments being explored as therapeutic agents.[32,33]

As researchers we need to be more mindful of our *in vitro* cell conditions by considering substrate concentrations (*e.g.* glucose, glutamine, fatty acids and lactate) in addition to oxygen levels. As we are starting to observe with glucose, the conditions we work in can have a significant impact on cell metabolism[34] and the impact pharmacological agents.[35] Developing models which allow better control of environmental conditions maintaining substrates such as glucose at tissue levels (1–5 mM) in long-term experiments are essential if we are to make the relevant conclusions from our *in vitro* culture systems. We need to change our thinking from our current belief that leaving cells, in what is essentially glucose syrup, for many days will produce meaningful data. This is especially relevant in data produced from an experiment, which explores, for example, the long-term effects of hypoxia. Cells grown in non-physiological conditions will have significantly altered metabolic programs.[36]

One example of potential problems caused as a consequence of cell culture conditions relates to the Warburg Effect. For decades the dominating belief was that the Warburg Effect (also commonly known as aerobic glycolysis) was a primary feature of cancer. In 2011 it was even assigned as a hallmark of the disease.[37] However, in more recent times studies are emerging that lay challenge to this. Due to advances in the technologies that allow the study of cellular metabolism using metabolic tracers *in vivo* and a new-found realisation of the importance of controlling and accounting for non-physiologic conditions of *in vitro* testing cancer cells are proving to be a diverse mix of bioenergetic profiles which have a significant degree of plasticity depending upon environmental conditions. For instance, the well-documented Crabtree effect results in a reduction in respiration when glucose levels are elevated.[38] Glucose concentration can have a major impact on the bioenergetics of a cell and depending on the level of mitochondrial activity, cells may respond differently to xenobiotics designed to target the mitochondria. Marroquin *et al.* demonstrated that glycolytic HepG2 cells were resistant to some well-known mitochondrial toxicants like rotenone and antimycin A when cultured under standard (high glucose) conditions. However, when glucose media was replaced with media containing galactose cells derived ATP from oxidative metabolism and sensitivity to the mitochondrial toxicants was restored.[35] Using this approach, mild to moderate impairment of mitochondrial respiration can be identified; impairments that may have been masked when cultured in high glucose.[39] This highlights the potential importance of controlling cell culture conditions to better reflect the conditions in actual cells/tumours in order to gain a deeper insight into cellular bioenergetics. Oxygen can have similar effects to glucose as shown in recent study by Huang *et al.* using the PI3K inhibitor LY294002.[40] The drug displayed a differential effect on cell lines at 20% and 1% oxygen with significantly reduced levels of killing under low oxygen conditions. In experiments carried out in 25 mM glucose the two cell lines RD (high OxPhos) and U87MG (low OxPhos, high glycolysis) also showed different drug effects with the U87MG cells adopting a glycolytic phenotype being much more sensitive to PI3K inhibition. As U87MG cells switch to a more OxPhos phenotype under physiological glucose conditions (1 mM)[34] it is very likely the drug response to the PI3K inhibitor will be significantly different in this cell line under physiological glucose conditions.

Moving into the future being mindful of and adopting more relevant cell culture practices is imperative. There is an array of technology available that allows for in-depth investigation of cellular metabolism but their use and relevance are reliant on adopting more physiological *in vitro* culturing protocols. The plasticity of cells to adapt their metabolism based on substrate supply/availability and the inaccurate assumption that intracellular oxygen can be inferred from applied oxygen concentrations cannot be disregarded. To do so may lead to inaccurate results and flawed conclusions, which only serve to weaken our knowledge and understanding.

## Acknowledgements

Dr Michelle Potter is funded by the European Horizon-2020 project MetaCELL and Dr Tiffany Lodge by the ME association UK.

## References

1. J. W. Baynes, *Oxygen and Life*, MOSBY Elsevier, 2009, ch. 37.
2. X. Xu, Z. Li, X. Leng, M. He, J. Wang, S. Cheng and H. Wu, *Arch. Virol.*, 2016, **163**, 603–610.
3. M. Koopman, H. Michels, B. M. Dancy, R. Kamble, L. Mouchiroud, J. Auwerx, E. A. Nollen and R. H. Houtkooper, *Nat. Protoc.*, 2016, **11**, 1798–1816.
4. C. Diepart, J. Verrax, P. B. Calderon, O. Feron, B. F. Jordan and B. Gallez, *Anal. Biochem.*, 2010, **396**, 250–256.
5. J. Zhang, E. Nuebel, D. R. Wisidagama, K. Setoguchi, J. S. Hong, C. M. Van Horn, S. S. Imam, L. Vergnes, C. S. Malone, C. M. Koehler and M. A. Teitell, *Nat. Protoc.*, 2012, **7**, 1068–1085.
6. O. S. Wolfbeis, *BioEssays*, 2015, **37**, 921–928.
7. W. L. Rumsey, J. M. Vanderkooi and D. F. Wilson, *Science*, 1988, **241**, 1649–1651.
8. S. Chomicz, D. Ferrick and W. Min, *Genet. Eng. Biotechnol. News*, 2007, **27**, 19.
9. Y. Will, J. Hynes, V. I. Ogurtsov and D. B. Papkovsky, *Nat. Protoc.*, 2006, **1**, 2563–2572.
10. J. Hynes, E. Natoli Jr and Y. Will, *Curr. Protoc. Toxicol.*, 2009, ch. 2, Unit 2 16.
11. J. Aragones, P. Fraisl, M. Baes and P. Carmeliet, *Cell Metab.*, 2009, **9**, 11–22.
12. G. L. Semenza, *Science*, 2007, **318**, 62–64.
13. P. Vaupel, M. Hockel and A. Mayer, *Antioxid. Redox Signaling*, 2007, **9**, 1221–1235.

14. J. T. Chi, Z. Wang, D. S. Nuyten, E. H. Rodriguez, M. E. Schaner, A. Salim, Y. Wang, G. B. Kristensen, A. Helland, A. L. Borresen-Dale, A. Giaccia, M. T. Longaker, T. Hastie, G. P. Yang, M. J. van de Vijver and P. O. Brown, *PLoS Med.*, 2006, **3**, e47.
15. M. Takahashi, *J. Reprod. Dev.*, 2012, **58**, 1–9.
16. N. Haque, M. T. Rahman, N. H. Abu Kasim and A. M. Alabsi, *Sci. World J.*, 2013, **2013**, 632972.
17. K. Fynes, R. Tostoes, L. Ruban, B. Weil, C. Mason and F. S. Veraitch, *Stem Cells Dev.*, 2014, **23**, 1910–1922.
18. S. Parrinello, E. Samper, A. Krtolica, J. Goldstein, S. Melov and J. Campisi, *Nat. Cell Biol.*, 2003, **5**, 741–747.
19. D. H. Betts, S. D. Perrault and W. A. King, *Biogerontology*, 2008, **9**, 19–31.
20. M. Benhar, *Biochim. Biophys. Acta*, 2015, **1850**, 2476–2484.
21. D. P. Jones and H. Sies, *Antioxid. Redox Signaling*, 2015, **23**, 734–746.
22. A. Fercher, S. M. Borisov, A. V. Zhdanov, I. Klimant and D. B. Papkovsky, *ACS Nano*, 2011, **5**, 5499–5508.
23. D. B. Papkovsky, *Methods Enzymol.*, 2004, **381**, 715–735.
24. E. Schmalzlin, J. T. van Dongen, I. Klimant, B. Marmodee, M. Steup, J. Fisahn, P. Geigenberger and H. G. Lohmannsroben, *Biophys. J.*, 2005, **89**, 1339–1345.
25. T. C. O'Riordan, K. Fitzgerald, G. V. Ponomarev, J. Mackrill, J. Hynes, C. Taylor and D. B. Papkovsky, *Am. J. Physiol.: Regul., Integr. Comp. Physiol.*, 2007, **292**, R1613–R1620.
26. A. V. Kondrashina, D. B. Papkovsky and R. I. Dmitriev, *Analyst*, 2013, **138**, 4915–4921.
27. M. Potter, L. Badder, Y. Hoade, I. G. Johnston and K. J. Morten, *Adv. Exp. Med. Biol.*, 2016, **876**, 257–263.
28. I. Papandreou, T. Goliasova and N. C. Denko, *Int. J. Cancer*, 2011, **128**, 1001–1008.
29. K. J. Morten, M. Potter, L. Badder, P. Sivathondan and R. Dragovic, *et al.*, *Wellcome Open Res.*, 2017, **2**, 14.
30. R. I. Dmitriev, A. V. Zhdanov, Y. M. Nolan and D. B. Papkovsky, *Biomaterials*, 2013, **34**, 9307–9317.
31. A. Carreau, B. El Hafny-Rahbi, A. Matejuk, C. Grillon and C. Kieda, *J. Cell. Mol. Med.*, 2011, **15**, 1239–1253.
32. J. A. Marwick, D. A. Dorward, C. D. Lucas, K. O. Jones, T. A. Sheldrake, S. Fox, C. Ward, J. Murray, M. Brittan, N. Hirani, R. Duffin, I. Dransfield, C. Haslett and A. G. Rossi, *J. Leukocyte Biol.*, 2013, **94**, 1285–1292.
33. A. G. Vij, K. Kishore and J. Dey, *Indian J. Med. Res.*, 2012, **135**, 211–216.
34. M. Potter, E. Newport and K. J. Morten, *Biochem. Soc. Trans.*, 2016, **44**, 1499–1505.
35. L. D. Marroquin, J. Hynes, J. A. Dykens, J. D. Jamieson and Y. Will, *Toxicol. Sci.*, 2007, **97**, 539–547.
36. G. Gstraunthaler, T. Seppi and W. Pfaller, *Cell. Physiol. Biochem.*, 1999, **9**, 150–172.
37. D. Hanahan and R. A. Weinberg, *Cell*, 2011, **144**, 646–674.
38. H. G. Crabtree, *Biochem. J.*, 1929, **23**, 536–545.

39. C. Aguer, D. Gambarotta, R. J. Mailloux, C. Moffat, R. Dent, R. McPherson and M. E. Harper, *PLoS One*, 2011, **6**, e28536.
40. X. Huang, M. Potter, B. Pilgrim, R. Ardkhean, M. Kabeshov, T. Claridge, M. Wiseman, K. J. Morten, T. Donohoe and H. Townley, *Int. J. Cancer Ther. Oncol.*, 2017, **5**, 1.

CHAPTER 9

# *Monitoring Parameters of Oxygen Transport to Cells in the Microcirculation*

ALEKSANDER S. GOLUB AND ROLAND N. PITTMAN*

Virginia Commonwealth University, Department of Physiology and Biophysics, Richmond, VA 23298-0551, USA
*E-mail: Roland.Pittman@vcuhealth.org

## 9.1 Introduction

Oxygen is a key element for oxidative metabolism, providing energy for living animal cells. $O_2$ is a freely available resource, but its utilization by an organism requires its conversion to a transportable form (binding to hemoglobin (Hb) as $HbO_2$), transportation from the lungs to tissue capillaries and then diffusion to mitochondria in somatic cells.

The convective transport of $O_2$ by circulating blood is combined with the distribution of arterial blood flow between organs through a system of branching arteries. The distribution of blood flow within an organ and reducing the characteristic size of the flowing blood to fine channels provides the coupling of convective and diffusive $O_2$ transport to cells and mitochondria, all taking place in the microcirculation. The arteriolar network also plays an important role in active regulation of local blood flow. Arterial blood entering an organ has high $O_2$ saturation and oxygen losses in large arterioles are

Detection Science Series No. 11
Quenched-phosphorescence Detection of Molecular Oxygen: Applications in Life Sciences
Edited by Dmitri B. Papkovsky and Ruslan I. Dmitriev

Published by the Royal Society of Chemistry, www.rsc.org

small. For the study of $O_2$ transport phenomena in the microcirculation the oxygen tension, $PO_2$, is a convenient variable and its level in arterial blood is close to 100 mmHg.

Diffusional losses of oxygen through the vascular wall become significant in pre-capillary arterioles and especially in the capillary network, due to the small diameter, thin wall, slowest linear velocity and proximity to $O_2$-consuming cells in these microvessels. The connection between convective and diffusive $O_2$ transport takes place on a scale of tens of microns, or even microns, and the $O_2$ transport chain terminates in the mitochondria. Constant refreshment of oxygenated blood in capillaries and its consumption in mitochondria forms a very steep local $PO_2$ gradient in tissues, on the order of tens of mmHg $\mu m^{-1}$.

Even a microscopic size sensor of oxygen is surrounded by an environment having a heterogeneous $PO_2$. Fundamental problems in $O_2$ transport are how to characterize the state of tissue oxygenation in an organ, how to evaluate rates of $O_2$ delivery and consumption, and how to characterize their balance in dynamic situations during changing functional activity.

The first attempts of microscopic $PO_2$ measurements were made during the last century with polarographic microelectrodes, using a needle inserted into the tissue.[1] This approach averages the $PO_2$ over a diffusion-limited volume around the electrode tip, thus combining the $O_2$ tensions within small microvessels, interstitial fluid (about 10% of the volume) and cells, occupying the largest fraction of the volume and having the lowest $PO_2$. The measured value was used as a measure of "tissue $PO_2$"; it was low, and contained little information about $O_2$ transport. This term is still used in the literature, sometimes without an appreciation of the complexity of local $PO_2$ gradients and oxygen fluxes.

The invention of the phosphorescence quenching method, PQM, for measurements of $[O_2]$ and $PO_2$ by Vanderkooi and Wilson gave biologists a new non-invasive and accurate technique to study oxygen processing in living tissues.[2] These pioneers developed biologically compatible oxygen probes, based on albumin-bound metalloporphyrins, and provided methods for their calibration, with consideration of dependences on pH and temperature. They also pointed to the inevitable photo-consumption of $O_2$ by the method itself, and formulated basic design principles for biological phosphorometers.

Development of the microscopic PQM technique gave a new instrument to measure $PO_2$ on a scale of microns, in different biological compartments, and opened a window into the fine details of oxygen delivery, consumption and regulation of their balance on a cellular level.[3–5] This great opportunity was accompanied with a multitude of hidden problems, that required many technical innovations and a better understanding of processes limiting oxygen measurements in living tissues with the PQM. In this chapter we will present our own experience in the microscopic application of PQM to evaluation of $PO_2$ in different compartments in the microcirculation and technical approaches to evaluation of the rate of oxygen consumption.

*Objects of study*. Only a small number of tissues are available for biomicroscopy using transmitted and reflected illumination, based on the image quality needed to accurately localize the signal and establish its anatomical correlations. Thin skeletal muscles, like the spinotrapezius muscle in rats, are suitable for the study of respiration rate, due to the possibility of varying that rate over a wide range.[6] However, they are not ideal in an optical sense, because only vessels in the upper layers of the tissue are available for observation. The mesentery is a thin, sparsely vascularized tissue, in which the whole vascular network is available for microscopic observation and $PO_2$ measurements.[7] However, this tissue has a very low rate of oxygen consumption, except in the vicinity of working lymphatic pumps.

*Animal anesthesia*. Pentobarbital and similar anesthetics are known to depress cardiovascular function, especially reducing perfusion of peripheral organs. In most of our own experiments we used a controlled intravenous infusion of Alfaxan, while monitoring heart rate with a veterinary hemoximeter and blood pressure sensor in a catheterized femoral artery.[8]

*Compartments for oxygen measurements*. Oxygen is a lipid-soluble, gaseous molecule and cannot be confined to a specific compartment. Albumin-conjugated oxygen probes have a large molecular size and can be retained within a compartment, whether blood plasma, interstitial space, cytosol and even mitochondria. This property allows the reporting of the dynamics of oxygen tension in every step of the $PO_2$ ladder from erythrocyte to mitochondrion. For intravascular loading, the probe is injected as a bolus into the circulation and remains for an extended time in the plasma. In some tissues (*e.g.*, mesentery and tumors) the conjugated probe may be extravasated into the interstitial space; it takes about 30 min to accumulate an extravascular concentration of probe necessary for $PO_2$ measurements.[4] The volume of extravascular and extracellular fluid in a muscle depends on the age and functional state of the animal; generally, the fractional volume of the interstitial space is about 10–15%. In skeletal muscle, due to the dense capillary network with high surface area, the interstitial $PO_2$ mainly reflects the pericapillary space, the interface between the capillary network and muscle fibers (*i.e.*, the $PO_2$ on the surface of cells).

*Spectral characteristics of excitation*. The oxygen probe used in our measurements was a Pd(II) meso-tetra(4-carboxyphenyl) porphine (PdT790, www.frontiersci.com), known as PdTCCP, conjugated with BSA (bovine serum albumin). It forms a red solution having two absorbance peaks: a Soret band in the violet and a Q-band in the green region of the spectrum. Reported peaks from the literature and our own measurements at 37 °C and pH 7 are very similar: the Soret band peak is within 410–415 nm, and the Q-band peak is within 520–525 nm. The Q-band absorbance peak magnitude is about 12 times smaller than that for the Soret band. That difference provides an opportunity to use both excitation bands to optimize the excitation light according to experimental conditions. Emitted phosphorescence light is red; a wide emission band starts at 650 nm and extends to the infra-red (IR) spectral zone. The second generation dendritic shell oxygen probe R2 (Oxygen Enterprises, Philadelphia, PA) has the same spectral characteristics.[9]

*Thermostabilization of animal and organ.* Maintaining normal body temperature in an anesthetized small rodent requires a heating pad. Thin organs used for bio-microscopy and PQM oximetry also require accurate thermostabilization, not only for maintaining metabolic activity, but also to avoid errors caused by the temperature dependence of the quenching coefficients of oxygen probes.[9] We have designed a microscope stage for small rodents, providing independent temperature control for the body and for tissues under observation.[10] The temperature of the animal heating pad was set at 39 °C and tissue temperature on the transparent sapphire window was set at 37 °C.

*Dark room conditions.* The phosphorescence in tissue and microvessels is rather weak, so that for the best signal quality, the PQM measurements require a darkened room and shielding the tissue under microscopic viewing from stray light of computer monitors, LED displays and indicators.

*Phosphorescence microscope.* Any fluorescence microscope with sufficient space for an animal and tissue under the objective lens can be used. A microscope stage and its mechanical support have to be strong enough for the weight of the laboratory animal platform and animal body on it.

The PQM employs intense light sources that produce microsecond duration light pulses, which do not appear bright or dangerous, but they are. That is why no direct visual access of the operator through the eyepieces is recommended. A video camera provides a microscopic image of the tissue, required for targeting the excitation light spot and for measurements of anatomical variables. Using a high voltage in the photomultiplier tube (PMT) and a Xenon flash lamp requires the microscope to be grounded according to high voltage regulations.

*Excitation light source.* In the first PQM instruments the only available source of an intense light pulse was a Xenon flash lamp which emitted light in a wide band, from UV to IR, so that the required narrow spectral band was obtained with an exciter optical filter, mounted in a filter cube together with a dichroic mirror beam-splitter and the emission filter. Disadvantages of the Xe flash lamp were: the high voltage power supply of the discharge capacitor and trigger electrode, leading to a very high current during the main capacitor discharge. Altogether it generates a strong electromagnetic pulse, EMP, producing an electromagnetic pickup in the photo-detector circuits. The light pulse from a Xe flash lamp has a non-ideal shape with a sharp front and then a slower exponential decay; it also requires special measures to suppress the "ringing in the tail."

During the last decade high power laser diodes, LD, for both excitation bands of Pd-TCCP became available. Continuous work, CW, high power LDs are employed in order to put out a short rectangular light pulse of sufficient flash energy. For example, to produce a 1 μs duration pulse having 1 μJ energy requires a 1 W LD, specified for CW mode. However, these pulses may be generated at any required rate. LDs use a low voltage, they do not produce EMPs, and the control of flash duration and flash rate is simple and flexible. The LDs have a narrow band that makes unnecessary an expensive exciter interference filter. Generally, application of laser diodes is cost effective, compared to Xe flash lamps. In the highly improbable accident of a LD's driver failure in the "on" position, even reflected light of that intensity could harm the eyes of the operator through an eyepiece. That is why no eyepieces should be allowed in the setup, only a video camera. LDs and drivers are available from many sellers (*e.g.*, www.roithner-laser.com); they are available as focusable laser modules, with wavelength bins 410–415 nm and 520–525 nm, and may also be ordered for an exact wavelength, if necessary. Considering the difference in absorbance and applications for two excitation bands, the recommended CW powers are different; for a green LD the CW power is below 1 W, but the power for a violet LD should be less than 100 mW.

A simple electronic circuit to control pulse duration and specific patterns of excitation pulses are usually homemade. We use the Arduino microcontroller board (www.arduino.cc) to control the flash rate and synchronize flashes and electrical stimulation of the muscle, as well as other experimental events. The pulse for the LD driver is formed by a precision one-shot IC, usually, 0.5, 1, and 2 µs. To facilitate the beam alignment, the laser module is mounted on a simple kinematic platform at the upper illuminator port of the microscope. Due to the small size of laser modules, they both can be mounted together with a beam combiner, thus enhancing the experimental capability of the instrument.

A high-power light emission diode, LED, is a cost-efficient light source, useful for measurements with a wide excitation area, but it is not as good when a small excitation spot is required. Thus a 525 nm, 1–3 W LED may be used to produce a microsecond flash, using over-current for short pulses in LEDs.

*Pulse energy density*. Absorbed pulse energy depends on the delivered light pulse energy and the volume within which it is absorbed. It is directly related to the artifact of oxygen consumption by the method itself, so that it should be minimized to a level, which provides a good phosphorescence decay curve without substantial distortion of local $PO_2$. A high flash rate may aggravate the photo-consumption error, if the time interval between flashes is shorter than the time necessary to replenish the oxygen consumed by the flash. For interstitial fluid, it is the time of diffusion from nearby tissue and vessels into the excited region. In flowing blood, it is the time interval needed to refresh the excitation volume by flowing blood.

In the detection volume of a tissue having heterogeneous $PO_2$, a long pulse duration produces a disproportionate excitation in sub-volumes with different oxygen tensions during the initial phosphorescence decay.[11] For the Pd-TCCP + BSA probe the excitation pulse duration has to be shorter than 5 µs; 1 µs pulse duration is recommended.

An epi-illumination field diaphragm, circular or rectangular in shape, is used to define the shape of the excitation region on the surface of the tissue. The width of a small spot can be changed with a laser module focusing attachment, and shaped with a diaphragm. For measurements with a scanning excitation spot, a galvano- or resonant mirror can be added to the light source, and operated in a synchronized (or not) mode with the excitation flash.[12] A neutral density filter wheel is helpful to quickly adjust the excitation energy when the size of the light spot is not constant, because of the switch to a different magnification or when refocusing the laser module.

For adjustment of the shape and size of the excitation spot, its scanning direction, amplitude and pattern, a fluorescent flat screen (*i.e.*, fluorescent film or glass slide painted with a fluorescent marker) can be used instead of the object. A dried drop of probe on the slide is also a useful tool to tune up the detection circuit of the setup.

*Optical filter cube.* If LDs are employed as excitation light sources, the optical filter cube does not require an exciter filter and the dichroic mirror beam splitter may be selected that is equally effective for both excitation bands. The beam of excitation light is reflected by a dichroic mirror through the objective lens onto the living tissue or vessel, loaded with the oxygen probe. Emitted NIR phosphorescence light is collected by the objective lens and the beam is directed through the dichroic mirror to the emitter filter. The quality of this filter is important for fidelity of the collected signal. Colored glass filters are not good for this purpose because the reflected excitation light pulse, present in the detection beam, evokes a red luminescence in the colored glass with a lifetime of about 30 μs. An optical interference filter, oriented with its reflective side toward the incoming light substantially reduces this unwanted artifact. Long-pass interference filters also use red Schott glass as a substrate to improve the spectral quality, therefore a weak glass luminescence, caused by a residual fraction of reflected excitation light, is still possible. However, glass luminescence is a widely scattered light, while the collected phosphorescence light is projected as a confined beam. Thus, it is good practice to use a relatively long tube with a black internal wall to absorb the glass luminescence component before the photo detector.

*Objective lens.* Microscope objectives and condensers are made of relatively thick glass lenses that may also add fluorescence contamination to the phosphorescent signal. Violet light produces higher intensity fluorescence than green light and special objectives used in fluorescence microscopy are preferable. The numerical aperture (NA) is the most important characteristic of an objective lens, because it determines the sensitivity of the instrument. Objectives with high NA also have high magnification and, unfortunately, a short working distance. Considering the spatial extent of the object under study, 40× magnification is recommended.

*Phosphorescence emission.* The beam of excitation light is focused onto a small spot on the tissue or vessel in a thin organ, so the window in the observation pedestal is also illuminated and may emit luminescence. The best material for this window is sapphire or quartz. It is not important for work with skeletal muscle (*e.g.*, rat spinotrapezius muscle with thickness about 0.5 mm), but for experiments with microvessels in the mesentery, some luminescence may come even from the glass of the condenser. An additional interference filter reflecting light above 600 nm and placed under the thermostatic window blocks this parasitic emission from the condenser and reflects to the objective an additional 40–60% of the phosphorescent light.

*Transillumination limitations*. When an oxygen probe is injected into the bloodstream or loaded into the interstitial space, the traditional continuous white light illumination becomes damaging for the tissue. A Pd-porphyrin probe is a photosensitizer, generating singlet oxygen, a very strong oxidative agent. In this process the metalloporphyrin consumes oxygen, the substance of interest, and "burns" the living cells with the singlet form. To avoid this problem, transillumination is recommended with a light beam in which the excitation bands are excluded from the spectrum. Transillumination with a combination of two LEDs, royal blue and yellow, may be safe for a tissue with loaded probe. A yellow or orange glass filter is recommended for transillumination with an incandescent lamp; even in that case the duration of illumination needs to be minimal.

A shutter in the transillumination train is required to block possible light contamination from the LED light source, which may produce a weak glow when turned off. The photodetector must be protected from the transmitted light, so that the sliding mirror of the microscope must be switched to the video camera position before turning on the transillumination.

*Photodetector*. A photoelectron multiplier tube, PMT or a PMT module, are the most widely used photodetectors. We use a side-on NIR-sensitive PMT (model R9110, www.hamamatsu.com). A variety of sockets are offered for these PMTs, with a built-in high voltage power supply and trans-impedance amplifier. The sensitivity of the PMT is regulated by high voltage, controlled *via* an external potentiometer. The photocathode size in side-on PMTs is smaller than the maximal width of the emitted light beam, so a light collector lens focused on the photocathode is necessary when a wide excitation area is used. All the recommendations on amplification and processing of the phosphorescence signal are related to the time-domain method of lifetime evaluation. The frequency-domain technique has a long excitation time period and is not suitable for microscopic measurement of rapid changes in tissue oxygenation, in our opinion.

*Trans-impedance amplifier*. A short excitation flash evokes in the oxygen probe a relatively long phosphorescence decay, which contains information about the dissolved oxygen concentration. The PMT converts the emitted light into an electrical signal, representing a phosphorescence decay curve. The PMT has a current output, so that a trans-impedance amplifier is required to convert it into a voltage signal. This may be a unit built into the PMT socket or an external operational amplifier. It is important to tune the amplifier in order to save information contained in the decay curve and at the same time prevent aliasing of high frequency PMT noise and sampling rate in an analog-to-digital converter (ADC). Usually, the frequency response of the amplifier is limited to 160–200 kHz, which prevents aliasing at sampling rates of 200–500 kS $s^{-1}$, but does not affect the observed lifetimes for the highest levels of physiological $PO_2$ (about 100 mmHg).

*Gating of signal*. In order to eliminate artifacts in the phosphorescence decay curve produced during the excitation flash by electromagnetic pickup from a Xe flash lamp and autofluorescence of the tissue and optics, it is useful to disable the amplifier for the duration of the excitation light pulse. However, this is not necessary if a green LD or LED and short (1–2 μs) light pulses are used. A violet LD excitation may produce an artifact when applied to a wide tissue area, but not to intravascular $PO_2$ measurements. Gating may be performed with a special PMT socket, using an operational amplifier with a "disable" option, or with an additional analog switch circuit, synchronized with the excitation flash. Gating may be realized digitally in the data acquisition program by skipping the first data point after triggering the curve recording.

*Data acquisition*. The oxygen probe Pd-TCCP + BSA has a phosphorescence decay time constant of about 600 μs for zero-$PO_2$ and about 30 μs for the highest physiological $PO_2$. For a high level $PO_2$ a 2 μs sampling period (500 kS $s^{-1}$) is used. In experiments using barbiturate anesthesia, when $PO_2$ in tissue is under 40 mm Hg, using a 200 kS $s^{-1}$ (5 μs sampling period) is justified. There are requirements for parameters of data collection, such as sampling rate, recorded curve duration, and accuracy of digitization, for an accurate exponential analysis of the decay curves.[13] The ADC has to have at least 12 bit resolution and the duration of the curves should not be shorter than 5–7 characteristic lifetimes.

*Data analysis*. Analysis of phosphorescence decays, obtained in homogeneous detection volumes (*e.g.*, *in vitro* well-stirred fluid samples) is relatively simple. A semi-logarithmic plot of the decay curve is transformed into a straight line, and a monoexponential fitting of the experimental signal returns the lifetime value. For an *in vivo* situation the $PO_2$ in the detection volume of a tissue is always heterogeneous, so that polyexponential phosphorescence decays are generated by the probe in sub-volumes having different $PO_2$.

A monoexponential fitting of heterogeneous curves is valid for the initial segment of a curve and returns a value close to the mean $PO_2$ in the detection volume.[14] A more complicated fitting to a "rectangular" model represents the $PO_2$ in the detection volume as a flat distribution, characterized by statistical moments.[14] The first moment (mean) can be recovered accurately, but the accuracy of evaluation of higher moments progressively decreases. The choice of method depends on the quality of the signal; however, it is recommended that the same data analysis approach be used for measurements as for calibration.[13,15]

*Signal localization*. The reference (or detection) volume of a vessel or tissue is formed by the intersection of the excitation light beam with the probe, distributed in a specified compartment. The excitation region is formed by the light beam focused by optics and is shaped with a field diaphragm, iris or rectangular slit. The boundary of the excitation region can be visualized with a fluorescent screen under the objective lens and marked on a video monitor. The depth of the excitation volume depends on the light absorbance in the tissue that is different for violet (shallow) and green (deeper) spectral bands. Another factor determining the limitations on light pulse energy and excitation area is the oxygen consumption by the method itself.

*Oxygen consumption by PQM.* Quenching the phosphorescence of excited metalloporphyrin with oxygen leads to the production of singlet oxygen, which is a highly reactive substance, consumed in the oxidation of local organic substrates. Each excitation flash consumes some fraction of the dissolved oxygen, depending on the probe concentration, excitation band, delivered energy density and flash rate. Authors of the PQM demonstrated that in *in vitro* measurements the effect is negligible;[2] however, in the microscopic application of the method, all factors influencing the photo-consumption of oxygen were increased several-fold due to pulse energy, flash rate and probe concentration. As a result, the photo-consumption error may become significant in microscopic applications and must be minimized, evaluated and corrected, if necessary.[16]

A simple way to evaluate the oxygen consumption by PQM is measuring the decrease of $PO_2$ in a dissected tissue, killed by freezing and thawing, under a high flash rate of the excitation light.[17] A recorded oxygen disappearance curve (ODC) provides the rate of $O_2$ photo-consumption from flash to flash. The *in vivo* estimation of the artifact in measurements of interstitial $PO_2$ is calculated from two ODC's collected for the same detection volume at two different flash rates.[18] If the photo-consumption per flash exceeds 1% of the $PO_2$, the flash energy must be attenuated using current control for a LD or a ND filter wheel for a flash lamp. In some past microcirculation studies, the quality of collected decay curves was modest; to improve the quality of the analyzed decay curves, accumulation and averaging of multiple curves obtained in a series of high frequency flashes was used.[3–5] In this case, the cumulative effects of photo-consumption in a motionless detection volume led to a large reduction in "tissue $PO_2$"; this did not occur in the $PO_2$ data collected from moving blood in a nearby microvessel. As a result, an exaggerated $PO_2$ gradient appeared between the blood and tissue, which was interpreted as high $O_2$ uptake by the vascular wall.[16]

*Intravascular $PO_2$ measurements.* The oxygen probe is infused intravenously to achieve a Pd-TCCP concentration in plasma of 0.5 mg ml$^{-1}$ for microvessels of 20–100 μm in diameter, and 1 mg ml$^{-1}$ for smaller vessels. The best excitation band for vessels smaller than 30 μm is violet and for the larger vessels it is green. The excitation spot has to be limited by the vessel boundary, in order to avoid strong tissue autofluorescence caused by a violet light pulse. That band is absorbed well by the probe and hemoglobin, thus reducing the autofluorescence artifact from underlying tissue. The half-penetration depth was 7 μm for violet excitation and 140 μm for green excitation, calculated for a homogeneous solution of hemoglobin, equivalent to 42% hematocrit.

The flash energy density is adjusted to minimize the oxygen photo-consumption in plasma (<1% of $PO_2$ per flash). For a low flash rate (<10 Hz) the $O_2$ lost in plasma is compensated by diffusion from red blood cells. In normal circulatory conditions blood flow refreshes the oxygen in samples from flash to flash. However, a large excitation spot and high flash rate may cause accumulation of an error in $PO_2$ measurements, due to multiple excitations of the same volume of plasma.[16]

*Interstitial $PO_2$ measurements.* A motionless detection volume in the tissue with multiple excitations is the most vulnerable for accumulated error related to oxygen photo-consumption. The oxygen probe is loaded into thin tissues (*e.g.*, mesentery or spinotrapezius muscle) by topical application of the probe solution to the tissue surface and allowing 15–30 min for diffusion into the interstitial space.[17,18] The resulting interstitial concentration of the probe is low and, in order to measure $PO_2$ in the interstitial fluid, the size of the excitation region needs to be much wider than for intravascular measurements. For measurements in muscle tissue it is also important to use the green excitation band, as it does not produce significant tissue autofluorescence; but it requires higher pulse energy from the light source. That is why for a violet LD the CW power is measured by tens (~20–60) of mW, while for a green LD it is hundreds of mW, depending on the diameter of the excitation area, and only a small fraction of the light is absorbed by the interstitial probe. Green light penetrates deeper into the muscle, providing a representative signal from a larger volume. There is a strong restriction on flash rate in interstitial $PO_2$ measurements, determined by the slow diffusion of oxygen from capillaries to the interstitium; it cannot be much higher than 5 flashes per second in a normally perfused muscle.[17–19]

*$PO_2$ in arrested blood and extravascular $PO_2$ profile.* A special case of motionless blood takes place when blood flow in an arteriole is rapidly arrested in order to study the rate of oxygen depletion.[12] Accumulation of oxygen photo-consumption can be prevented by scanning the excitation spot along a vessel, to avoid multiple excitations of the same plasma volume and to provide enough time for $O_2$ diffusion from erythrocyte to plasma.

*Scanning excitation spot.* A scanning mirror that moves the excitation spot can be used for a linear scan along vessels with arrested flow and a linear scan to obtain the $PO_2$ profile in a thin tissue (mesentery) around a microvessel.[12] It can also form multiple excitation spots for simultaneous measurements of $PO_2$ in arterioles and venules, and the longitudinal $PO_2$ gradient in perfused capillaries. Data collection and binning must be synchronized with the position of the excitation spot. Scanning is an efficient method to reduce the artifact of accumulated photo-consumption in stationary media when a small and intense excitation spot is employed.

*Gas barrier film and organ compression with an "airbag".* In order to prevent tissue desiccation and to separate the tissue from ambient oxygen, a gas barrier film is used to wrap the surgically exposed tissue. In our experiments we use the polyvinylidene chloride film, Krehalon (www.kureha.com) both of 12 and 20 μm thickness. A thin film 12 μm thick is used to wrap the tissue or organ, while a stronger film 20 μm thick is used to fabricate a flexible membrane for the pneumatic compression device, "airbag," mounted on the objective lens barrel.[12,17–19] A thin layer of fluid containing the oxygen probe always forms between the covering film and muscle tissue, forming some small pockets. This solution does not consume oxygen, but has a small inflow of oxygen from the air through the film, sometimes creating a high $PO_2$ component in the signal. Low air pressure (4–12 mmHg) in the airbag membrane, attached to the objective, is used for gently pressing the film to the tissue surface, eliminating this problem and also improving image quality and fixation of the tissue in place. That low pressure is safe for the circulation; small changes in blood flow and tissue oxygenation start to appear when the pneumatic pressure exceeds 45 mmHg. To prevent sticking between the two sheets of film (*i.e.*, barrier film on tissue and airbag), a drop of water-based lubricant is applied. The addition of fluorescent ink to the lubricant helps to visualize the excitation spot and the scanning pattern, using a fluorescein optical cube in the microscope.

*Oxygen disappearance curve and* $VO_2$ *evaluation.* When, during continuous recording of the interstitial $PO_2$, the pressure in the airbag is rapidly elevated to a supra-systolic level (140 mm Hg), flowing blood is immediately arrested and extruded from the compressed region of muscle. From that moment the tissue volume is isolated from oxygen inflow and outflow through microvessels, so that the local $PO_2$ reflects the level of oxygen dissolved in the muscle. The $PO_2$ decrease due to local oxygen consumption forms the oxygen disappearance curve, ODC.[19] The oxygen consumption, $VO_2$, of the muscle is equal to the negative first derivative of the ODC, with a proportionality coefficient equal to the oxygen solubility in the muscle. This is correct until reaching a low $PO_2$ level, when myoglobin-bound oxygen may distort the curve. To obtain muscle $VO_2$, it is enough to obtain only the initial slope of the ODC at a high level of oxygenation by applying a brief compression of the airbag and then release to low air pressure for recovery.[17] This procedure can be repeated after a period of $PO_2$ restoration, thereby continuously sampling the tissue $PO_2$ and the respiration rate.[18] An empirically found optimal cycle comprises 5 s of compression and 15 s of low pressure recovery, provided by an automatic pressure regulator.

*Probe calibration.* Application of a new formulation for an oxygen probe or using a well-known probe at different temperatures and pH, requires an accurate calibration method, suitable for the laboratory. We have developed a method with variable barometric pressure in a test vial, controlled with a digital manometer.[15]

## References

1. W. J. Whalen, P. Nair, D. Buerk and C. A. Thuning, *Am. J. Physiol.*, 1974, **227**(6), 1221–1225.
2. J. M. Vanderkooi, G. Maniara, T. J. Green and D. F. Wilson, *J. Biol. Chem.*, 1987, **262**(12), 5476–5482.
3. I. P. Torres Filho and M. Intaglietta, *Am. J. Physiol.*, 1993, **265**(4), H1434–H1438.
4. I. P. Torres Filho, M. Leunig, F. Yuan, M. Intaglietta and R. K. Jain, *Proc. Natl. Acad. Sci. U. S. A.*, 1994, **91**(6), 2081–2085.
5. L. Zheng, A. S. Golub and R. N. Pittman, *Am. J. Physiol.*, 1996, **271**(1), H365–H372.
6. S. D. Gray, *Microvasc. Res.*, 1973, **5**(3), 395–400.

7. B. W. Zweifach, *Anat. Rec.*, 1954, **120**(1), 277–291.
8. A. Brammer, C. D. West and S. L. Allen, *Lab. Anim.*, 1993, **27**(3), 250–257.
9. I. Dunphy, S. A. Vinogradov and D. F. Wilson, *Anal. Biochem.*, 2002, **310**(2), 191–198.
10. A. S. Golub and R. N. Pittman, *Microvasc. Res.*, 2003, **66**(3), 213–217.
11. A. S. Golub, A. S. Popel, L. Zheng and R. N. Pittman, *Photochem. Photobiol.*, 1999, **69**(6), 624–632.
12. A. S. Golub, B. K. Song and R. N. Pittman, *Am. J. Physiol.: Heart Circ. Physiol.*, 2011, **301**(3), H737–H745.
13. A. A. Istratov and O. F. Vyvenko, *Rev. Sci. Instrum.*, 1999, **70**(2), 1233–1257.
14. A. S. Golub, A. S. Popel, L. Zheng and R. N. Pittman, *Biophys. J.*, 1997, **73**(1), 452–465.
15. A. S. Golub and R. N. Pittman, *J. Appl. Physiol.*, 2016, **120**(7), 809–816.
16. A. S. Golub and R. N. Pittman, *Am. J. Physiol.: Heart Circ. Physiol.*, 2008, **294**(6), H2905–H2916.
17. A. S. Golub, M. A. Tevald and R. N. Pittman, *Am. J. Physiol.: Heart Circ. Physiol.*, 2011, **300**(1), H135–H143.
18. W. H. Nugent, B. K. Song, R. N. Pittman and A. S. Golub, *Microvasc. Res.*, 2016, **105**, 15–22.
19. A. S. Golub and R. N. Pittman, *Am. J. Physiol.: Heart Circ. Physiol.*, 2012, **303**(1), H47–H56.

CHAPTER 10

# *Photoacoustic Imaging of Oxygen*

CHANG H. LEE[a], JANGGUN JO[b], XUEDING WANG[b] AND RAOUL KOPELMAN*[a,b]

[a]The University of Michigan, Department of Chemistry, Ann Arbor, MI, USA;
[b]The University of Michigan, Department of Biomedical Engineering, Ann Arbor, MI, USA
*E-mail: kopelman@umich.edu

## 10.1 Introduction

Oxygen ($O_2$) is inevitable for living organisms. Humans, like other mammals, have evolved to precisely control the oxygen levels in their body. Imbalances or decreases in oxygen levels are expressed in various pathological states and can be fatal.[1] Especially in tumors, oxygen levels are known to be significantly lower than in normal tissue. This cancer hallmark of *hypoxia* has been discussed and explained by the Warburg effect.[2,3] Hypoxia is observed not only in cancer but also in other inflammatory diseases, as well as in relation to the growth of stem cells.[4,5]

Photoacoustic (PA) imaging is a hybrid imaging modality combining in a unique way traditional optical imaging and ultrasound imaging (Figure 10.1).[6] An optical contrast agent absorbs pulsed non-ionizing radiation, and converts the excited energy into thermoelastic expansion, which induces a

Detection Science Series No. 11
Quenched-phosphorescence Detection of Molecular Oxygen: Applications in Life Sciences
Edited by Dmitri B. Papkovsky and Ruslan I. Dmitriev

Published by the Royal Society of Chemistry, www.rsc.org

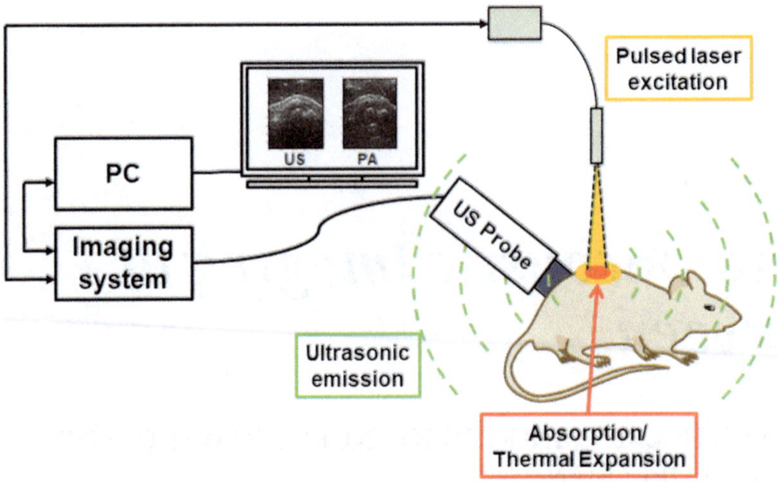


**Figure 10.1** Photoacoustic (PA) imaging scheme. Pulsed laser is used to excite the optical contrast agent. Absorption by the laser excitation creates thermal expansion causing ultrasound emission. The ultrasonic emission (from both laser pulses and ultrasound pulses) is detected by an ultrasound (US) probe and the acquired information can be converted into both US and PA images.

slight temperature increase and ultrasonic emission. The initial pressure rise due to a pulsed optical absorption ($p_0$) can be estimated from

$$p_0 = \Gamma \mu_a F, \tag{10.1}$$

where $\Gamma$ is the Grüneisen parameter of the tissue, $\mu_a$ is the absorption coefficient, and $F$ is the local light fluence. As shown in eqn (10.1), the PA signal generated correlates linearly with the absorbance of the optical contrast agent, and thus with the optical absorption spectrum. Because ultrasound travels far better than light in tissues, *i.e.* with little scattering, its imaging depth is mostly limited by the excitation light, thus providing a deeper tissue imaging ability than traditional optical imaging. Here we concentrate on functional/chemical imaging of oxygen. Although label-free PA imaging has been widely explored for both structural and functional imaging, well-established optical sensors have also been coupled with PA imaging.[7–49]

This chapter is intended to provide an overview of PA imaging of tissue oxygen; however, it is inconceivable to discuss PA oxygen sensing without mentioning PA blood oxygenation measurements. The PA blood oxygenation measurement is acquired through direct ratiometric measurements of oxygen carrying proteins, such as hemoglobin and myoglobin.[7–32] Hemoglobin (as well as myoglobin) changes its optical property, or simply color, with varying oxygen concentration. This change in optical property follows the standard principle of pulse oximetry.[50,51] The ratio of the PA signals at two or more light wavelengths allows a non-invasive indirect oxygen measurement *in vivo*.

More recently, the method of *photoacoustic lifetime* (PALT) measurement-based oxygen monitoring was introduced so as to measure directly tissue oxygenation levels. This method is an extension of time-resolved (or transient) absorption spectroscopy.[33–43] The sensing mechanism is similar to that of phosphorescence quenching based oxygen sensors, but it is measured by PA imaging/spectroscopy, *i.e.* it is based on optical absorption rather than emission. Theories and examples for both methods will be discussed in the subsequent sections.

## 10.2 Photoacoustic Oxygen Monitoring Using Biological Chromophores (*i.e.* Hemoglobin)

Photoacoustic (PA) imaging has been initially applied for functional hemodynamic monitoring. Hemoglobin is an oxygen carrying protein associated with red blood cells. It is well known that hemoglobin changes its optical absorption (color) when bound with $O_2$ molecules (referred to as oxy- and deoxy-hemoglobin).[50,51] The change in optical characteristics of hemoglobin have long been applied clinically for non-invasive blood oxygenation measurements. Oxy- and deoxy-hemoglobin ($HbO_2$ and Hb, respectively) are natural chromophores, which can be applied as PA contrast agents for blood oxygenation mapping.[7–31] A similar principle has been applied for the other oxygen carrying protein in muscle tissue, myoglobin.[32]

The method described below is one of the most common algorithms for measuring $O_2$, based on computing ratios between two PA signals. Assuming that the main chromophores in the PA signal are $HbO_2$ and Hb, the PA intensity at a single exciting wavelength $\lambda$ can be expressed as:

$$\mathrm{PA}_\lambda = k(\varepsilon_{\mathrm{Hb}}{}^{\lambda}[\mathrm{Hb}] + \varepsilon_{\mathrm{HbO_2}}{}^{\lambda}[\mathrm{HbO_2}]), \tag{10.2}$$

where $k$ is a PA signal constant including the Grüneisen parameter of the tissue, the light fluence, and the sensitivity of the imaging system; the constant $k$ is independent of the wavelength when we assume the light fluencies as being the same for all the optical wavelengths; $\varepsilon_{\mathrm{Hb}}{}^{\lambda}$ and $\varepsilon_{\mathrm{HbO_2}}{}^{\lambda}$ are the known optical absorption coefficients at the excitation wavelength, $\lambda$, of the Hb and $HbO_2$, respectively; [Hb] and [$HbO_2$] are concentrations of deoxy- and oxy-hemoglobin, respectively. A similar equation can be drawn when using another excitation wavelength. With two different wavelengths ($\lambda_1$ and $\lambda_2$), the PA signal can be expressed as:

$$\mathrm{PA}_{\lambda_1} = k\left(\varepsilon_{\mathrm{Hb}}{}^{\lambda_1}[\mathrm{Hb}] + \varepsilon_{\mathrm{HbO_2}}{}^{\lambda_1}[\mathrm{HbO_2}]\right), \tag{10.3}$$

$$\mathrm{PA}_{\lambda_2} = k\left(\varepsilon_{\mathrm{Hb}}{}^{\lambda_2}[\mathrm{Hb}] + \varepsilon_{\mathrm{HbO_2}}{}^{\lambda_2}[\mathrm{HbO_2}]\right). \tag{10.4}$$

Eqn (10.3) and (10.4) can be converted into matrix form:

$$k \cdot \begin{bmatrix} \varepsilon_{\mathrm{Hb}}{}^{\lambda_1} & \varepsilon_{\mathrm{HbO_2}}{}^{\lambda_1} \\ \varepsilon_{\mathrm{Hb}}{}^{\lambda_2} & \varepsilon_{\mathrm{HbO_2}}{}^{\lambda_2} \end{bmatrix} \begin{bmatrix} [\mathrm{Hb}] \\ [\mathrm{HbO_2}] \end{bmatrix} = \begin{bmatrix} \mathrm{PA}_{\lambda_1} \\ \mathrm{PA}_{\lambda_2} \end{bmatrix}, \tag{10.5}$$

$$k\cdot\begin{bmatrix}[\mathrm{Hb}]\\ [\mathrm{HbO_2}]\end{bmatrix}=\begin{bmatrix}\varepsilon_{\mathrm{Hb}}{}^{\lambda_1} & \varepsilon_{\mathrm{HbO_2}}{}^{\lambda_1}\\ \varepsilon_{\mathrm{Hb}}{}^{\lambda_2} & \varepsilon_{\mathrm{HbO_2}}{}^{\lambda_2}\end{bmatrix}^{-1}\begin{bmatrix}\mathrm{PA}_{\lambda_1}\\ \mathrm{PA}_{\lambda_2}\end{bmatrix}. \tag{10.6}$$

$PA_{\lambda_1}$ and $PA_{\lambda_2}$ are experimentally determined values and the optical absorption coefficients ($\varepsilon$) are known constants. Thus, the concentrations of Hb and $HbO_2$ can be calculated. The blood oxygenation level ($sO_2$) can be computed by

$$\mathrm{sO_2}=\frac{k[\mathrm{HbO_2}]}{k([\mathrm{HbO_2}]+[\mathrm{Hb}])}. \tag{10.7}$$

This algorithm allows for approximate blood oxygenation using PA spectroscopy or imaging.

Figure 10.2 is an example of *in vivo* PA hemoglobin oxygen saturation ($sO_2$) image of a tumor bearing mouse using a similar algorithm to that described above. The images were taken using an ultrasound (US) and PA dual imaging system built on a commercially available research US platform (V1, Verasonics) with a linear array probe working at central frequency of 11.25 MHz (CL15-7, Philips) excited with a tunable optical parametric oscillator (OPO) laser working at a 10 Hz pulse repetition rate with a pulse width of 5 ns. Four wavelengths (565 nm, 576 nm, 584 nm, and 600 nm) were used. As shown in Figure 10.2, blood oxygenation in the peripheral tissue is higher than in the subsurface, due to tumor hypoxia.

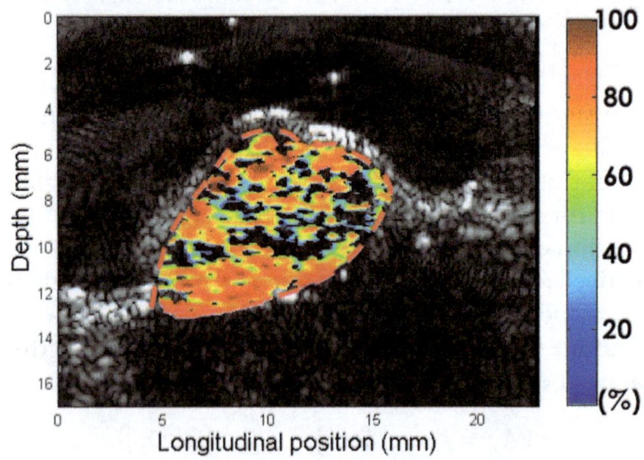


**Figure 10.2** PA hemoglobin oxygen saturation ($sO_2$) image of the tumor *in vivo*. The PA image in pseudo-color is superimposed on the gray-scale ultrasound (US) image. US image is used to delineate the tumor boundary (red dashed line).[49] Reprinted from ref. 49, https://dx.doi.org/10.1038/s41467-017-00598-1, © The Author(s) 2017. Published under the terms of the CC BY 4.0 licence, https://creativecommons.org/licenses/by/4.0/.

However, there have been numerous approximations made in this algorithm, such as excluding variations in key biologically dependent parameters, *e.g.*, light fluence. This, in fact, is a concern for all optical based imaging modalities used for sub-surface medical imaging.[52] The propagation of visible or near-infrared photons in biological tissues suffers from attenuation as a function of depth and wavelength. Without compensation for the light attenuation, the $O_2$ quantification from PA $sO_2$ images results in errors.

Several groups have shown examples of successfully overcoming the local light fluence difference. Yao *et al.* reported a single-wavelength pulse-width-based method (PW-$sO_2$) by employing two of the same wavelength lasers with different pulse widths (3 ns and 3 ps).[16] $HbO_2$ and Hb have different saturation intensities, which results in reduction of the absorption coefficients, depending on the different excitation intensities. An initial picosecond pulse excitation of $HbO_2$ results in more saturation, and thus lower PA intensity, compared to a following nanosecond pulse excitation. Hb, on the other hand, is not influenced by a change in pulse widths. Because only one wavelength of light is used, light fluence can easily be compensated. More recently, Tzoumas *et al.* reported Eigenspectra multispectral optoacoustic tomography. They utilized 21 different wavelengths and a vector corresponding to the light fluence spectrum so as to accurately quantify high resolution $sO_2$ images *in vivo*, up to 1 cm in depth.[14]

## 10.3 Photoacoustic Lifetime Measurements for Oxygen Sensing (PALT)

Blood oxygenation provides indirect measurements of oxygen through oxygen binding blood proteins, not by measuring tissue oxygen. It is important to note that tissue hypoxia is not always accompanied by ischemia.[53] Moreover, tissue oxygen monitoring can be as valuable a diagnostics tool as blood oxygen monitoring.[54] To directly measure tissue oxygenation, photoacoustic lifetime measurements for oxygen sensing (PALT) was introduced.[33–43]

The mechanism behind PALT is equivalent to that of phosphorescence quenching-based $O_2$ sensing (Figure 10.3), but the signal is detected photoacoustically rather than optically. The triplet state has a relatively long lifetime (typically in the μs range) due to the "forbidden" nature of the transition (requiring a change of spin, *i.e.* a relativistic effect). The lifetime of the sensor dye molecule excited triplet state can be shortened by its interactions with the local environment, such as molecular $O_2$. Energy from the $T_1$ state is transferred into the paramagnetic triplet ground state of a neighboring oxygen molecule, creating reactive oxygen species (ROS, including singlet oxygen, superoxide, $H_2O_2$, *etc.*) as side products. Thus, the phosphorescence is "quenched" (red arrow) by oxygen, and the lifetime decay rate of the phosphorescence is dependent upon the probability of colliding with $O_2$ molecules, which is linear with the local $O_2$ concentration. The dashed lines represent non-radiative relaxation, *i.e.* local heating that creates ultrasound,

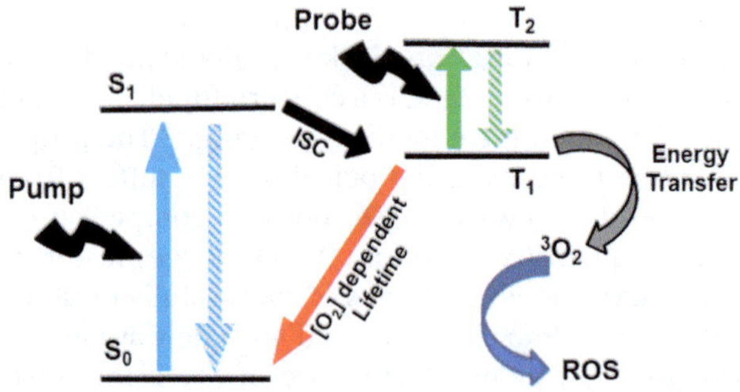


**Figure 10.3** Basic mechanism of PALT. An oxygen sensing dye (a dye molecule with long-lived phosphorescence) at its ground state ($S_0$) absorbs the initial "pump" pulse (solid blue arrow). The excited molecule, through internal conversion (not shown) and intersystem crossing (ISC), quickly relaxes from the first excited singlet state ($S_1$) to the first (lowest) triplet state ($T_1$). Then, the "probe" pulse (solid green arrow) excites the dye molecules from the first triplet state to a higher lying triplet state.

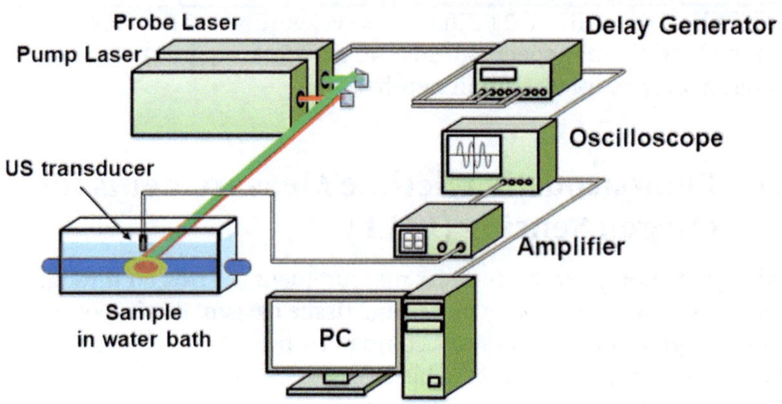


**Figure 10.4** Exemplary PALT setup. The "pump" and "probe" laser beams are guided into the target (sample) region, triggered by a delay generator. PA signals are collected by an ultrasound (US) transducer.

which can be detected by the PA signal. The lifetime of the excited state of the dye (as well as its phosphorescence) correlates non-linearly with the local $O_2$ concentrations, as shown below in the Stern–Volmer relationship.

$$\frac{T^0}{T} = 1 + k_Q T^0 \text{pO}_2. \tag{10.8}$$

where $T$ is the lifetime, $\text{pO}_2$ is the partial pressure of oxygen, $T^0$ is the lifetime at $\text{pO}_2 = 0$, and $k_Q$ is the quenching rate constant.

The PALT method employs two laser pulses, called "pump" and "probe" beams (Figure 10.4). The pump beam excites the oxygen sensing dye molecules into their first excited singlet state, which quickly transfers the energy

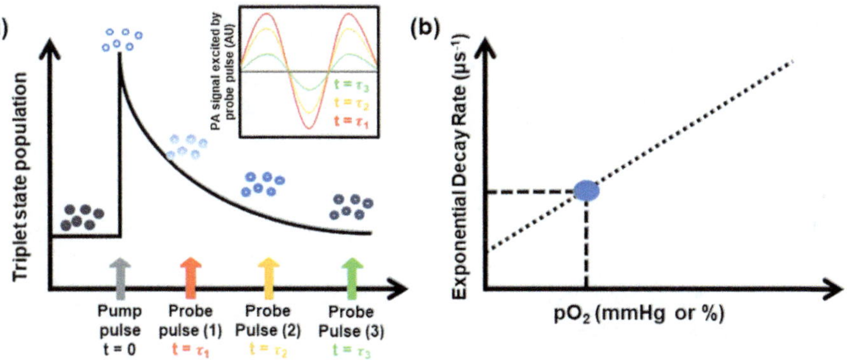


**Figure 10.5** PALT schematic (a) The phosphorescence always decays exponentially and the exponential decay rate is known to correlate linearly with the oxygen concentration, typical of a first order reaction. The delay time between the pump and probe pulse ($t = \tau$) can be controlled by a delay generator. Multiple PA signals from probe pulses are collected at varying delay times (*i.e.* at $t = \tau_1$, $t = \tau_2$, and $t = \tau_3$). The PA signals from probe pulses at different delay times are used to compute the exponential decay rate at a given oxygen concentration. (b) The exponential decay rate is calculated from the exponential decay curve fitting. The decay rate correlates linearly with oxygen concentration ($pO_2$).

into the first triplet state. The PA signal acquired from the "pump" excitation can be detected through an ultrasound transducer. The probe beam allows measuring the transient absorption of the dye triplet state, which can also be detected through the ultrasound transducer. By varying the delay times between the pump and probe beams (*i.e.* $\tau = 1$ µs, 2 µs, *etc.*), the exponential decay of the dye triplet state can be calculated (Figure 10.5a). The decay rate and lifetime vary at different $O_2$ concentrations. A calibration curve can be drawn with the decay rate and $pO_2$ (Figure 10.5b).

It is important to note that PALT requires a secondary sensing dye. Three dyes have previously been reported (Figure 10.6). Both PtOEP and Oxyphor G2 are known phosphorescence quenching-based $O_2$ sensing dyes.[33,43] PALT was performed, not only *via* phosphorescence based quenching $O_2$ sensing dyes, but also *via* a well-known PDT photosensitizer, Methylene Blue (MB).[34–40]

Ashkenazi *et al.* first reported the PALT using a standard phosphorescence quenching based $O_2$ sensing dye, PtOEP, with a 532 nm pump beam and a 740 nm probe beam.[33] The PA based transient absorption measurements were used to fit exponentially, in analogy to the emission lifetime of the dye, varying with low and high $O_2$ concentrations. The calculated lifetimes of the PtOEP dye, whether using PALT or emission based measurements, have been shown to be comparable in both high and low $O_2$ concentrations.

Ray *et al.* reported the PALT *in vivo* using a different phosphorescence quenching based $O_2$ indicator, Oxyphor G2.[43] Its monitoring employed a 630 nm pump beam and a 950 nm probe beam. Having two near-infrared light sources is optimal for biomedical applications, by minimizing light

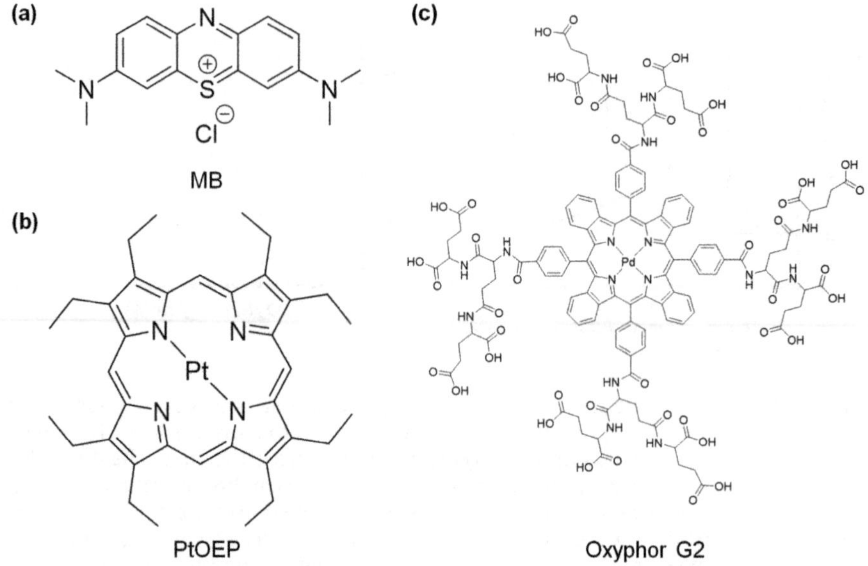


**Figure 10.6** Chemical structures of known PA $O_2$ sensing dyes. (a) Methylene blue (MB); (b) Pt(II) octaethylporphine (PtOEP); (c) Oxyphor G2.

scattering and absorptions from tissues. Exponential decay fits were shown for varying oxygen concentrations, *in vivo*.

Methylene Blue (MB) was also shown to be a potential PALT indicator for *in vivo* $O_2$ imaging.[34–40] Photons with 650 nm and 810 nm wavelength (pump and probe beam, respectively) were employed. A series of PA images, with varying pump-probe delays (0.25 μs, 0.5 μs, 1 μs, 2 μs, 4 μs, 8 μs, and 100 ms), were taken and computed so as to map $O_2$ in an *in vivo* mouse model.

To process the PALT using MB,[37] a transient PA signal from the probe beam ($PA_{810}$), at a specific pump-probe delay ($\tau$), can be extracted by subtracting the PA signal generated by the pump beam ($PA_{650}$) from the PA signal generated by both beams ($PA_{650+810}$). This is expressed as:

$$PA_{810,\tau} = PA_{650+810,\tau} - PA_{650,\tau}. \tag{10.9}$$

However, the $PA_{810,\tau}$ is not a pure transient PA signal because it also contains background tissue absorption at the 810 nm excitation, which mostly comes from blood tissue. The background PA signals with 810 nm excitation ($PA_{810,\text{background}}$) can be extracted by subtracting the PA signal generated by the pump beam ($PA_{650}$) from the PA signal generated by both beams ($PA_{650+810}$) at a long $\tau$ (100 ms). Since the lifetime of MB is known to be much shorter (<100 μs), $PA_{650+810,\tau=100\text{ ms}}$ does not contain any transient PA signals.

$$PA_{810,\text{background}} = PA_{650+810,\tau=100\text{ ms}} - PA_{650,\tau=100\text{ms}}, \tag{10.10}$$

$$PA_{\text{transient},\tau} = PA_{810,\tau} - PA_{810,\text{background}}. \tag{10.11}$$

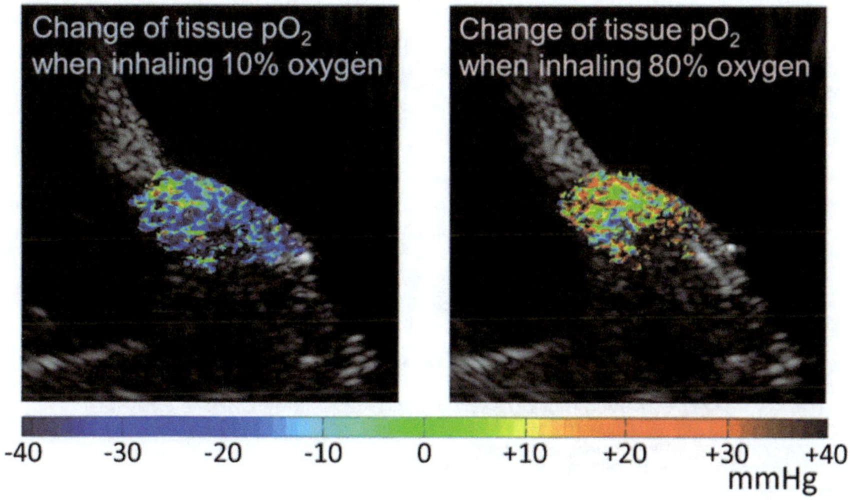


**Figure 10.7** Relative changes in tissue $pO_2$ during inhalation of low oxygen (10%) and high oxygen (80%) contents. Changes in $pO_2$ are calculated from PALT image with normal air breathing (21%). Reprinted with permission from Qi Shao and Shai Ashkenazi, Photoacoustic lifetime imaging for direct *in vivo* tissue oxygen monitoring, *J. Biomed. Opt.*, 2015, **20**(3), 036004.[34]

A series of transient PA signals, at different pump-probe delays ($\tau$), can be collected and fit into the exponential decay function expressed as:

$$PA_{transient} = PA_{transient,0}\, e^{-\frac{1}{T}\tau}, \tag{10.12}$$

where $T$ is the lifetime and $PA_{transient,0}$ is computed from the exponential decay function.

Exemplary *in vivo* PALT images using MB are shown in Figure 10.7.[34] MB was injected subcutaneously into a normal mouse at the hindlimb. The mouse was allowed to inhale different mixtures of oxygen, nitrogen, and air by a lab-made breathing apparatus. The $pO_2$ of the hindlimb was continuously monitored by a needle type oxygen probe. As shown in the figure, the PALT images show clear differences in tissue $pO_2$ during inhalation of low oxygen (10%) and high oxygen (80%). High oxygen inhalation resulted in higher tissue $pO_2$, while low oxygen inhalation resulted in hypoxic tissue. The tissue $pO_2$ changes also match with the oxygen probe single-point measurement.

## 10.4 Applying Nanotechnology to PALT

It is important to note that PALT requires a secondary oxygen indicator, which can be easily coupled with nanotechnology.[55–59] Nanotechnology for biomedical applications has made significant advances over the past few decades. The main advantages relate to the delivery (*i.e.* by targeting) and

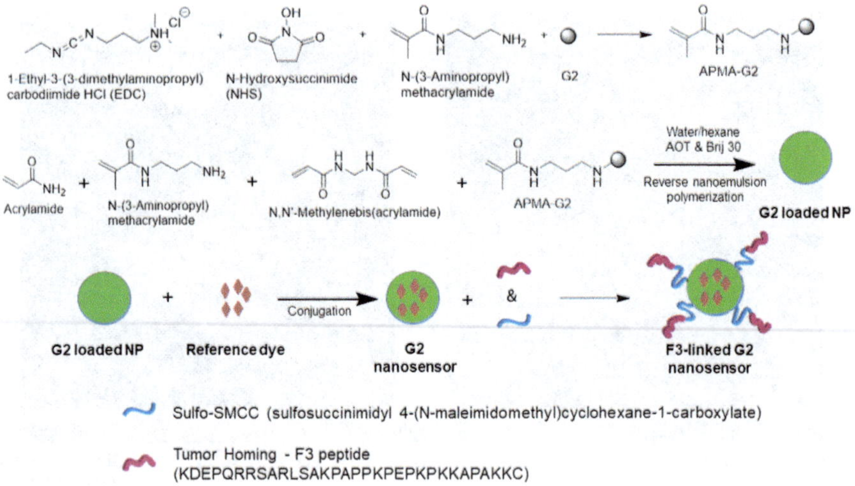


**Figure 10.8** Schematic of Oxyphor G2 conjugated polyacrylamide nanoparticles.

protection (*i.e.*, from blood enzymes).[49,60] The surface of the nanoparticles can be engineered to conjugate targeting moieties, such as peptides and antibodies. The nanoparticle (NP) matrix can physically encapsulate and protect the therapeutic cargoes as well as the indicator dyes so as to protect them from interactions with any larger biomolecules, such as proteins (including enzymes). Such interactions may cause degradation of dyes or alterations in their optical properties. The NP can also include additional calibration dyes required for reliable chemical analysis *in vivo*. Several chemical nanosensors have been developed for multiple analytes, including oxygen.[44–49]

We have already reported an oxygen nanosensor using a known PALT indicator, Oxyphor G2, and tested its oxygen sensitivity *in vitro* using fluorescence spectroscopy and imaging (Figure 10.8).[44] The nanosensor is composed of base monomer, acrylamide, with addition of a secondary monomer, *N*-(3-aminopropyl)methacrylamide, and a crosslinker, *N*,*N*′-methylene-bis(acrylamide). The crosslinker allows conjugating linear polyacrylamide chains into 3D nanoparticles. The secondary monomer, *N*-(3-aminopropyl)methacrylamide, provides primary amine groups allowing flexibility to functionalize the NP with amine-reactive crosslinking chemistry. Oxyphor G2 contains 16 carboxylic acid groups, which were conjugated with primary amines on the secondary monomer prior to NP formation. The prepared G2 conjugated NP was modified with a reference dye and a targeting moiety, tumor homing F3-peptide, to give the final nanosensor.

The diameter of the nanosensor was reported to be around 30 nm, determined by scanning electron microscope (SEM), and 51 ± 5 nm, determined by dynamic light scattering (DLS).[44] Hydrogel polyacrylamide NP is known to swell in aqueous solutions, explaining a large difference between two methods of size analyses (SEM for dried NP and DLS for NP in aqueous solutions). Other than the innate advantages of NP, *i.e.* targetability, G2 nanosensor had greatly improved photostability over the free Oxyphor G2 dye. Under continuous illumination by 630 nm light (1.5 mW), the photobleaching half-life of G2 nanosensor was found to be 39 min while that of the free dye was only 3 min. Using the G2 nanosensor, the average intracellular oxygen concentrations were reported to be 110 ± 27 mmHg in C6 glioma cells and 99 ± 22 mmHg in the A549 cells.

The same nanosensor can also be applied to PALT. PALT can be much further advanced by combination with nanotechnology. Other than the previously described advantages, the NP can also be incorporated with other therapeutic agents (*i.e.*, chemotherapy and/or photo-activated therapy) on top of the sensing capability turning them into multifunctional theranostic nanoparticles.

## 10.5 Summary and Discussion

PA imaging is well-suited for measuring blood oxygenation, based on the different optical properties of oxy- and deoxy-hemoglobin. The technique is advantageous by being non-invasive and with no need for secondary indicators. The biggest challenge is to increase the imaging depth while retaining quantitative $sO_2$ precision of measurement. Optical fluence at different tissue depths varies significantly with wavelength, as well as with the nature of the tissue. The aim now for researchers is to accurately measure $sO_2$ in deeper tissues (>1 cm).

In order to provide the tissue oxygenation level ($pO_2$) directly, PALT has been introduced. Unlike the PA blood oxygenation imaging, PALT requires a secondary indicator dye, typically identical to indicators for quenched-phosphorescence based oxygen sensing. A well-known common photosensitizer, methylene blue (MB), and other optical oxygen indicators, *i.e.*, Oxyphor G2 and PtOEP, have so far been applied for PALT.

A side product of PALT is the production of reactive oxygen species (ROS, including singlet oxygen, superoxide, $H_2O_2$, *etc.*). This mechanism is actually also shared with photodynamic therapy (PDT). PDT is a therapeutic modality, which was originally developed for cancer treatment and later extended to different diseases such as bacterial and viral infections as well as cardiac arrhythmia.[59–66] Photosensitizers receive light energy, which is transferred into nearby $O_2$ molecules, so as to produce ROS. The excess ROS causes oxidative stress within the cell, leading to cell death. PDT is a highly selective therapeutic modality whereas it requires the presence of three components (light, photosensitizer, and $O_2$). During PDT, the molecular $O_2$ is essentially consumed. PALT has the potential to be combined with PDT so as to enable real-time therapeutic monitoring, thereby acquiring theranostic operation for therapy guidance (Figure 10.9).[35,39,42] On the other hand, it is obvious that

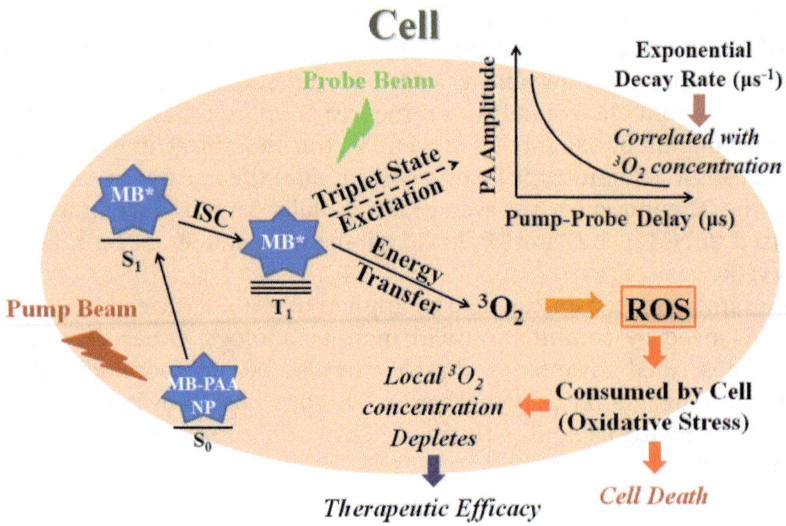


**Figure 10.9** Schematic of monitoring photodynamic therapy through PALT.

any diagnostic use of PALT should be done with minimal laser power, so as to minimize any potential photodamage to the tissue.

In this context we also note that hypoxia is liable to suppress any radiation therapy that is based on the production of ROS, whether using photons in the optical, X-ray, or gamma ray domain. Thus, the potential utility of such therapy may be predicted by mapping the tumor oxygen. In contrast, electrode measurements only give a spatially local value, while biopsies cannot provide any information on oxygen content. Due to the notorious heterogeneity of tumors, getting such a global time resolved description of its oxygen distribution may turn out to be highly useful for planning correctly the individual patient's treatment plan.

PALT can also be applied to different analytes, such as glucose. We have previously used optical oxygen sensing nanoparticles for intracellular glucose measurements.[48] The nanosensor was composed of polyacrylamide based nanoparticles containing an oxygen-sensitive fluorescence indicator $(Ru[dpp(SO_3Na)_2I_3)Cl_2$, an oxygen-insensitive reference fluorescence dye, and the glucose-selective enzyme, glucose oxidase. Glucose oxidase catalyzes the oxidation of glucose to produce D-gluconolactone, which hydrolyzes into D-gluconic acid. The oxidation results in the nanosensor's depletion of oxygen, which can be detected by the oxygen indicator, thus indirectly, but reliably, detecting and quantitatively measuring the glucose level.

In conclusion, some of the most recent advances in PA $sO_2$ imaging and some initial studies of PALT based PA $pO_2$ imaging were introduced here. Although PA $pO_2$ imaging is still in its infancy, the initial studies suggest its promising potential and the possibilities for a variety of applications. Finally, we emphasize that subsurface *in vivo* oxygen analysis can now be done non-invasively, in both spatially and time-resolved mode.

## Acknowledgements

We acknowledge the permission to reproduce a figure from Dr Qi Shao, Dr Shai Ashkenazi, and SPIE (ref. 34). This work was supported by NIH/NCI under the grant number R01CA186769 (R. K. and X. W.).

## References

1. D. B. Papkovsky and R. I. Dmitriev, *Chem. Soc. Rev.*, 2013, **42**, 8700.
2. W. R. Wilson and M. P. Hay, *Nat. Rev. Cancer*, 2011, **11**, 393.
3. O. Warburg, *Science*, 1956, **123**, 309.
4. H. K. Eltzschig and P. Carmeliet, *N. Engl. J. Med.*, 2011, **364**, 656.
5. H. Abdollahi, L. J. Harris, P. Zhang, S. McIlhenny, T. Tulenko and P. J. DiMuzio, *J. Surg. Res.*, 2011, **165**, 112.
6. M. Xu and L. V. Wang, *Rev. Sci. Instrum.*, 2006, **77**, 041101.
7. X. Wang, G. Ku, X. Xie, Y. Wang, G. Stoica and L. V. Wang, *Proc. SPIE*, 2004, **5320**, 69.
8. X. Wang, X. Xie, G. Ku, L. V. Wang and G. Stoica, *J. Biomed. Opt.*, 2006, **11**, 024015.
9. J. Yao, K. I. Maslov, Y. Zhang, Y. Xia and L. V. Wang, *J. Biomed. Opt.*, 2011, **16**, 076003.
10. J. Jo and X. Yang, *J. Biomed. Opt.*, 2011, **16**, 090506.
11. C.-S. Friedrich, M. P. Mienkina, C. Brenner, N. C. Gerhardt, M. Jörger, A. Strauß, M. F. Beckmann, G. Schmitz and M. R. Hofmann, *Proc. SPIE*, 2011, **8088**, 808814.
12. C. L. Bayer, B. J. Wlodarczyk, R. H. Finnell and S. Y. Emelianov, *Biomed. Opt. Express*, 2017, **8**, 757.
13. S. Mathangi, M. Konstantin, F. Z. Hao, S. George and V. W. Lihong, *Phys. Med. Biol.*, 2007, **52**, 1349.
14. S. Tzoumas, A. Nunes, I. Olefir, S. Stangl, P. Symvoulidis, S. Glasl, C. Bayer, G. Multhoff and V. Ntziachristos, *Nat. Commun.*, 2016, **7**, 12121.
15. E. W. Stein, K. Maslov and L. V. Wang, *J. Biomed. Opt.*, 2009, **14**, 020502.
16. J. Yao, L. Wang, J.-M. Yang, K. I. Maslov, T. T. W. Wong, L. Li, C.-H. Huang, J. Zou and L. V. Wang, *Nat. Methods*, 2015, **12**, 407.
17. A. Hussain, W. Petersen, J. Staley, E. Hondebrink and W. Steenbergen, *Opt. Lett.*, 2016, **41**, 1720.
18. G. Yin, D. Xing and S. Yang, *J. Appl. Phys.*, 2009, **106**, 013109.
19. C. J. Arthuis, A. Novell, F. Raes, J.-M. Escoffre, S. Lerondel, A. Le Pape, A. Bouakaz and F. Perrotin, *PLoS One*, 2017, **12**, e0169850.
20. L. J. Rich and M. Seshadri, *Radiology*, 2015, **275**, 110.
21. R. G. M. Kolkman, W. Steenbergen and T. G. van Leeuwen, *Laser Med. Sci.*, 2006, **21**, 134.
22. R. Fainchtein, B. J. Stoyanov, J. C. Murphy, D. A. Wilson and D. F. Hanley, *Proc. SPIE*, 2000, **3916**, 19.
23. H. F. Zhang, K. Maslov, M. Sivaramakrishnan, G. Stoica and L. V. Wang, *Appl. Phys. Lett.*, 2007, **90**, 053901.

24. R. K. Saha, S. Karmakar and M. Roy, *J. Opt.*, 2013, **42**, 141.
25. M. Petri, I. Stoffels, J. Jose, J. Leyh, A. Schulz, J. Dissemond, D. Schadendorf and J. Klode, *J. Wound Care*, 2016, **25**, 87.
26. K. S. Valluru and J. K. Willmann, *Ultrasonography*, 2016, **35**, 267.
27. S. N. Hennen, W. Xing, Y.-B. Shui, Y. Zhou, J. Kalishman, L. B. Andrews-Kaminsky, M. A. Kass, D. C. Beebe, K. I. Maslov and L. V. Wang, *Exp. Eye Res.*, 2015, **138**, 153.
28. X. Luis Dean-Ben and D. Razansky, *Light: Sci. Appl.*, 2014, **3**, e137.
29. X. Wang, Y. Pang, G. Ku, X. Xie, G. Stoica and L. V. Wang, *Nat. Biotechnol.*, 2003, **21**, 803.
30. H. F. Zhang, K. Maslov, G. Stoica and L. V. Wang, *Nat. Biotechnol.*, 2006, **24**, 848.
31. E. Herzog, A. Taruttis, N. Beziere, A. A. Lutich, D. Razansky and V. Ntziachristos, *Radiology*, 2012, **263**, 461.
32. L. Lin, J. Yao, L. Li and L. V. Wang, *J. Biomed. Opt.*, 2016, **21**, 061002.
33. S. Ashkenazi, S.-W. Huang, T. Horvath, Y.-E. L. Koo and R. Kopelman, *J. Biomed. Opt.*, 2008, **13**, 034023.
34. Q. Shao and S. Ashkenazi, *J. Biomed. Opt.*, 2015, **20**, 036004.
35. Q. Shao, M. A. Biel and S. Ashkenazi, *Proc. SPIE*, 2014, **8931**, 89310H.
36. Q. Shao, E. Morgounova and S. Ashkenazi, *Proc. SPIE*, 2014, **8943**, 89430F.
37. Q. Shao, E. Morgounova, C. Jiang, J. Choi, J. Bischof and S. Ashkenazi, *J. Biomed. Opt.*, 2013, **18**, 076019.
38. S. Ashkenazi, *J. Biomed. Opt.*, 2010, **15**, 040501.
39. Q. Shao, E. Morgounova and S. Ashkenazi, *Proc. SPIE*, 2015, **9323**, 932320.
40. E. Morgounova, Q. Shao, B. J. Hackel, D. D. Thomas and S. Ashkenazi, *J. Biomed. Opt.*, 2013, **18**, 056004.
41. A. Forbrich, P. Shao, W. Shi and J. Z. Roger, *J. Opt.*, 2016, **18**, 124001.
42. J. Jo, C. H. Lee, R. Kopelman and X. Wang, *Proc. SPIE*, 2016, **9708**, 97081L.
43. A. Ray, J. R. Rajian, Y.-E. K. Lee, X. Wang and R. Kopelman, *J. Biomed. Opt.*, 2012, **17**, 0570041.
44. Y.-E. Koo Lee, E. E. Ulbrich, G. Kim, H. Hah, C. Strollo, W. Fan, R. Gurjar, S. Koo and R. Kopelman, *Anal. Chem.*, 2010, **82**, 8446.
45. C. H. Lee, J. Folz, W. Zhang, J. Jo, J. W. Y. Tan, X. Wang and R. Kopelman, *Anal. Chem.*, 2017, **89**, 7943.
46. K. Sasaki, Z.-Y. Shi, R. Kopelman and H. Masuhara, *Chem. Lett.*, 1996, **25**, 141.
47. Y.-E. L. Koo, Y. Cao, R. Kopelman, S. M. Koo, M. Brasuel and M. A. Philbert, *Anal. Chem.*, 2004, **76**, 2498.
48. H. Xu, J. W. Aylott and R. Kopelman, *Analyst*, 2002, **127**, 1471.
49. J. Jo, C. H. Lee, R. Kopelman and X. Wang, *Nat. Commun.*, 2017, **8**, 471.
50. A. Jubran, *Crit. Care*, 2015, **19**, 272.
51. F. Jobsis, *Science*, 1977, **198**, 1264.
52. W. F. Cheong, S. A. Prahl and A. J. Welch, *IEEE J. Quantum Electron.*, 1990, **26**, 2166.
53. J. G. Rajendran and K. A. Krohn, *Radiol. Clin.*, 2005, **43**, 169.

54. M. Potter, L. Badder, Y. Hoade, I. G. Johnston and K. J. Morten, in *Oxygen Transport to Tissue XXXVII*, ed. C. E. Elwell, T. S. Leung and D. K. Harrison, Springer New York, New York, NY, 2016, pp. 257–263.
55. Y.-E. L. Koo, G. R. Reddy, M. Bhojani, R. Schneider, M. A. Philbert, A. Rehemtulla, B. D. Ross and R. Kopelman, *Adv. Drug Delivery Rev.*, 2006, **58**, 1556.
56. H. Clark, R. Kopelman and S. Barker, in *Biomedical Diagnostic Science*, ed. N. Akmal, W. T. Law and A. M. Usmani, CRC Press, Boca Raton, FL, 2002, ch. 7, pp. 139–164.
57. M. Brasuel, R. Kopelman, M. A. Philbert, J. W. Aylott, H. Clark, I. Kasman, M. King, E. Monson, J. Sumner, H. Xu, M. Hoyer, T. J. Miller and R. Tjalkens, in *Optical Biosensors: Present and Future*, ed. F. S. Ligler and C. A. Rowe Taitt, Elsevier Science, Amsterdam, 2002, ch. 16, pp. 497–536.
58. Y.-E. Koo Lee, E. Monson, M. Brasuel, M. A. Philbert and R. Kopelman, in *Biomedical Photonics Handbook*, ed. T. Vo-Dinh, CRC Press, Boca Raton, FL, 2nd edn, 2014, vol. III, ch. 22, pp. 555–574.
59. A. Gupta, S. Wang, P. Pera, K. V. R. Rao, N. Patel, T. Y. Ohulchanskyy, J. Missert, J. Morgan, Y.-E. Koo-Lee, R. Kopelman and R. K. Pandey, *Nanomedicine*, 2012, **8**, 941.
60. U. M. R. Avula, H. K. Yoon, C. H. Lee, K. Kaur, R. J. Ramirez, Y. Takemoto, S. R. Ennis, F. Morady, T. Herron, O. Berenfeld, R. Kopelman and J. Kalifa, *Sci. Transl. Med.*, 2015, **7**, 311ra172.
61. D. E. J. G. J. Dolmans, D. Fukumura and R. K. Jain, *Nat. Rev. Cancer*, 2003, **3**, 380–387.
62. H. K. Yoon, X. Lou, Y.-C. Chen, Y.-E. Koo Lee, E. Yoon and R. Kopelman, *Chem. Mater.*, 2014, **26**, 1592.
63. W. Tang, H. Xu, R. Kopelman and M. A. Philbert, *Photochem. Photobiol.*, 2005, **81**, 242.
64. M. Qin, H. J. Hah, G. Kim, G. Nie, Y.-E. K. Lee and R. Kopelman, *Photochem. Photobiol. Sci.*, 2011, **10**, 832.
65. T. J. Dougherty and C. J. Gomer, *Lasers Surg. Med.*, 2011, **43**, 541.
66. Z. P. Bernstein, T. Dougherty, S. Gollnick, S. A. Schwartz, S. D. Mahajan, J. Kepner, A. Sumlin, C. Stewart, P. Wallace, A. Adal, H. Walder and B. Poiesz, *Curr. HIV Res.*, 2008, **6**, 152.

CHAPTER 11

# *Imaging of Tissue Oxygen* Ex vivo

ALEXANDER V. ZHDANOV* AND DMITRI B. PAPKOVSKY

University College Cork, School of Biochemistry and Cell Biology, Cavanagh Pharmacy Building, College Road, Cork, Ireland
*E-mail: a.zhdanov@ucc.ie

## 11.1 Introduction

One of the central molecules in living systems, molecular oxygen ($O_2$) is a valuable functional biomarker and reporter for aerobic organisms. Analysis of mitochondrial, cellular and tissue oxygenation and $O_2$ consumption is used in cell biology and toxicology, as these parameters can report on the activity of the electron transport chain, numerous oxidases, oxygenases and their substrates. The levels of cell oxygenation can also report on the contribution of the different energy producing pathways (*e.g.* OxPhos, Krebs cycle), their resting and spare capacities, substrate utilisation, cell and context dependent preferences in energy source (*e.g.* glucose, glutamine, short chain fatty acids), metabolic rewiring and drug toxicity. Common cell cultures are widely used in such studies,[1,2] however, they are too simplistic and often irrelevant, because cells are usually maintained in an unnatural 2D micro-environment and at non-physiologically high levels of $O_2$ (~21 kPa in air or ~200 µM in the medium) and metabolites (20 mM glucose).

Changes in $O_2$ availability, due to altered balance between $O_2$ supply and consumption, sensed at the level of the whole organism and individual cells,

Detection Science Series No. 11
Quenched-phosphorescence Detection of Molecular Oxygen: Applications in Life Sciences
Edited by Dmitri B. Papkovsky and Ruslan I. Dmitriev

Published by the Royal Society of Chemistry, www.rsc.org

trigger adaptive and maladaptive responses, which can be missed or misinterpreted in the above simplistic models. To account for the complex roles of $O_2$ in the regulation of cell life, function and fate under various conditions and in different types of cells and tissues, more sophisticated approaches are needed.

To this end, current cell biology and biophysics tends to move away from analyses of average or 'bulk' oxygenation of cells and tissue samples towards more detailed imaging studies and mapping of $O_2$ concentration at a subcellular level. While the former task can be performed with relatively simple commercial instruments (TR-F fluorescence readers[3] or dedicated platforms[4,5]), the latter group of analytical tasks requires considerably more sophisticated optical set-ups. The use of laser-scanning fluorescence microscopy, particularly in phosphorescence lifetime imaging (PLIM) mode and in conjunction with dedicated cell-penetrating $O_2$ sensing probes, enables researchers to implement precise control and 2D/3D mapping of $O_2$ concentration in live tissue samples and parallel measurement of other functional markers.

From this prospective, 3D tissue models, such as multicellular spheroids, organoid structures, bioengineered tissue and scaffolds, can help to reproduce more closely the *in vivo* conditions in normal and diseased animal tissue and thus significantly improve the physiological relevance of the data and facilitate their mathematical modelling.[6–9] The recent studies with intestinal organoids as heterocellular mimetics of mammalian tissue have demonstrated their high applicability for $O_2$ imaging.[10] Still, these *in vitro* tissue models need to be developed to a level where they can replace the existing *in vivo* and *ex vivo* models. The long culturing time, high price of organoid cultures, limited number of tissues enriched with stem cells, lack of vascularisation and relatively small size are currently the main bottlenecks.

On the other hand, *ex vivo* models represented by tissue sections (*e.g.* brain slices), organs and even organ systems allow rapid and detailed functional analysis of live tissue by $O_2$ microscopy. The characteristic spectra and lifetimes of phosphorescent $O_2$ sensing probes are well suited for live multi-parametric *ex vivo* imaging of tissue samples, especially when combined with other fluorescence based metabolic reporters.[11–13] *Ex vivo* imaging has several advantages over *in vitro* and *in vivo* $O_2$ microscopy. (i) Simple, controllable delivery of the $O_2$ probe to targeted tissue and good staining efficiency ensure high quality and reproducibility of results. (ii) The excised tissue can be immobilised, thus reducing its movement and contraction and notably increasing the resolution of imaging. (iii) Unidirectional $O_2$ supply to the tissue (only from the surrounding medium) simplifies interpretation of $O_2$ imaging data and physical modelling of $O_2$ dynamics in the tissue. (iv) The tissue sample is disconnected from the whole organism, therefore direct cell and tissue specific effects of drugs and other treatments can be analysed.

Below we provide three representative examples of *ex vivo* $O_2$ imaging studies conducted with the following tissue models: (1) colon, including sections of rodent intestine and human biopsies, (2) whole mouse urinary bladders and (3) sections of porcine and mouse carotid artery.

# 11.2 Experimental

## 11.2.1 Materials

Cell-permeable $O_2$-sensitive probes, Pt-Glc[14] and IR-Glc[15] were kindly provided by Dr R. Dmitriev (University College Cork, Cork, Ireland). Tetramethyl rhodamine methyl ester (TMRM), bis-(1,3-diethylthiobarbituric acid) trimethine oxonol (DiSBAC$_2$(3)), Hoechst 33342 (HXT) and Calcein Green (calcein) were from Thermo Fisher Scientific (Waltham, MA, USA). ROSstar 550 was from LI-COR Biotechnology (Lincoln, NE). Stainless Steel Minuten Pins (0.2 mm) were from Fine Science Tools (Foster City, CA). Sylgard ® 184 silicon elastomer kit was from Dow Corning (Midland, MI). Antimycin A (AntA), carbonyl cyanide 4-(trifluoromethoxy) phenylhydrazone (FCCP), atropine, nifedipine and other reagents were from Sigma-Aldrich. Plasticware was from Sarstedt (Wexford, Republic of Ireland) and Corning Life Sciences (NY). Extracellular $O_2$-sensitive probe MitoXpress® Xtra was from Luxcel Biosciences (Cork, Ireland).

## 11.2.2 Animals and Tissue Staining

Experiments with animals were conducted under the licence from the Irish Government, in accordance with national and EU legislation (European Directive 2010/63/EU) and the protocols approved by the University College Cork Animal Experimentation Ethics Committee. C57BL/6OlaHsD male mice (Harlan, UK) aged 10–12 weeks were used for staining and imaging experiments with colon (induced colitis and NADPH oxidase deficiency models) and urinary bladder tissue, as described.[12,13] Mice deficient in NADPH oxidases in the intestine were generated as described previously.[16] Similar experiments were performed on fragments of colon of male Sprague-Dawley rats (Harlan) aged 3–4 months, and human colon biopsies. Fragments of porcine and mouse carotid arteries were kindly provided by Dr K. Martin and Prof. N. Caplice (CRVB, Cork, Ireland). Chronic vaso-occlusive disease was mimicked by ligating porcine carotid artery for 6 weeks as previously described.[17]

For *ex vivo* staining, tissues of euthanised animals were dissected, cleaned, stretched on Sylgard-coated 3.5 cm$^2$ Petri dishes using Minuten Pins and incubated with 10–20 μM Pt-Glc in 2 ml of Krebs buffer supplemented with 5 mM glucose and 2% BSA for 1–3 h at 37 °C in a humidified atmosphere containing 5% $CO_2$. Mouse carotid artery was excised, perfused with PBS, filled with 20 μM IR-Glc probe[15] in Krebs buffer (formulated as above) and incubated for 1 h at 37 °C.

For *in vivo* staining of bladder epithelium, C57BL/6 mice were anesthetised with isoflurane,[12] Pt-Glc probe (200 μl) was instilled into the bladder by means of a 24G Angiocath catheter (BD, Franklin Lakes, NJ). Mice were left on a heated pad for 3 h with inserted catheter and then sacrificed by cervical dislocation.

Co-staining with ROSstar 550 (free radical probe), TMRM (mitochondrial membrane potential indicator), calcein (cell viability probe), Hoechst 33342 (nuclear stain), or DiSBAC$_2$(3) (plasma membrane potential indicator) was performed as described.[12,13,16,18] After washing, fresh Krebs buffer supplemented with glucose and 2 mM L-glutamine was added to the tissue; nifedipine or atropine were used to prevent spontaneous tissue movements due to muscle contraction during the microscopy.[12,16]

### 11.2.3 Live Tissue Imaging by Confocal FLIM/PLIM Microscopy

Live confocal measurements by PLIM (for Pt-Glc and IR-Glc), FLIM and intensity imaging (for all the other probes) were performed on a custom-built system based on an upright fluorescent microscope AxioExaminer Z1 with 20×/1.0 W-Plan-Apochromat water dipping objective (Carl Zeiss), integrated *T* and *Z*-axis control, DCS-120 confocal scanner and time correlated single photon counting (TCSPC) module (Becker & Hickl), and R10467U-40 photon counting detector (Hamamatsu Photonics K.K.).[14]

Mini Petri dishes with tissue samples were equilibrated for 10 min on a heated stage at 37 °C. Pt-Glc was excited with a 405 nm picosecond diode laser BDL-SMC (B&H) and emission was collected at 635–675 nm in PLIM mode using SPCM software (B & H). IR-Glc was excited at 570 nm (supercontinuum laser SC400-4, Fianium, UK) and emission was collected at 710–750 nm. Imaging of the other probes was conducted using a 405 nm diode laser or Fianium laser with the excitation/emission at: 405 nm/420–460 nm (HXT), 488 nm/512–536 nm (Calcein Green) and 543 nm/565–605 nm (ROSstar 550, TMRM and $DiSBAC_2(3)$). For 3D reconstructions, confocal images were collected from 5–10 focal planes (5–10 μm steps). When required, the samples were treated with activators and inhibitors of respiration.

Phosphorescence lifetime (LT) images were produced by single-exponential fitting of the PLIM–TCSPC data in SPCImage software (B & H). 2D matrices of the LT data (256 × 256 pixels) for each measurement were extracted to Excel, correlated with the intensity images and converted into $O_2$ concentration using calibration function (**1**) for the colon: $[O_2]$ (μM) = $10 \times e^{(39.27 - LT)/5.08}$,[13] and function (**2**) for the urinary bladder: $O_2$ (μM) = $10 \times e^{(40.172 - LT)/4.789}$.[12] 3D $O_2$ distribution maps were constructed using ImageJ and processed in Adobe Photoshop and Illustrator software.

### 11.2.4 Analysis of Oxygen Consumption Rate (OCR) in Colonic Mucosa Samples

OCR analysis was performed as described[1] with modifications.[13] First, muscular and serosal layers were gently stripped off freshly dissected fragments of mouse distal colon. Then, using a mini-puncher, regular size slices of epithelial tissue were prepared (2.5 mm in diameter, 0.135 ± 0.007 mg of protein per slice) and dispensed into wells of a 96-well plate (two slices per well, in triplicate). Wells were then filled with 120 μl of respiration medium containing MitoXpress®-Xtra probe (200 nM), sealed with mineral oil and monitored on a Victor 2 reader (PerkinElmer Life Sciences). OCRs were calculated as nmoles of $O_2$ consumed by 1 mg of tissue per minute.

### 11.2.5 Statistical Analysis and Data Presentation

SPSS and Excel software programs were used for statistical analysis of the results. $p$ values of $\leq 0.01$ calculated for 3–5 independent experiments were deemed as statistically significant. The differences in mean $O_2$ values between groups of animals and tissue samples, with and without metabolic treatments, were evaluated using Mann–Whitney U-test and dependent t-test. The associations between Pt-Glc intensity and LT values, TMRM intensity and $O_2$ concentrations were evaluated using nonparametric Spearman's rank correlation analysis. Co-localisation of low $O_2$ and high TMRM intensity areas was analysed after $10 \times 10$ binning of the corresponding values. Confocal $O_2$ images show LT of Pt-Glc as rainbow false colours.

In the DSS-colitis study, frequency histograms describing $O_2$ distribution in colonic mucosa were obtained for three individual focal planes within each field of view ($256 \times 256$ pixel matrices). For each focal plane in each tissue sample, data from three fields of view were averaged, thus giving seven data points per plane per group. $O_2$ values corresponding to the 25th, 50th and 75th percentiles on the cumulative frequency histograms were determined for both groups and presented as mean ± SEM. The differences between these $O_2$ values were analysed using Mann–Whitney U-test. Also, pooled average $O_2$ levels (7 + 7 animals) were calculated for each percentiles and each focal plane (nine values altogether). Then $O_2$ levels corresponding to 25th, 50th and 75th percentiles for each individual sample were scored as 'high $O_2$' or 'low $O_2$' (> or < the pooled average $O_2$ levels), and the scores were compared between groups with Fisher's exact test.

The differences between OCR values were analysed using one-sample Wilcoxon signed rank test. For this, OCRs in DSS samples were normalised to control values in each sample pair.

## 11.3 Results and Discussion

### 11.3.1 Oxygen Imaging in the Colon Tissue

Being exposed to almost anoxic lumen, normal colonic epithelium is adopted to low $O_2$ levels (down to 10 μM [19]). Fast proliferation and high physiological activity of colonocytes require large amounts of ATP. Substantial OxPhos flux is necessary for the maintenance of integrity of colonic epithelium,[20] therefore pathological processes that limit $O_2$ and nutrient supply to the tissue (*e.g.* inflammation) usually affect viability of colonocytes and functioning of the colon. On the other hand, intestinal inflammation, such as in ulcerative colitis, can trigger adaptive metabolic processes which counteract shortages in $O_2$ (*e.g.* *via* hypoxia inducible factor HIF-1 and other pathways). Similarly, deficiency in particular enzymes utilising $O_2$ (such as NOX) may also affect $O_2$ homeostasis in the tissue. We hypothesised that metabolic shifts associated with such conditions can be traced by monitoring changes in oxygenation of colonic mucosa in excised fragments of colonic tissue using *ex vivo* $O_2$ imaging techniques. When blood circulation through tissue is stopped, $O_2$ can be supplied to the epithelium only by diffusion from the atmosphere through the layer of tissue-supporting medium. This supply of $O_2$ is balanced by its consumption by the tissue, thus producing a steady state in which cell/tissue oxygenation reflects mostly the OCR.

#### *11.3.1.1 Staining of Colonic Tissue with $O_2$ Sensing Probes*

The main technical issue for staining the intestinal tissue with probes (both *in vivo* and *ex vivo*) is the presence of protective mucus. In mouse colon, two layers of mucus (inner, 50 µm thick, and outer, 100 µm thick) effectively shield the epithelium from the probe.[21] In our initial experiments, staining of the tissue with nanoparticle probes NanO2, MM2 and PA2 [11,22,23] was completely prevented by the mucus. However, when we tested a small molecule probe Pt-Glc, quite appropriate *ex vivo* staining of mouse colon was obtained. This probe worked even better in staining rat and human colon tissue samples. However, we failed to stain colonic tissue *in vivo*: delivered through a catheter Pt-Glc could not stain the colonocytes, even when mice were fasted overnight or their colon was washed by instillation with physiological solution. Therefore, in further experiments we used *ex vivo* staining of colon tissue with 12.5–20 µM Pt-Glc for 70–180 min,[13] which provided us strong intensity signals and stable phosphorescence LT during long imaging sessions (3 h). These staining conditions were not cytotoxic and allowed flexibility in the experiments with other disease models and metabolic probes.[13,16]

#### *11.3.1.2 Imaging of $O_2$ in Colonic Tissue*

Incubation of colon tissue samples with Pt-Glc probe resulted in efficient staining of mostly colonocytes (Figure 11.1A–C). The phosphorescence intensity signal decreased progressively with distance from the tissue surface, and was reliably measurable up to depths of approximately 50 µm. This loss of signal was due to reduced number of crypts with tissue depth, decreased staining efficiency in the inner cryptal regions, attenuation of excitation and emitted light by the tissue.

The surface of colonic mucosa formed by characteristic finger-like invaginations (crypts) is highly irregular. Therefore, we had to introduce a conventional coordinate system, in order to quantify in-depth distribution of the probe signal and properly compare tissue oxygenation in groups of healthy and diseased animals. Starting the imaging from the lumen, we selected the uppermost optical sections with >80% of the maximal phosphorescent signal as zero depth (*Z*-coordinate, 0 µm) for all ROI analysed. For LT

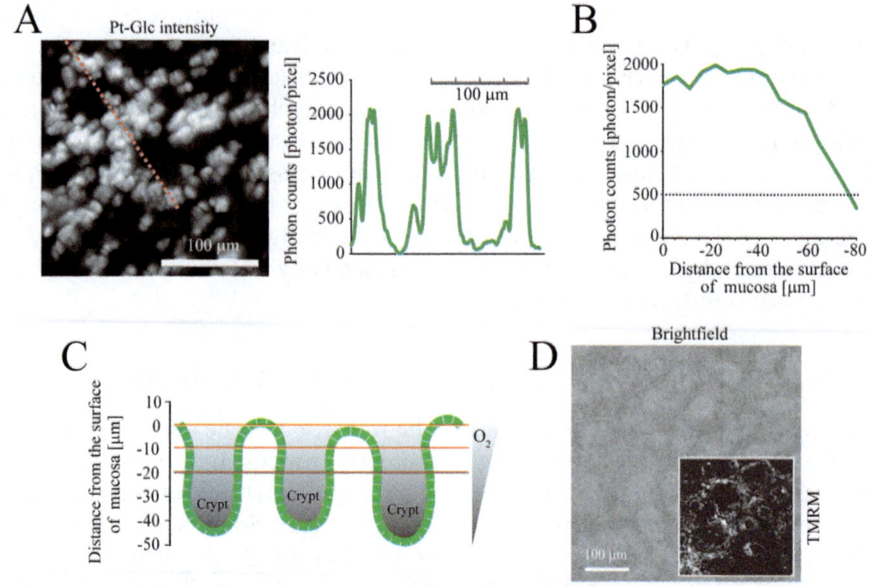


**Figure 11.1** *Ex vivo* staining of mouse colon epithelium with Pt-Glc probe. A. 3D reconstruction of signal intensity in the tissue incubated with 20 μM probe for 3 h. A stack of five focal planes taken with 10 μm steps (left panel) and intensity profile along the red dashed line (right panel) show localisation of Pt-Glc in crypt walls. B. Depth dependence of the probe intensity signals averaged for all epithelial cells in the tissue within a representative field of view. The dashed line shows the selected threshold of the signal. C. A scheme of mucosal architecture; lines show focal planes taken for the analysis of $O_2$ levels; gradients show the expected decrease in tissue $O_2$ with a distance from the surface of the mucosa. D. Bright field image of colonic mucosa. In the inset, TMRM staining shows polarised mitochondria. Reprinted from *Cellular and Molecular Life Sciences*, Quantitative analysis of mucosal oxygenation using *ex vivo* imaging of healthy and inflamed mammalian colon tissue, **74**, 2016, 141–151,[13] A. V. Zhdanov, I. A. Okkelman, A. V. Golubeva, B. Doerr, N. P. Hyland, S. Melgar, F. Shanahan, J. F. Cryan and D. B. Papkovsky, Copyright © 2016, Springer International Publishing, with permission of Springer.

calculations, we used 3 focal planes at 0, 10 and 20 μm below tissue surface. To increase the accuracy of $O_2$ determination, phosphorescence LT was calculated only for the regions having >10% of the maximal intensity signal. Approximately 90% of all LT values ranged 34–44 μs (Figure 11.2A and B) which corresponded to 35–5 μM $O_2$ concentration. $O_2$ values in the deeper mucosal layers tended to decrease (Figure 11.2B and C). Bright TMRM staining (mitochondrial membrane potential probe) suggested high viability and respiratory activity of the colonocytes (Figure 11.1D). This was confirmed by lower LT values of Pt-Glc upon tissue treatment with complex III inhibitor antimycin A (not shown).

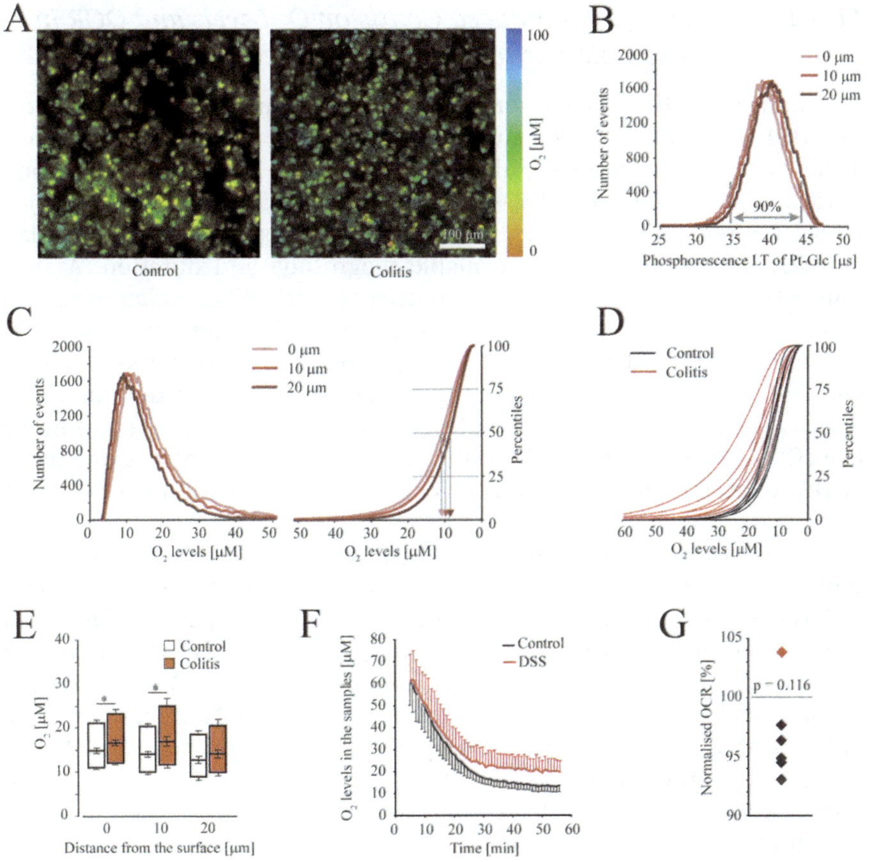


**Figure 11.2** Effect of DSS-induced colitis on oxygenation of excised mouse colon tissue. A. $O_2$ levels in colonic mucosa of mice with DSS-colitis and control mice, analysed by *ex vivo* imaging. Stacks of five focal planes (5 μm step) are shown in rainbow false colours. B. Representative frequency histogram demonstrates the distribution of phosphorescence LT values in the tissue (three focal planes). The range 34–45 μs covers >90% of LT values. C. $O_2$ frequency histogram (left) and cumulative $O_2$ frequency histogram (right), calculated from (B) according to the function (**1**); $O_2$ levels decrease in deeper tissue areas. D. Cumulative frequency histograms of $O_2$ concentration at 10 μm distance from the surface of the mucosa in DSS-colitis and control animals. E. $O_2$ values at 0, 10 and 20 μm depth, shown at 25th, 50th and 75th percentiles. F. Representative $O_2$ consumption profiles and (G) the difference in OCR values between control and DSS groups (one-sample Wilcoxon signed rank test). J. Asterisks indicate significant differences between $O_2$ levels for all percentiles ($p < 0.05$, Mann–Whitney U-test). Reprinted from *Cellular and Molecular Life Sciences*, Quantitative analysis of mucosal oxygenation using *ex vivo* imaging of healthy and inflamed mammalian colon tissue, **74**, 2016, 141–151,[13] A. V. Zhdanov, I. A. Okkelman, A. V. Golubeva, B. Doerr, N. P. Hyland, S. Melgar, F. Shanahan, J. F. Cryan and D. B. Papkovsky, Copyright © 2016, Springer International Publishing, with permission of Springer.

#### 11.3.1.3 *Effect of DSS-induced Colitis on $O_2$ Levels and OCR in Mouse Colonocytes*

DSS-induced colitis is a common model for human inflammatory bowel disease (IBD). According to the current classification, this chronic pathology includes two classical forms, ulcerative colitis and Crohn's disease, and a third undetermined form (IBD-U).[24] Colitis primarily affects colon mucosal layer, often with severe symptoms. DSS administration rapidly induces mucosal inflammation of controllable magnitude and duration. A short course of DSS treatment (4 days) induces an early-stage inflammation in mouse colon, with only moderate effects on the function and viability of the epithelium.[25] Indeed, we observed a profound increase in tissue IL6 and mKC mRNA levels, colon shortening, a slight reduction of the number of Goblet cells (not shown), but no effect on gross architecture of the mucosa. The efficiency of tissue staining with Pt-Glc (20 μM for 3 h) was unaffected by DSS treatment, and this allowed for non-biased comparison of $O_2$ levels in the two groups.

Oxygenation of the colonocytes in DSS-treated mice was higher than in control (Figure 11.2A). To accurately determine the differences, we compared $O_2$ levels in healthy and inflamed mucosa calculated separately for optical sections at 0, 10 and 20 μm depths according to equation (**1**). We analysed $O_2$ distribution histograms, which show the number of pixels (*Y*-axis) with particular $O_2$ values (*X*-axis) in the field of view (Figure 11.2B). These numbers were transformed into percents and plotted as cumulative $O_2$ distribution histograms (Figure 11.2C and D). Comparing the $O_2$ values at 25th, 50th and 75th percentiles, we found that colonic mucosa in mice with DSS-colitis was significantly more oxygenated than in control animals ($p < 0.05$ for the 0 and 10 μm depth), with a pronounced trend ($p = 0.061$) at 20 μm depth (Figure 11.2E). The observed increases in $O_2$ levels suggested that respiration of colonocytes in inflamed tissue was reduced.

Indeed, when we measured respiration of freshly dissected colonic epithelium samples from the two groups on a plate reader using MitoXpress® Xtra probe, we observed a strong trend towards reduced OCR in the tissue from DSS-treated mice (Figure 11.2F and G). In agreement with this data and with the overall concept of HIF-dependent down-regulation of oxidative metabolism in inflamed tissue,[26] our Western blotting and qPCR data also demonstrated activation of transcription of HIF-1 target genes.[13]

#### 11.3.1.4 *Effect of NOX Deficiency on $O_2$ Levels in Mouse Colonic Epithelium*

Intestinal epithelium plays a vital role in shaping the composition of commensal microbiota, continuous 'education' of the immune system and protection of the host from bacterial invasion. One of the key antimicrobial defense systems is the production of free radicals by NADPH oxidases (NOX, DUOX) of immune cells and mucosal epithelium. NOX malfunction in the

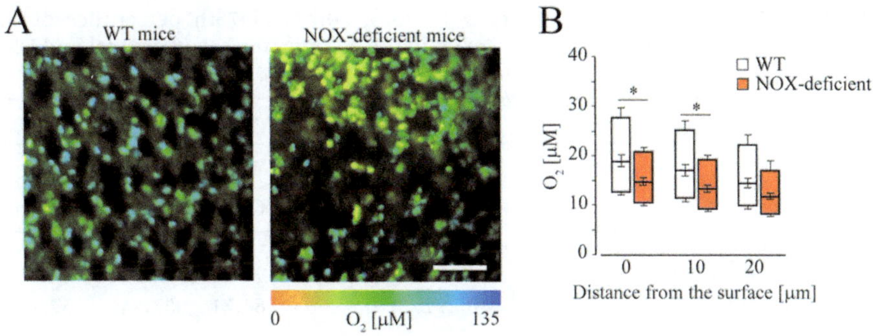


**Figure 11.3** Effect of NOX deficiency on oxygenation of mouse colonic epithelium. A. Confocal PLIM images demonstrate $O_2$ levels in the mucosa of WT and NOX-deficient mice. Stacks of three focal planes taken with 10 μm step are shown in rainbow false colours. B. $O_2$ values at 0, 10 and 20 μm depth, shown as medians and the interquartile range. Error bars represent SEM. Scale bar is 100 μm. Asterisks indicate significant difference ($p < 0.05$, Mann–Whitney U-test).

epithelium can therefore increase susceptibility to infections and cause dysregulation of intestinal homeostasis, including oxygenation.

In this study we modified the above staining protocol in order to co-stain the colon tissue with Pt-Glc and free radical sensitive probe ROSstar 550: both probes were applied to the tissue at 15 μM for 70 min. As expected, ROSstar 550 fluorescence signals were significantly lower in the tissue of NOX-deficient animals.[16,18] The deficit of free radicals in the epithelium was coupled with an enriched content of $H_2O_2$-producing bacterial strains (particularly lactobacilli) in the intestine.

Although Pt-Glc concentration and staining time were reduced compared to the colitis study, phosphorescence intensity signals were sufficient for accurate determination of LT and $O_2$ concentration in the mucosa ($>5 \times 10^2$ photons per pixel). Using the above algorithm (partitioning the results into three groups with 0, 10 and 20 μm depth followed by plotting a series of cumulative histograms), we found that median $O_2$ values in the colonic mucosa of NOX deficient animals were substantially lower than in the control group ($p < 0.05$, Mann–Whitney U-test, Figure 11.3A and B). Frequency distribution analysis also showed a significantly higher abundance of 'low $O_2$' medians in the samples from NOX-deficient mice at 0 and 10 μm depth ($p < 0.05$, Fisher's exact test), with a similar trend at 20 μm depth (Table 11.1).

### *11.3.1.5 $O_2$ Imaging in Rat and Human Colonic Epithelium*

In order to move this experimental model closer to clinical applications such as diagnostics of GI disorders (*e.g.* intestinal ischemia, diabetes, stress-induced impairment of gut barrier function, IBD), we also performed pilot $O_2$ imaging experiments with rat distal colon and human colon biopsies.

**Table 11.1** $O_2$ values corresponding to 25th, 50th and 75th percentiles of the frequency distribution of $O_2$ in the colonic mucosa of control and DSS-treated mice (Fisher's exact test).[a]

| | | Distance from the surface (μm) | | | | | |
|---|---|---|---|---|---|---|---|
| | | 0 | | 10 | | 20 | |
| | Parameters | Control | Colitis | Control | Colitis | Control | Colitis |
| 25th | Average pooled $O_2$ values (μM) | 22.8 | | 22.7 | | 19.3 | |
| | 'High $O_2$' frequency (%) | 14.29 | 85.71 | 0 | 85.71 | 57.14 | 57.14 |
| | *p*-value | 0.03* | | 0.005* | | 1 | |
| 50th | Average pooled $O_2$ values (μM) | 15.8 | | 15.6 | | 13.5 | |
| | 'High $O_2$' frequency (%) | 14.29 | 71.43 | 0 | 85.71 | 28.57 | 57.14 |
| | *p*-value | 0.1 | | 0.005* | | 0.59 | |
| 75th | Average pooled $O_2$ values (μM) | 11.3 | | 11 | | 9.5 | |
| | 'High $O_2$' frequency (%) | 28.57 | 71.43 | 14.29 | 71.43 | 57.14 | 14.29 |
| | *p*-value | 0.29 | | 0.1 | | 0.27 | |

[a]Asterisks indicate significant difference ($p < 0.05$).

The initial results with rat colon tissue suggested that the quality of $O_2$ imaging was significantly better than with mouse tissue. Pt-Glc signals were bright and evenly distributed in the mucosal epithelium. Image quality allowed accurate reconstruction of mucosal architecture, clearly visualising the crypts (Figure 11.4A–C) and showing decreased oxygenation in deep tissue layers, with approximately 90% of $O_2$ values ranging 7–55 μM (not shown). Interestingly, the serosal surface of rat colon samples could also be efficiently stained with Pt-Glc probe (Figure 11.4D); this opens a possibility of probing $O_2$ gradients between the mucosal and serosal sides of the colon.[27]

The results of $O_2$ imaging in human colon biopsies, although promising and showing $O_2$ distribution patterns similar to those seen in rodents (Figure 11.4F and G), point to the need for the development of standard tissue sampling and staining protocols, which so far limit the possibilities of such analyses.

### *11.3.1.6 Main Findings and Conclusions*

Several independent studies performed so far[13,16,28] have demonstrated the capabilities of *ex vivo* $O_2$ imaging in colonic mucosa by confocal PLIM and detection of minor changes in tissue oxygen homeostasis associated with DSS-induced colitis, NOX deficiency or metabolic therapy. Quantitative, high-resolution $O_2$ imaging is a valuable instrument for studies of mechanisms of pathogenesis in gastrointestinal diseases. Development of dedicated imaging hardware and cell-penetrating probes such as Pt-Glc, helps to quantify

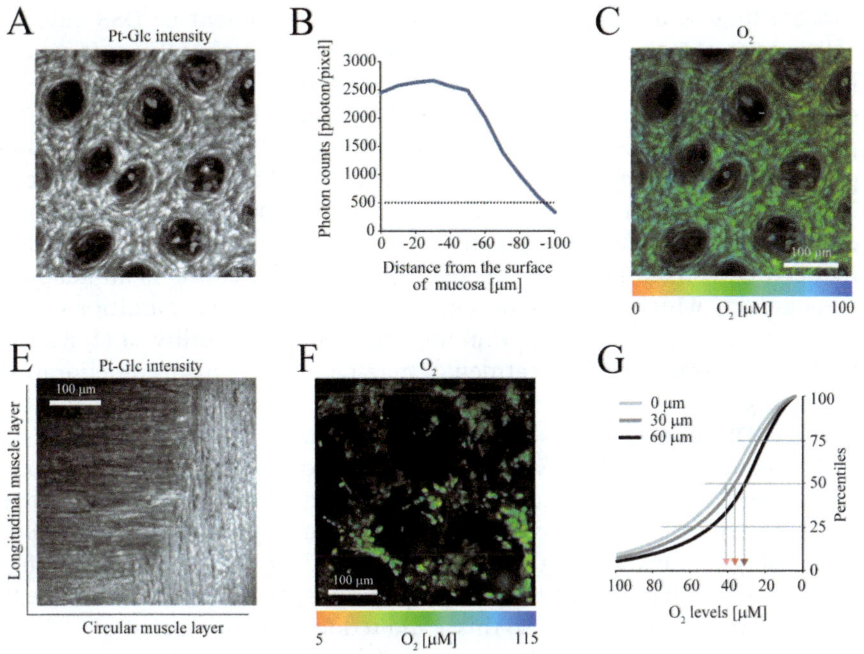


**Figure 11.4** *Ex vivo* analysis of $O_2$ levels in rat (A–E) and human (F, G) colon tissue. A. 3D reconstruction of Pt-Glc signal intensity. B. Depth dependence of Pt-Glc intensity signals averaged for the entire mucosa within a representative field of view. The dashed line shows the selected signal threshold. C. $O_2$ map of rat colonic mucosa. E. Pt-Glc intensity signal on the surface of colonic serosa; a single confocal plane. Tissue was stained with 20 μM probe for 2 h. In (A, C) stacks of seven focal planes taken with 10 μm step are shown. F. 3D reconstruction of $O_2$ levels in colonic epithelium in a healthy biopsy stained with 20 μM of Pt-Glc probe for 3 h; stack of 5 focal planes with 10 μm step. G. Typical cumulative frequency histograms of $O_2$ levels in healthy colonic biopsy. A depth-associated decrease in tissue oxygenation is marked by arrows. Reprinted from *Cellular and Molecular Life Sciences*, Quantitative analysis of mucosal oxygenation using *ex vivo* imaging of healthy and inflamed mammalian colon tissue, **74**, 2016, 141–151,[13] A. V. Zhdanov, I. A. Okkelman, A. V. Golubeva, B. Doerr, N. P. Hyland, S. Melgar, F. Shanahan, J. F. Cryan and D. B. Papkovsky, Copyright © 2016, Springer International Publishing, with permission of Springer.

$O_2$ levels in cells and even at subcellular level (see also Section 11.3 of this chapter). Currently, spatial resolution of this method is better than for microscopy with intravascular $O_2$ probes.[29] Deep deoxygenation of inflamed intestinal tissue has been already reported, however changes in respiratory activity of colonic mucosa in colitis have not been studied. The minute reduction in cell respiration could be overlooked in *in vivo* measurements (*e.g.* if masked by decreased $O_2$ supply through blood circulation or other factors[30,31]), but were reliably detectable by the current *ex vivo* imaging technique.

According to our proposed mechanism, development of DSS-induced colitis causes deep deoxygenation of colonocytes, which, acting *via* canonical HIF pathway, triggers the reconciling decrease in cellular respiration, seen *ex vivo* as an increase in epithelial $O_2$ levels. It is worth noting that plate reader-based oxygen consumption rate (OCR) analysis of small samples of mucosal tissue was unable to clearly reveal the difference between DSS-colitis and control groups. This shows the superiority of more reproducible and accurate data provided by PLIM technique. We think that reduced $O_2$ consumption by mucosal epithelium in colitis is an adaptive mechanism, which aids tissue survival under hypoxic conditions.[31] In agreement, reduced consumption and increased availability of $O_2$ caused by dimethyloxalylglycine treatment[28] may be a mechanism, ameliorating menacing effects of colon inflammation and aids the repair of injured tissue in DSS-colitis.[32]

The reduced $O_2$ levels in NOX-deficient tissue observed *ex vivo* still require explanation. Seeing no difference in the intestinal HIF-dependent signalling between the two groups (not shown), we can suggest that a burst in $O_2$ consuming lactobacilli in the mucus surrounding the colonocytes[16] may decrease $O_2$ availability. For *ex vivo* preparations, $O_2$ is delivered to the epithelium through the mucus, so this could reduce tissue oxygenation.

Fully cognisant that $O_2$ values measured *ex vivo* do not adequately reflect oxygenation of the mucosa *in vivo*,[30] we still emphasise that *ex vivo* $O_2$ imaging can help more easily discriminate between normal and abnormal colonic mucosa, and that it has potential for studying various pathological conditions in the intestine.

### 11.3.2 *Ex vivo* $O_2$ Imaging in the Urinary Bladder

Similar to the intestine, epithelial tissue of the urinary bladder is metabolically active in order to provide with ATP various energy demanding processes supporting tissue integrity, proliferation and ion transport.[33] In the urothelium, flat giant (up to 200 μm in diameter) umbrella cells interconnected with tight junctions form a low permeability shield which protects the inner tissue from contact with urine.[34] Large size, superficial location on the surface of the epithelium, extensive contact area with the liquid phase (urine or medium) and high energy demand make umbrella cells a good model for studying intracellular $O_2$ gradients and their role in cell physiology. So far, tissue oxygenation in the urinary bladder *in vivo* has not been studied. The data on $O_2$ levels in urine from human bladder is controversial (30–100 μM) and dependent on many factors such as lung ventilation, exercise and renal blood circulation.[35,36] This variability makes it unclear whether *in vivo* $O_2$ diffuses to bladder urothelium from urine or *vice versa*. In the *ex vivo* setup, $O_2$ is supplied to umbrella cells by simple diffusion from the surrounding medium, which helps data interpretation.

#### *11.3.2.1* Ex vivo *and* in vivo *Staining of the Urinary Bladder Epithelium with Pt-Glc and Other Probes*

The protocol for tissue staining with Pt-Glc in this study was similar to the colon. The dissected bladder was stretched on Sylgard with the urothelial layer facing up and was incubated with 20 µM Pt-Glc for 3 h. This resulted in a very strong staining of the uppermost epithelial cells on the surface of rugae, with almost no signal detectable in the underlying epithelium (Figures 11.5 and 11.6). Most of the stained cells were identified as giant umbrella cells by their specific hexagonal shape, polynuclearity and size (>100 µm in diameter, 15–25 µm thick).

Furthermore, we also managed to achieve similar efficacy with *in vivo* staining, when Pt-Glc was delivered to the bladder of anesthetised mice *via* a catheter. This method is advantageous, as tissue viability was preserved better. For both staining procedures, a fraction of umbrella cells tended to shed, as a result of manipulations with the tissue (Figure 11.5). On the other hand, in most of the tissue preparations, co-staining with fluorescent markers of the plasma membrane potential (DiSBAC$_2$(3)), mitochondrial membrane potential (TMRM), cell viability (Calcein Green) and nuclei (Hoechst 33342) demonstrated well-preserved integrity of the urothelium (Figure 11.6). This suggested that incomplete staining with Pt-Glc of a proportion of umbrella cells was due to the differences in their differentiation stage, which affected

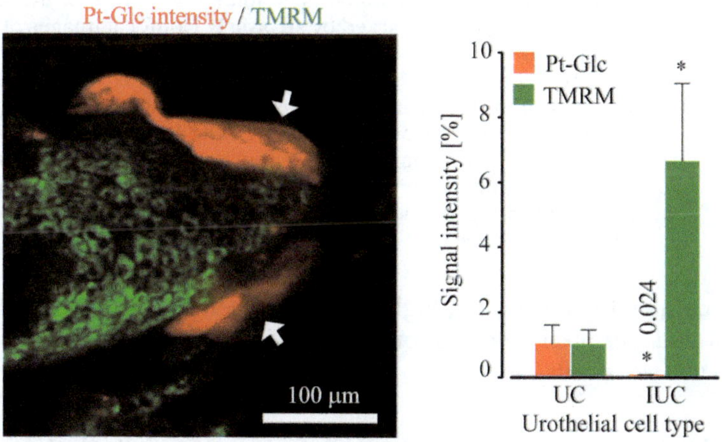


**Figure 11.5** Staining of giant umbrella cells with Pt-Glc probe: specificity and intensity. Co-staining with TMRM dye shows that Pt-Glc penetrates preferentially into the uppermost layer of the urothelium (umbrella cells, UC). In contrast, TMRM staining is noticeably brighter in the underlying intermediate urothelial cells (IUC). Arrows show umbrella cells, which are about to shed from the surface of the tissue. Asterisks indicate significant difference between the two cell types ($p < 0.01$, t-test).

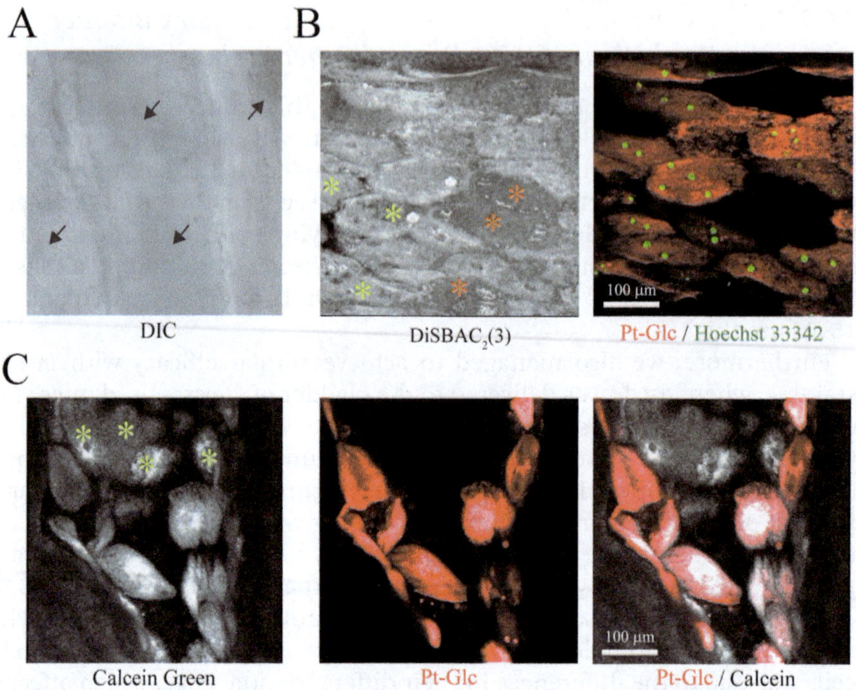


**Figure 11.6** Staining of umbrella cells with Pt-Glc probe: multiplexing and cell viability. A. DIC image demonstrates 'folded' structure of urinary bladder tissue, rugae are marked by arrows. B. Confocal images of bladder tissue stained *ex vivo* with Pt-Glc (20 μM for 3 h) and Hoechst 33342/ DiSBAC$_2$(3) (1 μM for 30 min) show that Pt-Glc (red) penetrates in the majority of umbrella cells with measurable plasma membrane potential (viable cells). C. Co-staining with Pt-Glc and calcein (a cell viability marker) gives similar results. Yellow asterisks show live umbrella cells not stained with Pt-Glc, red asterisks – presumably dead cells. Reprinted with permission from ref. 12. Copyright © 2015 the American Physiological Society.

Pt-Glc uptake. In all samples, high phosphorescence intensity signals (>1 × $10^3$ photons per pixel) allowed for accurate determination of $O_2$ concentration in umbrella cells.

### 11.3.2.2 Ex vivo *Imaging of Intracellular $O_2$ Gradients in Umbrella Cells*

LT values of Pt-Glc probe (30–46 μs), measured *ex vivo* in the layer of umbrella cells, corresponded to 85–5 μM $O_2$ (according to the function 2). Data analysis revealed relatively high (>40 μM) and uniform $O_2$ levels in approximately 40% of umbrella cells (Figure 11.7A), most likely due to their low respiration or delocalised mitochondrial network (Figure 11.7D). However, in the majority of cells, $O_2$ levels were seen to decrease significantly in the central areas

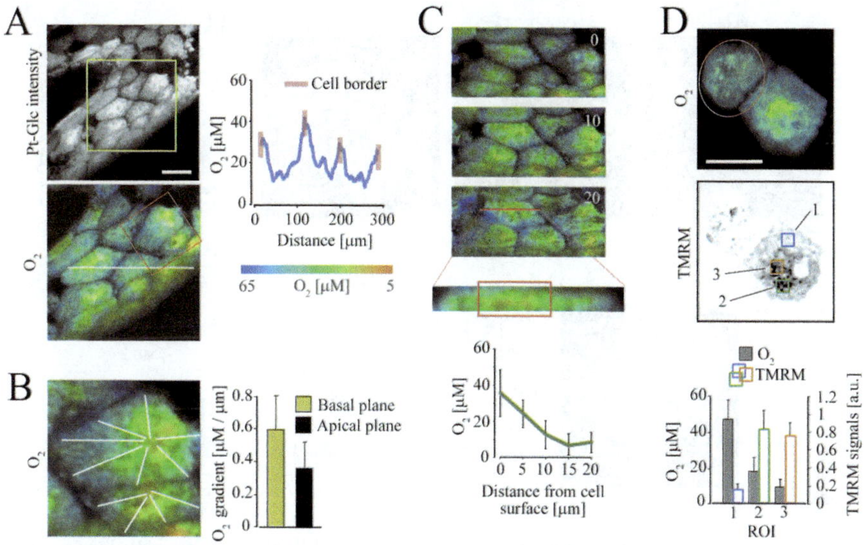


**Figure 11.7** $O_2$ gradients in the giant umbrella cells. A. Cell layer selected for the analysis. Pt-Glc intensity (top left panel) and calculated LT/$O_2$ concentration distribution in the ROI, highlighted by white quadrangle. Right panel demonstrates line profile analysis of $O_2$ gradients within cell layer (single focal plane). B. Quantification of $O_2$ gradients in individual cells (example ROI is shown by red quadrangle in (A)). Bars indicate average $O_2$ gradients along the lines connecting cell periphery and the most hypoxic interior regions within apical and basal parts of umbrella cells. C. 3D reconstruction and quantitative analysis of $O_2$ gradients along the apical–basal axis of a single cell (shown by red line). Five single plane $O_2$ concentration images, taken with a 5 μm step, were analysed (three planes are shown). D. $O_2$ distribution and mitochondrial localisation (TMRM probe) in the same cell. 3D images are reconstructed from three confocal planes (5 μm step). The red circle outlines the cell with low mitochondrial content and no $O_2$ gradients. Bottom panel presents $O_2$ levels and relative fluorescence intensities of TMRM in 3 ROI (highlighted with multicolour rectangles 1–3), demonstrating depletion of $O_2$ in the area enriched with polarised mitochondria. Error bars show SEM. Reprinted with permission from ref. 12. Copyright © 2015 the American Physiological Society.

(10–30 μM on average, with 40–60 μM $O_2$ on the periphery). Using line profile analysis of 30 large umbrella cells with clear $O_2$ heterogeneity, we found that average radial $O_2$ gradients are 0.6 ± 0.3 μM/μm (or 40 μM across the cell), reaching a maximum of 1.5 μM/μm due to non-linearity (Figure 11.7B). Importantly, both $O_2$ levels and local gradients were sustained during 1–3 h microscopy sessions. Approximately 5–10% of all analysed cells also tended to form $O_2$ gradients along their apical–basal axis (20 μm in average, Figure 11.7C), but due to the high inter-experimental variability, $O_2$ gradients between the apical area and more hypoxic core and basal areas could not be regarded as significant.

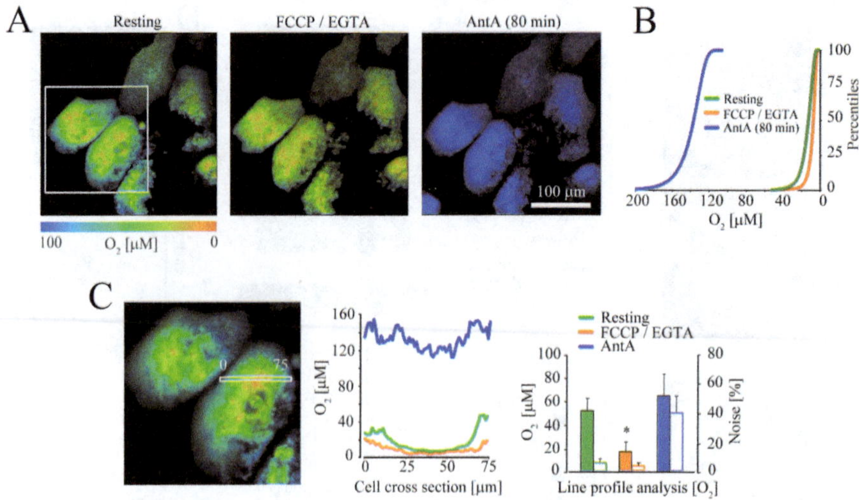


**Figure 11.8** Effects of mitochondrial activation and inhibition on cellular $O_2$ gradients. A. Changes in $O_2$ induced by double treatment of the tissue with FCCP (1 µM) and EGTA (2.5 mM) for 20 min, followed by AntA treatment (10 µM, 80 min). All images show stacks of five confocal planes (5 µm step). Quadrangle indicates ROI selected for single cell analyses illustrated in (B) and (C). B. Changes in cumulative $O_2$ frequency histogram upon activation (FCCP/EGTA) and inhibition (AntA) of mitochondrial respiration C. Single plane $O_2$ line profile analysis; $O_2$ values are averaged for six lines along the 75 µm axis of the rectangle, shown in white. Right panel: $O_2$ variability inside resting and stimulated cells (two cells shown in the left panel). Filled bars illustrate average $O_2$ range across the cell. Open bars demonstrate average noise, calculated for each pixel as: Noise (%) = (|$O_2$ measured − $O_2$ estimated|)/($O_2$ range) × 100. Estimated $O_2$ values were obtained from corresponding trendline equations; $O_2$ range was calculated as $({}_{max}O_2 - {}_{min}O_2)$ for each line profile. Asterisks indicate significant difference between treated and resting cells. Error bars show SEM. Reprinted with permission from ref. 12. Copyright © 2015 the American Physiological Society.

Superimposed oxygenation and TMRM maps revealed a pronounced correlation between low $O_2$ and high TMRM signals, suggesting $O_2$ depletion in the areas enriched with polarised mitochondria ($\rho = -0.65 \pm 0.13$, $p < 0.05$, Spearman's test) (Figure 11.7D). This agreed with the role of mitochondria as the main $O_2$ gradient-forming factor in the various models.[37–39]

This result was further confirmed by the changes in oxygenation and $O_2$ gradients in umbrella cells upon stimulation with activators and inhibitors of mitochondrial respiration. Monitoring actively respiring cells with pronounced $O_2$ gradients, we observed a decrease in $O_2$ levels upon double treatment with FCCP (a mitochondrial uncoupler) and EGTA (an activator of mitochondrial $Na^+/Ca^{2+}$ exchange[40]), shown as a shift of cumulative frequency histogram (Figure 11.8A and B). Radial $O_2$ gradient across the cell also decreased from >40 µM to ~20 µM (Figure 11.8C, left and middle

**Table 11.2** Oxygen levels in umbrella cells at rest and upon modulation of mitochondrial activity.

| | Resting | FCCP/EGTA | Bafilomycin A1 | Ant A |
|---|---|---|---|---|
| Percentile | $O_2$ levels (μM) | | | |
| 25th | 8.4 | 6.1 | 7.2 | 124 |
| 50th | 12.2 | 8.0 | 9.7 | 135 |
| 75th | 16.3 | 10.8 | 12.3 | 144 |

panels, Table 11.2). A similar reduction in oxygenation and $O_2$ gradients was observed upon treatment with bafilomycin A1, which also uncouples mitochondria (Table 11.2).[41]

Treatment with antimycin A (AntA) caused a large increase in tissue oxygenation over >80 min (Figure 11.8A–C). The response was delayed most likely due to slow penetration of AntA into the tissue (in monolayer 2D cell cultures AntA rapidly inhibits mitochondria). The increase in $O_2$ was coupled with the dissipation of $O_2$ gradients. At high $O_2$ concentrations small fluctuations in probe LT are seen as substantial changes on the $O_2$ scale due to their hyperbolic dependence.[42] Indeed, at $O_2$ >100 μM high noise interfered with the analysis (Figure 11.8C, right), but such high $O_2$ levels are not common for real tissue physiology.

Subcellular oxygenation maps were also used for computing theoretical activities of enzymes utilising $O_2$ and responsible for metabolic reprogramming and adaptation to low $O_2$.[43] Although enzymatic activity values are based on the data generated *in vitro*, they can still be important predicting factors. This is exemplified by the calculated perinuclear activities of prolyl hydroxylase domain-containing protein 2 (PHD2), which regulates HIF signalling.[12,44]

### 11.3.2.3 Ex vivo *Imaging of $O_2$ in Other Tissue Types of the Urinary Bladder*

To explore the bladder model further, we analysed Pt-Glc staining and oxygenation in its serosal and deeper mucosal tissue. The former was brightly stained and measured $O_2$ levels were found to be similar to that in the muscles of the intestine (see Figure 11.4E). The latter could only be stained in the regions where the layer of umbrella cells was completely shed. Figure 11.9 shows Pt-Glc staining and calculated $O_2$ levels in a blood micro-vessel located close to the surface of mucosal tissue. We observed intense staining of vessel wall (most likely endothelium) and surrounding tissue, with $O_2$ levels ranging 36–64 μM (interquartile range).

### 11.3.2.4 *Main Findings and Conclusions*

The *ex vivo* imaging demonstrated for the first time pronounced $O_2$ gradients (up to 40 μM) in individual umbrella cells residing in their native tissue environment (physiological context), in contrast to isolated cells commonly used

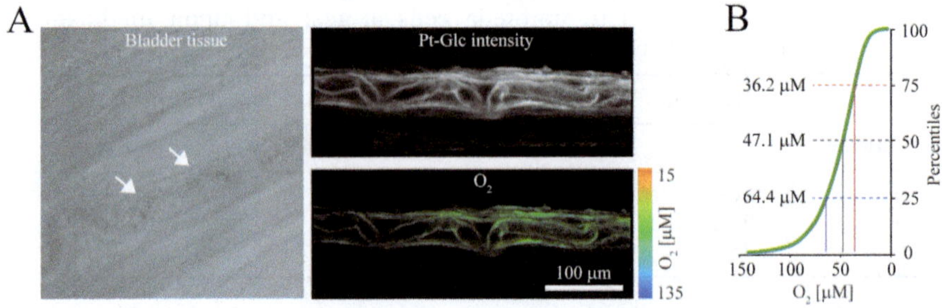


**Figure 11.9** *Ex vivo* $O_2$ imaging in mucosal blood vessel of the urinary bladder. A. Images of mucosal tissue, Pt-Glc phosphorescence intensity and $O_2$ levels. Arrows show the area of rugae, containing the vessel. In the right panel, images represent stacks of four sequential focal planes taken with 10 μm step produced using ImageJ program (Volume Viewer). B. Cumulative $O_2$ frequency histogram averaged for four focal planes selected for (A); $O_2$ values corresponding to 25th, 50th and 75th percentiles are shown.

for studying $O_2$ gradients.[45,46] We also conclude that prominent intracellular $O_2$ gradients are created by mitochondrial activity. However, extrapolation of the data obtained with giant umbrella cells to the cells of 'regular' size (10–20 μm) would reduce these $O_2$ gradients to values of approximately 4 μM, which can hardly be resolved by current $O_2$ imaging systems. Therefore, for selecting proper biological models for analysis of sub-cellular $O_2$ gradients, cell size is as important a parameter as mitochondrial respiration. Elaboration of functional roles of such gradients also requires sophisticated mathematical models and data processing algorithms. Apart from *ex vivo* $O_2$ analysis, successful *in vivo* staining with Pt-Glc permits the mapping of local $O_2$ levels *in vivo*, *e.g.* within normal and pathological urinary bladder (such as chronic cyclophosphamide model of bladder pain syndrome[33]).

### 11.3.3 *Ex vivo* $O_2$ Imaging in Carotid Arteries

The feasibility of $O_2$ imaging in blood vessels demonstrated in the previous study motivated us to perform a pilot *ex vivo* analysis of $O_2$ levels of mammalian carotid arteries (primarily in the *intima tunica* lined by endothelium), and their changes upon chronic occlusion. The access to a human scale porcine chronic vascular occlusion model was provided by Prof. N. Caplice (CRVB, Cork, Ireland).[17] Chronic vaso-occlusive disease is one of the major complications of cardiovascular medicine and surgery and novel strategies for prevention and treatment of coronary artery and carotid artery diseases are actively sought.[47,48] In particular, knowledge of metabolic rearrangements in arteries under pathological conditions are of major importance.

We found that staining efficiency of the ligated artery with Pt-Glc was somehow lower than for control (approximately $1.8 \times 10^3$ *vs* $3.2 \times 10^3$ photons per pixel), but still sufficient for the imaging (Figure 11.10A). Calculated $O_2$

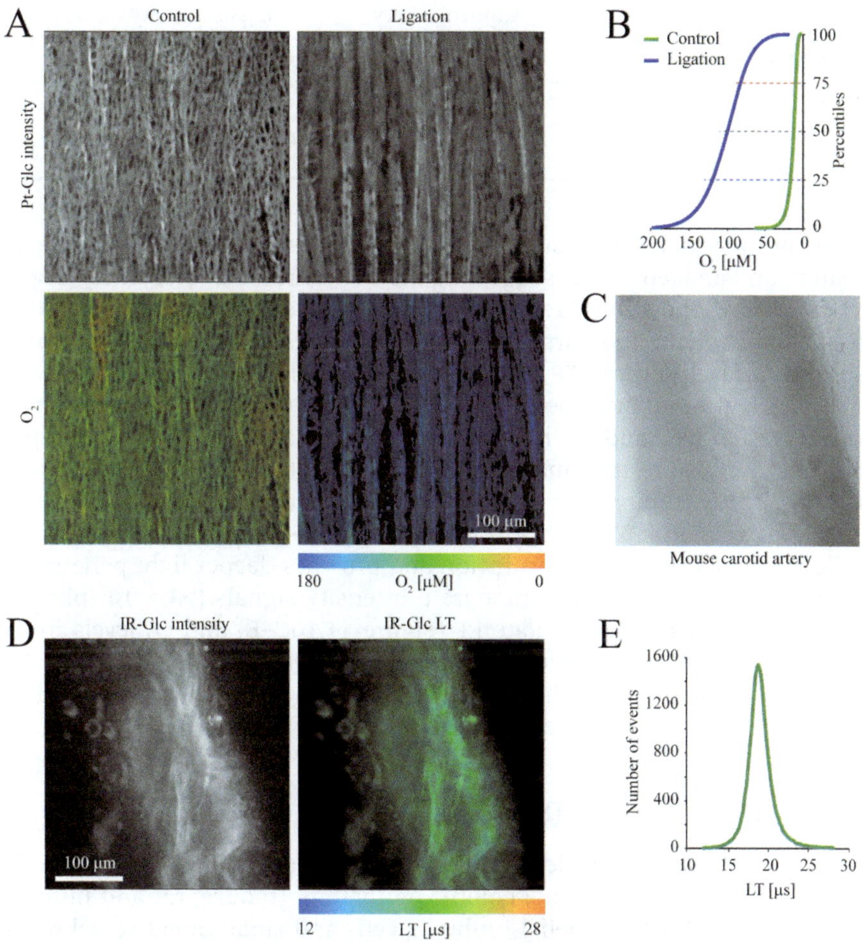


**Figure 11.10** *Ex vivo* $O_2$ imaging in carotid arteries of a pig (A, B) and a mouse (C–E). A. Intensity of Pt-Glc staining (upper panel) and tissue $O_2$ levels in the uppermost layers of the inner surface (endothelium and likely the tunica intima) of the control (left) and ligated for 6 weeks carotid arteries. Stacks of ten focal planes taken with 10 μm step are shown in a rainbow false colour scale (Volume Viewer, ImageJ). B. Cumulative $O_2$ frequency histogram averaged for four focal planes; $O_2$ values corresponding to 25th, 50th and 75th percentiles are shown in Table 11.3. C, D. Images of vessel tissue, IR-Glc phosphorescence intensity and LT. Imaging was conducted through the wall of the artery (50–100 μm); stacks of four focal planes taken with 5 μm step are shown. E. Frequency histograms demonstrate IR-Glc lifetime distribution in the tissue with an average value of approximately 18 μs. Note that IR-Glc and Pt-Glc probes have different LT ranges and $O_2$ calibrations.

**Table 11.3** Oxygen levels in the different samples of porcine carotid arteries.

| | Control artery | Ligated artery |
|---|---|---|
| Percentile | $O_2$ levels (μM) | |
| 25th | 14.5 | 120.3 |
| 50th | 10.4 | 100.5 |
| 75th | 7.6 | 82.7 |

levels in the innermost tissue layers of injured vessel were markedly higher than in undamaged tissue, suggesting their lower respiration activity (Figure 11.10A and B and Table 11.3). These data point to large rearrangements in tissue metabolism upon arterial occlusion, including but not limited to the shift towards glycolytic ATP production and HIF signalling.

This experiment was performed on a large thick-wall porcine artery, which was excised and cut longitudinally for the staining and imaging procedures. We also tested smaller animals (mice), trying to image intra-arterial $O_2$ directly through the wall of the carotid artery (50–100 μm thick) in a less invasive manner. For this model, we used IR-Glc probe,[15] which is less influenced by tissue light absorption and provides deeper light penetration depth. IR-Glc probe yielded moderate intensity signals (>3 × $10^2$ photons per pixel) sufficient to calculate LT (Figure 11.10C–E) and $O_2$ levels in the inner tissue layers of the carotid artery. This experiment can be regarded as a proof of concept of a high-resolution $O_2$ imaging of vascular tissue, which has potential for further use with *in vivo* models.

## 11.4 Concluding Remarks

In this chapter we briefly described the *ex vivo* imaging analysis of $O_2$ in different types of animal tissue: colonic epithelium (mouse, rat and human), colonic serosa (rat), epithelial umbrella cells and small blood vessel of the urinary bladder (mouse), the uppermost layers of the inner surface of carotid artery (pig and mouse); *ex vivo* $O_2$ imaging in brain tissue slices was described elsewhere.[11] All this data clearly demonstrates the utility of cell-penetrating phosphorescence based $O_2$ sensing probes, such as Pt-Glc and IR-Glc, and confocal one-photon PLIM microscopy for the generation of precise high-resolution maps of $O_2$ concentration in live tissues. The examples given here show that *ex vivo* $O_2$ imaging yields reproducible data, which allow for comparison of $O_2$ homeostasis and metabolism in the uppermost layers of different types of tissue, collected *post mortem* from normal and diseased animals. Importantly, the spectrum of disease models, for which *ex vivo* $O_2$ imaging can be used, is broad and includes cancer, inflammation and other pathological states characterised by mitochondrial dysfunction, redox imbalance, oxidative stress and hypoxia signalling. A more simple $O_2$ delivery to the tissue *ex vivo* (from medium rather than from circulating blood), which is often criticized as non-physiological, is also an advantage of the method, as it allows the detection of changes in $O_2$ consumption and metabolic shifts, which would be overseen by *in vivo* or *in vitro* approaches. Further progress in this technique is assured by development of powerful and flexible imaging platforms (confocal one- and two-photon PLIM) and phosphorescent $O_2$ sensing probes similar to Pt-Glc and IR-Glc with optimal photophysical and optical properties, improved bio-compatibility and tissue penetrating capability.

## Acknowledgements

Financial support of this work by the Science Foundation Ireland, grant number 12/RC/2276, is gratefully acknowledged. Authors would like to thank Dr R. Dmitriev (University College Cork, Cork, Ireland) for provision of Pt-Glc and IR-Glc probes; Prof. U. Knaus (University College Dublin, Ireland) and Dr G. Aviello (University of Aberdeen, UK) for sharing the NOX-deficient mouse model; Prof. N. Caplice (CRVB, Cork, Ireland) and Dr K. Martin (Dublin, Ireland) for the help with porcine and mouse artery tissue preparation. Permission to reproduce material published by Springer[13] is kindly acknowledged.

## References

1. A. Zhdanov, C. Favre, L. O'Flaherty, J. Adam, R. O'Connor, P. Pollard and D. Papkovsky, *Integr. Biol.*, 2011, **3**, 1135–1142.
2. J. Zhang, E. Nuebel, D. R. Wisidagama, K. Setoguchi, J. S. Hong, C. M. Van Horn, S. S. Imam, L. Vergnes, C. S. Malone, C. M. Koehler and M. A. Teitell, *Nat. Protoc.*, 2012, 7(6), 1068–1085.
3. T. C. O'Riordan, A. V. Zhdanov, G. V. Ponomarev and D. B. Papkovsky, *Anal. Chem.*, 2007, **79**, 9414–9419.
4. D. A. Ferrick, A. Neilson and C. Beeson, *Drug Discovery Today*, 2008, **13**, 268–274.
5. M. Wu, A. Neilson, A. L. Swift, R. Moran, J. Tamagnine, D. Parslow, S. Armistead, K. Lemire, J. Orrell and J. Teich, *Am. J. Physiol.: Cell Physiol.*, 2007, **292**, C125–C136.
6. R. I. Dmitriev, A. V. Zhdanov, Y. M. Nolan and D. B. Papkovsky, *Biomaterials*, 2013, **34**, 9307–9317.
7. D. R. Grimes, C. Kelly, K. Bloch and M. Partridge, *J. R. Soc., Interface*, 2014, **11**, 20131124.
8. J. Leedale, A. Herrmann, J. Bagnall, A. Fercher, D. Papkovsky, V. Sée and R. N. Bearon, *Math. Biosci.*, 2014, **258**, 33–43.
9. J. Jenkins, R. I. Dmitriev, K. Morten, K. W. McDermott and D. B. Papkovsky, *Acta Biomater.*, 2015, **16**, 126–135.
10. I. A. Okkelman, T. Foley, D. B. Papkovsky and R. I. Dmitriev, *Biomaterials*, 2017, **146**, 86–96.
11. R. I. Dmitriev, S. M. Borisov, A. V. Kondrashina, J. M. Pakan, U. Anilkumar, J. H. Prehn, A. V. Zhdanov, K. W. McDermott, I. Klimant and D. B. Papkovsky, *Cell. Mol. Life Sci.*, 2015, **72**, 367–381.

12. A. V. Zhdanov, A. V. Golubeva, I. A. Okkelman, J. F. Cryan and D. Papkovsky, *Am. J. Physiol.: Cell Physiol.*, 2015, **309**, C501–C509.
13. A. V. Zhdanov, I. A. Okkelman, A. V. Golubeva, B. Doerr, N. P. Hyland, S. Melgar, F. Shanahan, J. F. Cryan and D. B. Papkovsky, *Cell. Mol. Life Sci.*, 2017, **74**, 141–151.
14. R. I. Dmitriev, A. V. Kondrashina, K. Koren, I. Klimant, A. V. Zhdanov, J. M. Pakan, K. W. McDermott and D. B. Papkovsky, *Biomater. Sci.*, 2014, **2**, 853–866.
15. D. B. Papkovsky, R. I. Dmitriev and S. Borisov, *Proc. of SPIE*, 2015, **9329**, 93290B1–93290B6.
16. G. Pircalabioru, G. Aviello, M. Kubica, A. Zhdanov, M.-H. Paclet, L. Brennan, R. Hertzberger, D. Papkovsky, B. Bourke and U. G. Knaus, *Cell Host Microbe*, 2016, **19**, 651–663.
17. A. H. S. Kumar, K. Martin, B. Doyle, C.-L. Huang, G.-K. M. Pillai, M. T. Ali, K. A. Skelding, S. Wang, B. M. Gleeson, S. Jahangeer, E. L. Ritman, S. J. Russell and N. M. Caplice, *Biomaterials*, 2014, **35**, 9012–9022.
18. A. V. Zhdanov, G. Aviello, U. G. Knaus and D. B. Papkovsky, *Biochim. Biophys. Acta, Gen. Subj.*, 2017, **1861**, 198–204.
19. S. P. Colgan and E. L. Campbell, *J. Appl. Physiol.*, 2017, **123**(5), 1321–1327.
20. S. Heller, H. M. Penrose, C. Cable, D. Biswas, H. Nakhoul, M. Baddoo, E. Flemington, S. E. Crawford and S. D. Savkovic, *FASEB J.*, 2017, **31**, 2013–2025.
21. M. E. Johansson, J. M. H. Larsson and G. C. Hansson, *Proc. Natl. Acad. Sci. U. S. A.*, 2011, **108**, 4659–4665.
22. A. Fercher, S. M. Borisov, A. V. Zhdanov, I. Klimant and D. B. Papkovsky, *ACS Nano*, 2011, **5**, 5499–5508.
23. A. V. Kondrashina, R. I. Dmitriev, S. M. Borisov, I. Klimant, I. O'Brien, Y. M. Nolan, A. V. Zhdanov and D. B. Papkovsky, *Adv. Funct. Mater.*, 2012, **22**, 4931–4939.
24. M. Aloi, F. Nuti, L. Stronati and S. Cucchiara, *Nat. Rev. Gastroenterol. Hepatol.*, 2014, **11**, 99–108.
25. L. J. Hall, E. Faivre, A. Quinlan, F. Shanahan, K. Nally and S. Melgar, *Dig. Dis. Sci.*, 2011, **56**, 79–89.
26. S. P. Colgan, E. L. Campbell and D. J. Kominsky, *Annu. Rev. Pathol.: Mech. Dis.*, 2016, **11**, 77–100.
27. R. I. Dmitriev and D. B. Papkovsky, *Methods Appl. Fluoresc.*, 2015, **3**, 034001.
28. A. V. Zhdanov, I. A. Okkelman, F. W. Collins, S. Melgar and D. B. Papkovsky, *Biochim. Biophys. Acta, Bioenerg.*, 2015, **1847**, 1254–1266.
29. L. Albenberg, T. V. Esipova, C. P. Judge, K. Bittinger, J. Chen, A. Laughlin, S. Grunberg, R. N. Baldassano, J. D. Lewis, H. Li, S. R. Thom, F. D. Bushman, S. A. Vinogradov and G. D. Wu, *Gastroenterology*, 2014, **147**, 1055–1063.
30. L. Zheng, C. J. Kelly and S. P. Colgan, *Am. J. Physiol.: Cell Physiol.*, 2015, **309**, C350–C360.
31. S. P. Colgan and C. T. Taylor, *Nat. Rev. Gastroenterol. Hepatol.*, 2010, **7**, 281–287.

32. E. P. Cummins, F. Seeballuck, S. J. Keely, N. E. Mangan, J. J. Callanan, P. G. Fallon and C. T. Taylor, *Gastroenterology*, 2008, **134**, 156–165.
33. A. V. Golubeva, A. V. Zhdanov, G. Mallel, T. G. Dinan and J. F. Cryan, *Physiol. Rep.*, 2014, **2**, e00260.
34. P. Khandelwal, S. N. Abraham and G. Apodaca, *Am. J. Physiol. Renal Physiol.*, 2009, **297**, F1477–F1501.
35. G. Zaharchuk, R. F. Busse, G. Rosenthal, G. T. Manley, O. A. Glenn and W. P. Dillon, *Acad. Radiol.*, 2006, **13**, 1016–1024.
36. X. Giannakopoulos, A. Evangelou, V. Kalfakakou, E. Grammeniatis, I. Papandropoulos and K. Charalambopoulos, *Int. Urol. Nephrol.*, 1997, **29**, 393–401.
37. D. T. Jones, *Am. J. Physiol.: Cell Physiol.*, 1986, **250**, C663–C675.
38. R. I. Dmitriev, A. V. Zhdanov, G. Jasionek and D. B. Papkovsky, *Anal. Chem.*, 2012, **84**, 2930–2938.
39. E. Takahashi and K. Doi, *Jpn. J. Physiol.*, 1998, **48**, 243–252.
40. A. V. Zhdanov, M. W. Ward, C. T. Taylor, E. A. Souslova, D. M. Chudakov, J. H. Prehn and D. B. Papkovsky, *Biochim. Biophys. Acta, Bioenerg.*, 2010, **1797**, 1627–1637.
41. A. V. Zhdanov, R. I. Dmitriev and D. B. Papkovsky, *Cell. Mol. Life Sci.*, 2011, **68**, 903–917.
42. R. I. Dmitriev and D. B. Papkovsky, *Cell. Mol. Life Sci.*, 2012, **69**, 2025–2039.
43. M. S. Nakazawa, B. Keith and M. C. Simon, *Nat. Rev. Cancer*, 2016, **16**, 663.
44. M. Hirsilä, P. Koivunen, V. Günzler, K. I. Kivirikko and J. Myllyharju, *J. Biol. Chem.*, 2003, **278**, 30772–30780.
45. E. Takahashi, K. Sato, H. Endoh, Z.-L. Xu and K. Doi, *Am. J. Physiol.: Heart Circ. Physiol.*, 1998, **275**, H225–H233.
46. W. J. Federspiel, *Biophys. J.*, 1986, **49**, 857–868.
47. K. Rosenfield, J. S. Matsumura, S. Chaturvedi, T. Riles, G. M. Ansel, D. C. Metzger, L. Wechsler, M. R. Jaff and W. Gray, *N. Engl. J. Med.*, 2016, **374**, 1011–1020.
48. G. N. Levine, E. R. Bates, J. A. Bittl, R. G. Brindis, S. D. Fihn, L. A. Fleisher, C. B. Granger, R. A. Lange, M. J. Mack and L. Mauri, *Circulation*, 2016, **134**, e123–e155.

CHAPTER 12

# *Tracking of Hypoxia and Cancer Metastasis with Iridium(III)-based $O_2$ Probes*

XIANCHUANG ZHENG AND XIQUN JIANG*

Nanjing University, Department of Polymer Science and Engineering, College of Chemistry and Chemical Engineering, 22 Hankou Road, Nanjing, 210093, China
*E-mail: jiangx@nju.edu.cn

## 12.1 Introduction

The oxygen-deprived condition, termed as hypoxia, is a characteristic feature of the microenvironment in most solid tumours, due to the imbalance between the oxygen consumption by the cancer cells and limited oxygen supply from the blood vessels.[1–4] Tumour hypoxia often leads to the overexpression of the regulatory transcription factors such as hypoxia inducible factor 1α (HIF-1α), which plays key roles in the tumor initiation, malignant progression, metastasis and resistance to therapy.[5–9] Therefore, imaging of tumor hypoxia is of great interest for the diagnosis of cancer, study of the tumor microenvironment and the determination of treatment strategies.[10,11]

Although great progress has been achieved in the treatment of primary tumors using surgical and adjuvant therapies, cancer metastasis is still characterized by poor prognosis and has become the principle cause of death

Detection Science Series No. 11
Quenched-phosphorescence Detection of Molecular Oxygen: Applications in Life Sciences
Edited by Dmitri B. Papkovsky and Ruslan I. Dmitriev

Published by the Royal Society of Chemistry, www.rsc.org

among cancer patients.[12,13] Hypoxia is a driving force of cancer metastasis promoting it *via* multiple mechanisms, such as metastatic niche formation, cancer cell phenotype switching and tumor vascularization.[14,15] The real-time monitoring of hypoxia during metastasis, *e.g.* using molecular imaging methods, can be of great help for the understanding of hidden mechanisms and effective diagnostics.[16]

While positron emission tomography (PET) imaging with hypoxia-targeted contrast agents such as fluoromisonidazole (FMISO) and Cu(II)-diacetyl-bis($N^4$-methylthiosemicarbazone) (Cu-ATSM) has been widely used for hypoxia imaging in the clinic,[17–19] optical imaging based on hypoxia-sensitive phosphorescent transition metal complexes is emerging as a new non-radiative method for hypoxia imaging, with significantly higher sensitivity and specificity.[20–24] Among these phosphorescent transition metal complexes, Iridium (Ir) (III) complexes have drawn great attentions due to their excellent brightness, large Stokes shift, tunable absorption and emission wavelengths and high sensitivity to molecular oxygen.[25–28] Recently, a series of phosphorescent probes developed based on the Ir(III) complexes have shown potential in the imaging of hypoxia in tumors, ranging from solid tumors to small numbers of cancer cells, and from primary tumors to metastases.[28–31]

## 12.2 Hypoxia Imaging with Poly(*N*-Vinylpyrrolidone) (PVP)-conjugated Ir(III) Complex Probe (Ir-PVP)

### 12.2.1 Design of the Ir-PVP Probe

A near-infrared (NIR) emitting macromolecular probe (Ir-PVP) has been developed for highly sensitive hypoxia imaging, based on the conjugation of Ir(III) complex with poly(*N*-vinylpyrrolidone) (PVP).[29] The chemical structure of the macromolecular Ir-PVP probe is shown in Figure 12.1a. The Ir(III) complex component contains a large conjugated ligand, extending its phosphorescence emission to the NIR region, beneficial for deep tissue imaging. The phosphorescence emission of the Ir(III) complex component is highly sensitive to molecular oxygen (Figure 12.1b). It is quenched in the normoxic tissues but turned ON in the hypoxic tumour tissues, providing specific imaging of hypoxia with minimal background interference. Another component of the Ir-PVP probe is the hydrophilic poly(*N*-vinylpyrrolidone) (PVP) polymer. The PVP component makes the probe water-soluble and improves the tumour retention of the probe *via* enhanced permeation and retention effects.[32,33] Compared with the widely used hydrophilic polymer, polyethylene glycol (PEG), PVP does not trigger the accelerated blood clearance phenomenon and also can be more easily functionalized for further applications.[34] The terminal group of the PVP component can be linked to an internal standard ('R' group in Figure 12.1a), allowing for ratiometric imaging of molecular oxygen (Figure 12.1b).

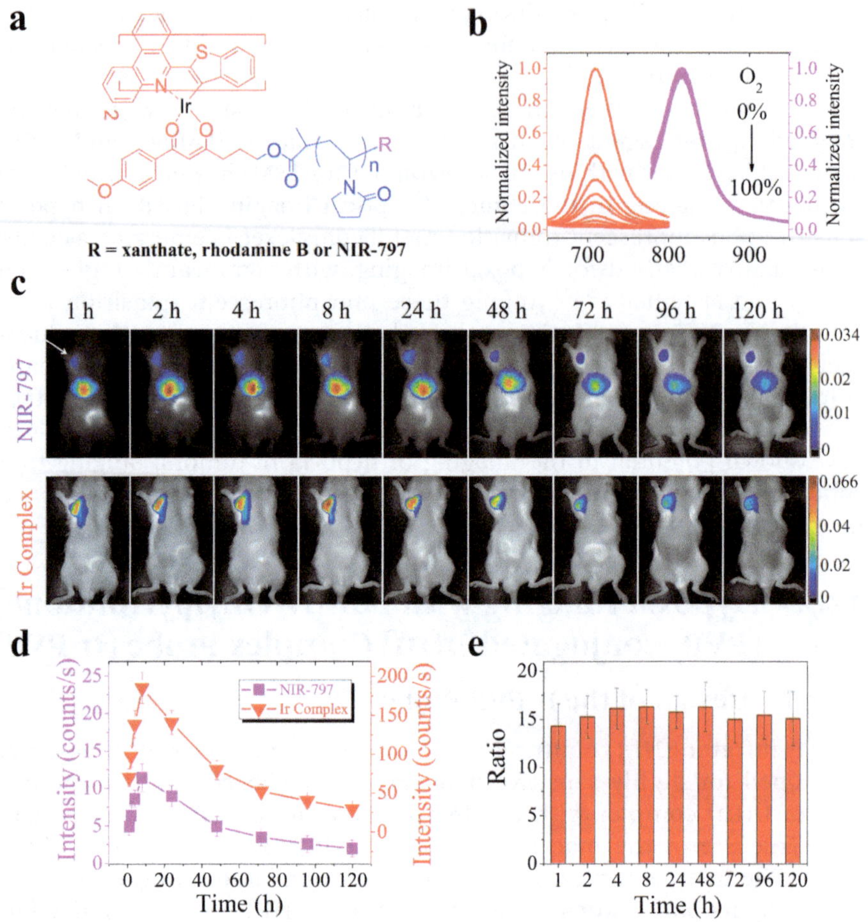


**Figure 12.1** Ratiometric imaging of tumour hypoxia with Ir-PVP probe. (a) Chemical structure of the Ir-PVP probe. 'R' represents the internal standard. (b) Emission spectra of Ir-PVP-NIR797 under increasing oxygen levels. (c) Whole-body optical images of ICR mice bearing H22 tumours in the armpit after i.v. injection of Ir-PVP-NIR797 (40 mg $kg^{-1}$). (d) The averaged signal intensities of Ir(III) complex (red) and NIR-797 component (purple) in the tumours at different time points after i.v. injection of the probe. (e) The ratio between the signal intensities of Ir complex component and NIR-797 component at different time points after i.v. injection of the probe. The results are presented as the mean ± s.d. ($n = 5$). Reprinted by permission from Macmillan Publishers Ltd: *Nature Communications* (ref. 29), copyright 2015.

### 12.2.2 Ratiometric Imaging of Hypoxia in Solid Tumours with Ir-PVP Probe

The ratiometric imaging of tumour hypoxia in living mice has been achieved by using the Ir-PVP probe with a NIR internal standard (NIR-797).[29] The probe was intravenously (i.v.) injected into the ICR mice implanted with murine hepatoma H22 tumours (40 mg $kg^{-1}$). The mice were scanned using a preclinical *in vivo* optical imaging system and the optical signals from the probe were collected in two spectral channels to detect the 710 nm phosphorescence emission from the oxygen-sensitive Ir(III) complex component and the 825 nm fluorescence emission from the oxygen-insensitive NIR internal standard (Figure 12.1c). The images in the fluorescence channel show the bio-distribution of the probe in the mouse. Most of the probe was taken up by the liver and subsequently cleared out from the body. The fluorescence signals detected from the tumour region show the accumulation and retention of the probe in this tissue. The images in the phosphorescence channel show the combined effect of the bio-distribution and hypoxia-specific response of the probe. Even though the uptake of the probe in the tumour was lower than in the liver, a much stronger phosphorescence signal was detected in the tumour, due to the response of the probe to hypoxia. The calculation of the ratio of the phosphorescence signal and the fluorescence signal provides a real-time semi-quantitative readout of the level of hypoxia in the tumour (Figure 12.1d–e).

### 12.2.3 Monitoring the Proliferation of Cancer Cells in Live Mice with Ir-PVP Probe

Due to the high sensitivity of the Ir-PVP to hypoxia, it is possible to track the proliferation of small numbers of cancer cells during the early stage of tumour initiation.[29] Mice subcutaneously implanted with different amounts of cancer cells ($10^7$ to $10^2$ H22 cells) were imaged over time after i.v. injection of the Ir-PVP probe (Figure 12.2a). Due to the absence of the extremely hypoxic solid tumours, the overall phosphorescence signals detected from the mice were relatively weak and some random non-specific signals were observed in normal tissues such as liver and bladder. However, strong probe signals were detected from the cancer cell implanted regions, especially at the higher cell numbers ($10^6$ to $10^7$ H22 cells). The probe signal detected from the cancer cell implanted region correlated to the number of cancer cells (Figure 12.2b). The detection limit of the probe with implanted H22 cells was determined to be $10^3$ cells at 24 h after the implantation of the cells. The macromolecular Ir-PVP probe also enabled long-term tracking of the hypoxia state of the implanted tumour tissue during the proliferation of the cancer cells post implantation. For example, continuous imaging of mice over the 96 h showed the gradual increase of hypoxia reaching saturated values of approx. $10^4$ H22 cells (Figure 12.2c).

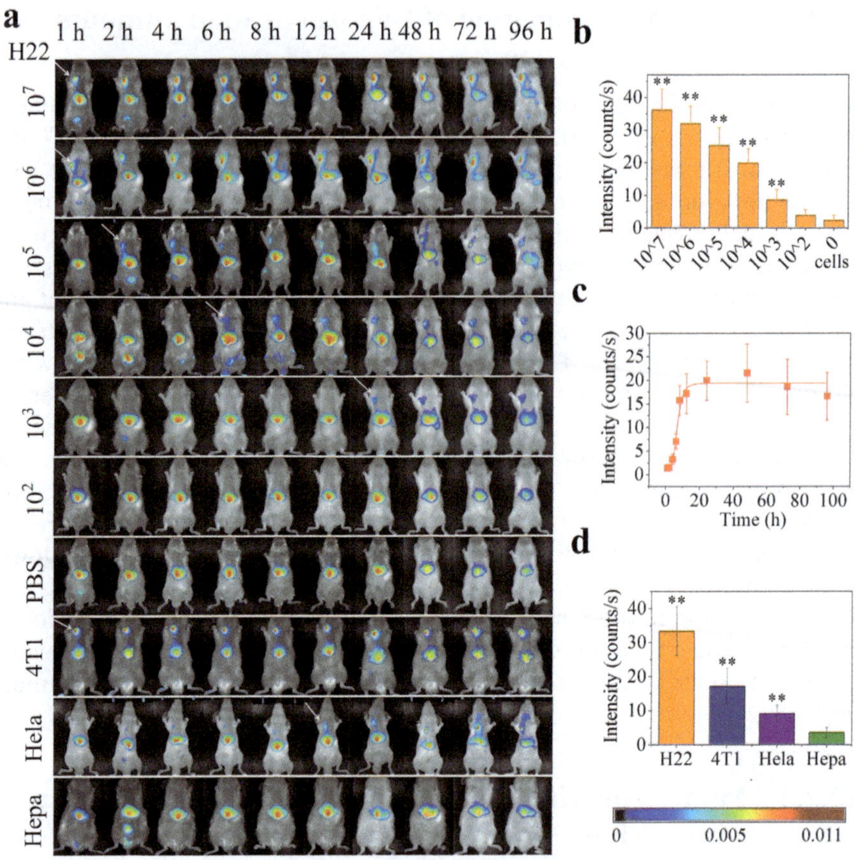


**Figure 12.2** Tracking cancer cell proliferation in living mice. (a) Optical images of mice subcutaneously implanted with different numbers of H22 cells ($10^7$ to $10^2$), 4T1 cells ($10^6$), HeLa cells ($10^6$) and murine hepatocytes (Hepa, $10^6$) after i.v. injection of Ir-PVP. (b) Comparison of probe signal intensities at different H22 cell numbers ($10^7$ to $10^2$) at 24 h after implantation. (c) Comparison of probe signal intensities at different time points after the implantation of $10^4$ H22 cells. (d) Comparison of probe signal intensities at different cell types (H22, 4T1, HeLa and Hepa) at the same cell number ($10^6$) at 12 h after implantation. The results are presented as the mean ± s.d. ($n$ = 5). Adapted by permission from Macmillan Publishers Ltd: *Nature Communications* (ref. 29), copyright 2015.

Comparison of the signal intensities detected from the implanted regions indicated the proliferation rate and degree of malignancy of the cancer cell types, reflected in increased probe signals from H22 cells, 4T1 cells, HeLa cells compared to hepatocytes (Figure 12.2d). These results show that the Ir-PVP probe is highly sensitive for tracking the hypoxia induced by the proliferation of cancer cells in small numbers and at the early stage of tumour initiation.

## 12.3 Tracking Cancer Metastasis with Ir(III)-based Oxygen Nanosensor

### 12.3.1 Design of the Ir(III)-based Oxygen Nanosensor (Ir-CM)

Based on the high sensitivity of the macromolecular Ir-PVP probe in the imaging of hypoxia in solid tumours and cancer cells, an Ir(III)-based oxygen nanosensor has been further developed for tracking cancer cell metastasis in live mice.[30] The nanosensor was prepared by the formation of co-micelles with poly(ε-caprolactone)-*b*-poly(*N*-vinylpyrrolidone) (PCL-PVP) and Ir-PVP, subsequently termed as Ir-CM (Figure 12.3a). The hydrophobic PCL chain and the Ir complex component constituted the core of the co-micelle, while the hydrophilic PVP chains formed the shell. The Ir-CM nanosensor has a diameter of about 30 nm in the dry state and a hydrodynamic diameter around 50 nm in the aqueous solution, and was stable over 7 days in PBS and serum (Figure 12.3b–d). Ir-CM has the similar NIR emission and sensitivity to molecular oxygen as Ir-PVP (Figure 12.3e), while its nanosize dimensions are beneficial for minimized uptake by the liver and higher accumulation and retention in the lymph node, making it suitable for the imaging of cancer metastasis.

### 12.3.2 Tracking Lung Cancer Metastasis with Ir-CM Nanosensor

The Ir-CM nanosensor has been applied for tracking lung metastasis in living mice.[30] ICR mice were i.v. injected though the bloodstream with H22 cells to establish the lung metastasis model. Whole-body optical imaging was performed to detect the NIR phosphorescence signals after administration of Ir-CM by i.v. injection. Strong phosphorescence signals were observed from the lungs bearing metastases compared to much lower non-specific signals from normal organs (Figure 12.4a–d). Ir-CM had relatively high accumulation and retention in the lung tissue, but the highest signals were confined to the spleen (Figure 12.4e), indicating that the phosphorescence signal of Ir-CM was selectively enhanced in the metastatic lungs due to hypoxia (Figure 12.4f–g). The phosphorescence signal of Ir-CM in metastatic lungs was about 60-fold higher than in normal lungs, which we attribute to significant differences in the level of hypoxia (Figure 12.4h–l), and by the high oxygen level in normal lungs (14–20% $O_2$). Thus, the metastatic cancer cells delivered through the bloodstream in the lungs can be accurately detected with high signal-to-background ratio by using optical imaging with the Ir-CM nanosensor.

### 12.3.3 Tracking Cancer Metastasis in the Lymph Node with Ir-CM Nanosensor

In addition to the bloodstream, the lymphatic system is another major metastatic route for cancer cells. The Ir-CM oxygen nanosensor has also been applied for tracking the lymph node metastasis.[30] The mouse lymph

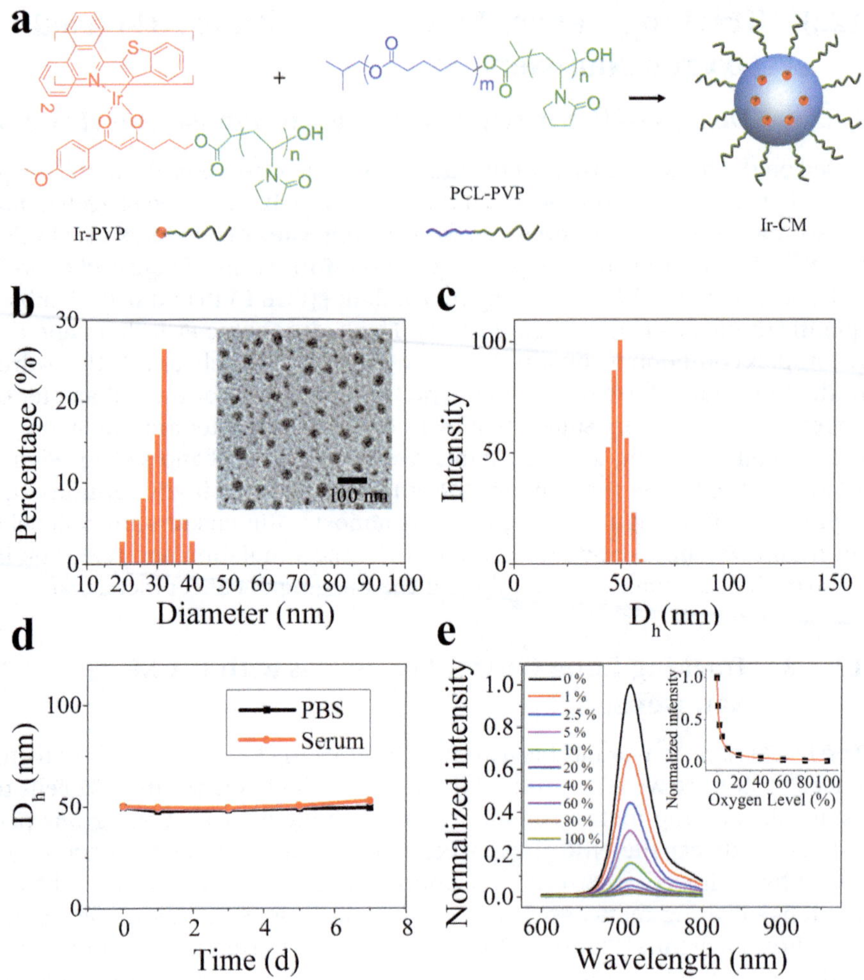


**Figure 12.3** The Ir(III)-based oxygen nanosensor (Ir-CM). (a) Synthesis of Ir-CM. (b) TEM image of Ir-CM (insert) and particle size distribution in the TEM image. (c) Hydrodynamic diameter of Ir-CM in distilled water. (d) Stability assessment of Ir-CM in PBS (1x) or fetal bovine serum buffer (serum/PBS 1:1) at room temperature. (e) Phosphorescence emission spectra of Ir-CM in distilled water at different oxygen levels. Excitation: 560 nm. Insert: Decay curve of peak intensities of Ir-CM with increasing oxygen level. Adapted with permission from ref. 30. © 2015 WILEY-VCH Verlag GmbH & Co. KGaA, Weinheim.

node metastasis model was established by implanting H22 cells into the right rear paw of the ICR mice. Whole-body optical imaging was performed after i.v. injection of Ir-CM to the mice. In addition to the primary tumour on the paw, strong phosphorescence signals were also detected in the right inguinal lymph node of the mice (Figure 12.5a and b). When comparing the Ir-CM signals in the metastatic lymph nodes and normal

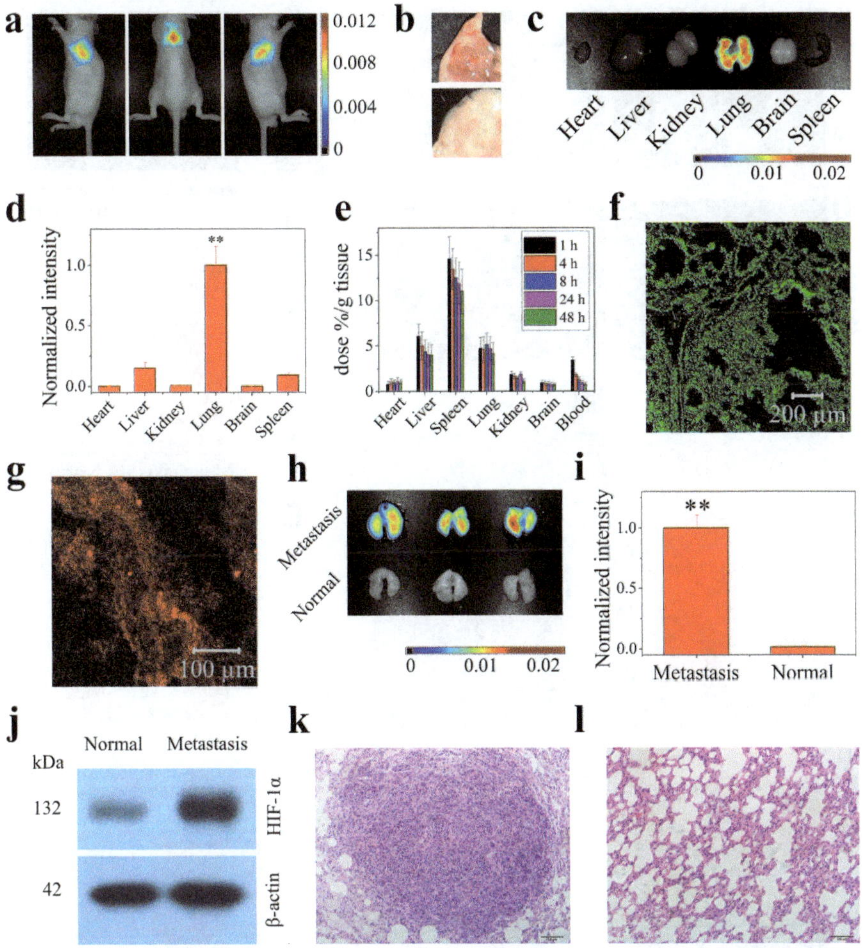


**Figure 12.4** Tracking lung metastasis with Ir-CM nanosensor. (a) Whole-body optical imaging of mice bearing lung metastasis at 1 h after i.v. injection of Ir-CM. (b) Bright-field images of lungs with metastasis (above) and normal lungs (below). (c) *Ex vivo* images of different organs dissected from ICR mice bearing lung metastasis. (d) Signal intensities detected from different organs ($n = 5$). (e) Bio-distribution assay of Ir-CM in different organs of mice bearing lung metastasis. (f,g) Frozen sections of the metastasis-bearing lungs after staining with (f) HIF-1α antibody and (g) pimonidazole antibody. (h) *Ex vivo* images of lungs bearing metastasis and normal lungs ($n = 3$). (i) Comparison of the signal intensities of the lung metastasis group and normal lungs group. (j) Western blots of HIF-1 expression in the normal lungs and metastasis-bearing lungs. (k,l) H&E stained slices of (k) metastasis-bearing lungs and (l) normal lungs. Scale bar: 100 mm. Adapted with permission from ref. 30. © 2015 WILEY-VCH Verlag GmbH & Co. KGaA, Weinheim.

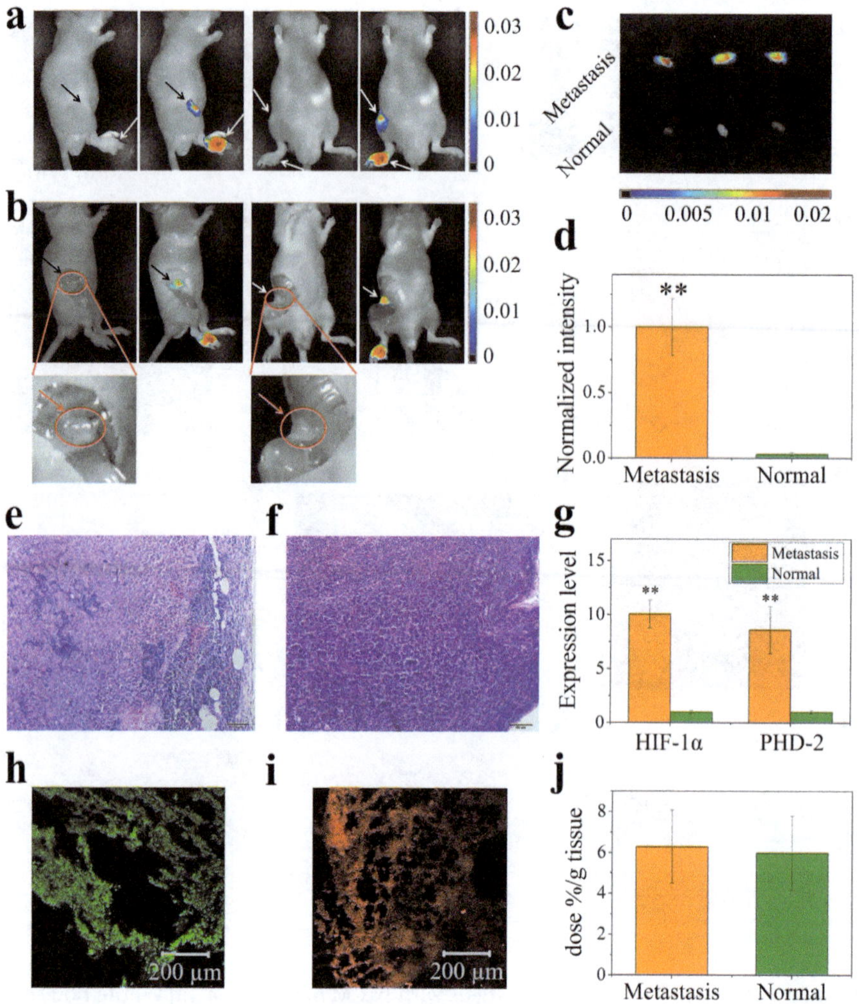


**Figure 12.5** Tracking lymph node metastasis with Ir-CM nanosensor. (a) Whole-body optical imaging of mice bearing H22 tumour in the right rear paw at 1 h after i.v. injection of Ir-CM. (b) The mice were imaged after removal of the local skin to reveal the location of the inguinal lymph node. (c) *Ex vivo* images of the right inguinal lymph nodes from mice bearing H22 tumour in the right rear paw and normal mice ($n = 3$). (d) Comparison between the signal intensities of the metastatic and normal right inguinal lymph nodes. (e,f) H&E stained slices of (e) metastatic and (f) normal lymph nodes. Scale bar: 100 mm. (g) The expression levels of HIF-1α mRNA and PHD-2 mRNA of metastatic and normal lymph nodes assayed by RT-PCR ($n = 3$). (h,i) Frozen sections of the metastatic lymph nodes were stained with (h) HIF-1α and (i) pimonidazole antibody. (j) The concentration of Ir-CM in the metastatic and normal lymph nodes ($n = 5$). Adapted with permission from ref. 30. © 2015 WILEY-VCH Verlag GmbH & Co. KGaA, Weinheim.

lymph nodes, we found that the Ir-CM signal intensity in the former was about 30-fold stronger than in the latter (Figure 12.5c–f). The metastatic lymph nodes were also found to be much more hypoxic than the normal lymph nodes, while the Ir-CM concentration was similar in metastatic and normal lymph nodes, indicating the selective enhancement of the phosphorescent signal of Ir-CM in the metastatic and hypoxic lymph nodes (Figure 12.5g–j). These results show that the Ir-CM nanosensor is not only feasible in the detection of primary tumors, but also sufficiently sensitive for tracking lymph node metastasis after systemic injection. Besides, it also provides high specificity to distinguish metastatic from normal lymph nodes.

## 12.4 Ir(III)-based Successively Activating Phosphorescent Probe for Ultrasensitive Hypoxia Imaging

### 12.4.1 Design of the Successively Activating Phosphorescent Probe (Ir-Im-PEG)

Insufficient image contrast and therefore the imaging sensitivity have always been major challenges in cancer and metastasis diagnostics, due to the limited expression levels of the biomarkers and limited signal amplification abilities of the contrast agents.[35,36] As a solution to this issue, an Ir(III)-based macromolecular phosphorescent probe has been developed which can successively respond to tumour acidifcation and hypoxia while amplifying detection sensitivity *via* signal propagation.[31] The primary or precursor form of the probe (named Ir-Im-PEG) consists of an Ir(III) complex chromophore and a PEG chain linked together *via* an pH-sensitive imine bond (Figure 12.6a). Under the neutral and normoxic condition, the precursor form displays weak phosphorescence with the excitation peak at 410 nm and emission peak at 610 nm. Once exposed to acidic conditions, the imine bond of the precursor breaks up, converting the Ir-Im-PEG probe from the precursor form to a secondary form, named the 'reporter form'. This conversion process red-shifts the excitation peak to 520 nm with a shoulder peak at 580 nm and an emission peak to 705 nm, falling within the NIR range suitable for *in vivo* imaging (Figure 12.6b–e). In addition, at low oxygenation conditions, the phosphorescence emission intensity of the reporter form at 705 nm will be significantly enhanced (Figure 12.6f–g). Therefore, the precursor should remain 'silent' (no NIR emission at 705 nm) in blood and normal tissues yet become activated with the switching ON of the NIR emission under the acidic and hypoxic tumour conditions. This propagation of signals leads to signal amplification in the hypoxic tumour microenvironment and thus improved detection sensitivity.

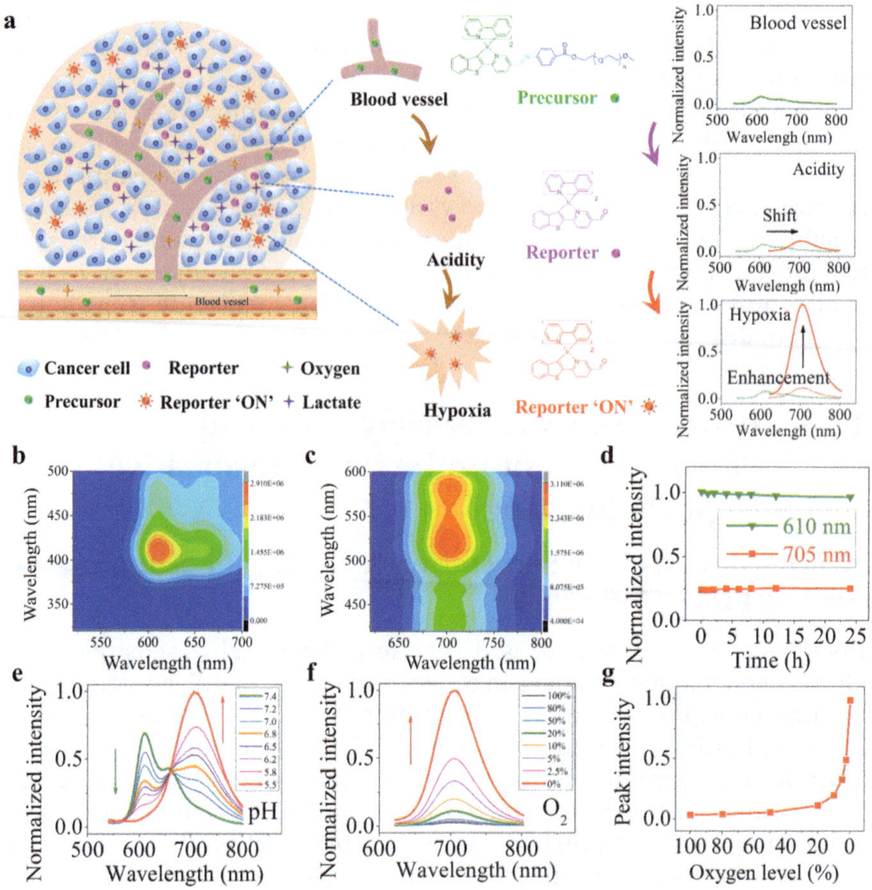


**Figure 12.6** The Ir(III)-based successively activatable phosphorescent probe (Ir-Im-PEG). (a) Chemical structures of the precursor and reporter forms of the probe and its two-step activation mechanism in the tumour microenvironment. (b,c) Three-dimensional fluorescence spectra of the (b) precursor and (c) reporter. (d) Serum stability test of the precursor at 37 °C. (e) Emission spectra of the probe within the pH range 5.5–7.4 when excited at 450 nm measured in deaerated PBS containing 10% DMSO (2 μM). (f) Emission spectra of the reporter at several oxygen levels when excited at 580 nm. (g) Emission intensities of the reporter at 705 nm at various oxygen levels. Adapted by permission from Macmillan Publishers Ltd: *Nature Biomedical Engineering* (ref. 31), copyright 2017.

### 12.4.2 Tracking Liver Cancer Metastasis with the Ir-Im-PEG Probe

On the basis of the successive response and signal propagation of the Ir-Im-PEG probe, tracking cancer metastasis in the liver has been achieved in mice models.[31] The liver metastasis model was established by the implantation of orthotopic human breast adenocarcinoma (MDA-MB-231) tumours in the female

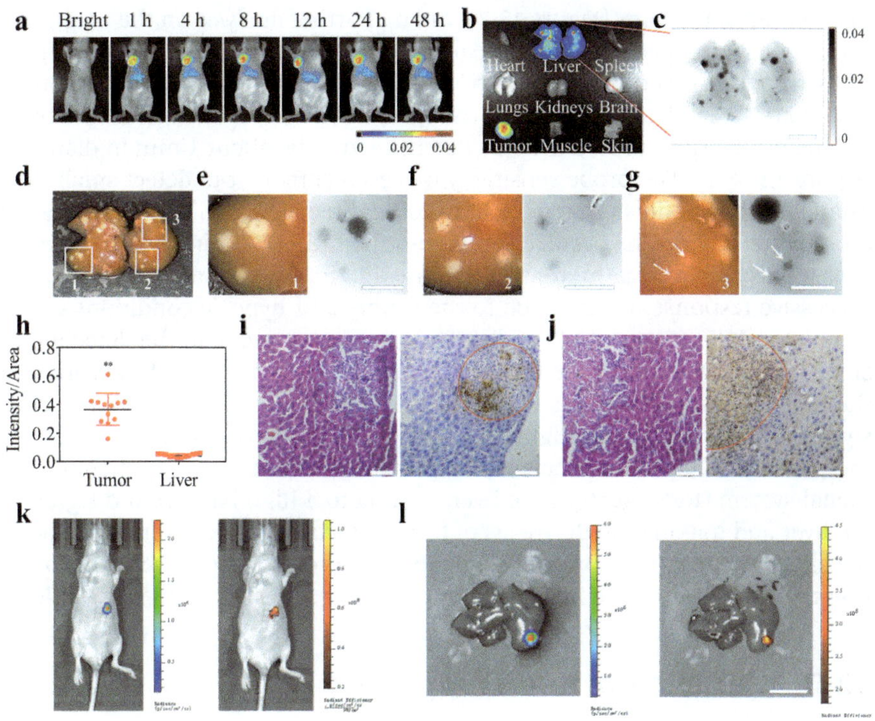


**Figure 12.7** Tracking liver metastasis with Ir-Im-PEG probe. (a) Whole-body NIR imaging of mice bearing an MDA-MB-231 breast tumour after i.v. injection of the precursor (40 mg $kg^{-1}$). (b) *Ex vivo* imaging of dissected organs at 48 h after injection of the precursor. (c) A zoomed-in, black-and-white contrast image of the *ex vivo* liver tissue showing the metastatic tumour nodules. Scale bar: 1 cm. (d) Photograph of the dissected liver. White boxes represent three regions of interest (ROIs). (e–g) Magnified bright-light and NIR images of ROI 1 – 3. Arrows in (g) point at two tumour nodules that were not easy to identify by the naked eye. Scale bars: 5 mm. (h) Signal intensities per area measured from the tumour nodules and the surrounding normal liver tissues (mean ± s.d., $n = 12$). (i,j) Immunohistochemical staining of liver tissue sections. Left: H&E staining. Right: antibody staining against (i) pimonidazole and (j) MCT-4. The red lines show the boundaries between the tumour and normal tissues. Scale bars: 100 µm. (k) Whole-body BLI (left) and fluorescence (right) imaging of mice bearing HCCLM3-luc tumour nodule in the liver at 24 h after i.v. injection of the precursor. (l) *Ex vivo* BLI (left) and NIR (right) imaging of the dissected liver tissue. Scale bar: 5 mm. Adapted by permission from Macmillan Publishers Ltd: *Nature Biomedical Engineering* (ref. 31), copyright 2017.

nude mice. The precursor form of the probe was i.v. injected into the tumour-bearing mice and whole-body optical imaging was performed to detect the NIR phosphorescence signal from the reporter form (Figure 12.7a). It was found that strong NIR phosphorescence signals were detected from the primary breast tumours as well as the metastatic livers, with low non-specific signals in other normal organs (Figure 12.7a and b). Further analysis on the magnified images of the metastatic livers revealed that the NIR phosphorescence signal of the reporter could precisely distinguish the small metastatic tumour nodules from the surrounding normal tissues. The detection limit of the probe to the metastatic tumour nodules was determined to be about 1 mm in diameter (Figure 12.7c–g). The probe sensitivity was even sufficient to detect small metastatic tumour nodules that were hard to be directly recognized by the naked eye (Figure 12.7g). The average signal-to-noise-ratio (SNR) in the detection of liver metastasis was as high as 10 (Figure 12.7h), which was ascribed to the successive response of the probe to the acidity and hypoxic conditions of the tumour nodules (Figure 12.7i and j). The accuracy of probe in the detection of small tumour nodules in the liver was also confirmed by the co-localization of the bioluminescence signals from the cancer cells and the phosphorescence signals from the reporter (Figure 12.7k and l). Many existing molecular imaging probes usually non-specifically accumulate in the organs of reticuloendothelial system (RES), such as the liver, leading to a high background signal in the liver and making the diagnosis of liver metastasis very challenging. Due to the signal propagation ability of the Ir-Im-PEG probe, the precise detection of liver metastasis with high imaging contrast and specificity becomes possible.

## 12.5 Conclusion and Perspective

In summary, a series of Ir(III) complex-based phosphorescent macromolecular or nanosized probes have been developed for tracking hypoxia and cancer metastasis. The combination of oxygen-sensitive phosphorescent Ir(III) complex, hydrophilic polymer and oxygen-insensitive fluorescent internal standard has made it possible to semi-quantitatively monitor hypoxia in solid tumours as well as cancer cell proliferation in the early stage of tumour initiation. In addition, an Ir(III) complex-based oxygen nanosensor with tailored design of size and surface properties has enabled the detection of cancer metastasis to the lungs *via* the bloodstream or to the lymph nodes providing high specificity and excellent imaging contrast. The precise detection of cancer metastasis in the liver has also been achieved by developing an Ir(III) complex-based phosphorescent probe successively responding to the acidity and low oxygenation in the hypoxic tumour microenvironment. This progress has shown the great potential of these kinds of phosphorescent materials in hypoxia imaging and the diagnostics of cancer metastases, which should be of great value for hypoxia-related preclinical mechanistic studies, drug development and surgical oncology applications such as precise intraoperative visualization of tumour margins and microscopic tumour invasion.

## Acknowledgement

This work was supported by the Natural Science Foundation of China (grant nos. 51690153, 21474045).

## References


1. M.-C. Hung, G. B. Mills and D. Yu, *Nat. Med.*, 2009, **15**, 246–247.
2. D. Hanahan and R. A. Weinberg, *Cell*, 2000, **100**, 57–70.
3. H. Harada, M. Inoue, S. Itasaka, K. Hirota, A. Morinibu, K. Shinomiya, L. Zeng, G. Ou, Y. Zhu, M. Yoshimura, W. G. McKenna, R. J. Muschel and M. Hiraoka, *Nat. Commun.*, 2012, **3**, 783.
4. P. Vaupel, F. Kallinowski and P. Okunieff, *Cancer Res.*, 1989, **49**, 6449–6465.
5. G. V. Thomas, C. Tran, I. K. Mellinghoff, D. S. Welsbie, E. Chan, B. Fueger, J. Czernin and C. L. Sawyers, *Nat. Med.*, 2006, **12**, 122–127.
6. B. J. Moeller, Y. Cao, C. Y. Li and M. W. Dewhirst, *Cancer Cell*, 2004, **5**, 429–441.
7. A. L. Kung, S. Wang, J. M. Klco, W. G. Kaelin and D. M. Livingston, *Nat. Med.*, 2000, **6**, 1335–1340.
8. H. Harada, *J. Radiat. Res.*, 2011, **52**, 545–556.
9. J. M. Brown and W. R. Wilson, *Nat. Rev. Cancer*, 2004, **4**, 437–447.
10. D. Thorwarth and M. Alber, *Radiother. Oncol.*, 2010, **97**, 172–175.
11. J. G. Rajendran, K. R. G. Hendrickson, A. M. Spence, M. Muzi, K. A. Krohn and D. A. Mankoff, *Eur. J. Nucl. Med. Mol. Imaging*, 2006, **33**(suppl. 1), 44–53.
12. T. N. Seyfried and L. C. Huysentruyt, *Crit. Rev. Oncog.*, 2013, **18**, 43–73.
13. P. Mehlen and A. Puisieux, *Nat. Rev. Cancer*, 2006, **6**, 449–458.
14. B. Muz, P. de la Puente, F. Azab and A. K. Azab, *Hypoxia*, 2015, 83.
15. E. B. Rankin and A. J. Giaccia, *Science*, 2016, **352**, 175–180.
16. R. Avni, B. Cohen and M. Neeman, *NMR Biomed.*, 2011, **24**, 569–581.
17. A. L. Vāvere and J. S. Lewis, *Dalton Trans.*, 2007, 4893–4902.
18. H. Kurihara, N. Honda, Y. Kono and Y. Arai, *Curr. Med. Chem.*, 2012, **19**, 3282–3289.
19. E. Lopci, I. Grassi, A. Chiti, C. Nanni, G. Cicoria, L. Toschi, C. Fonti, F. Lodi, S. Mattioli and S. Fanti, *Am. J. Nucl. Med. Mol. Imaging*, 2014, **4**, 365–384.
20. X. Wang, J. A. Stolwijk, T. Lang, M. Sperber, R. J. Meier, J. Wegener and O. S. Wolfbeis, *J. Am. Chem. Soc.*, 2012, **134**, 17011–17014.
21. T. Yoshihara, Y. Yamaguchi, M. Hosaka, T. Takeuchi and S. Tobita, *Angew. Chem., Int. Ed.*, 2012, **51**, 4148–4151.
22. A. J. Nichols, E. Roussakis, O. J. Klein and C. L. Evans, *Angew. Chem., Int. Ed.*, 2014, **53**, 3671–3674.
23. P. Babilas, G. Liebsch, V. Schacht, I. Klimant, O. S. Wolfbeis, R.-M. Szeimies and C. Abels, *Microcirculation*, 2005, **12**, 477–487.
24. X. Wang and O. S. Wolfbeis, *Chem. Soc. Rev.*, 2014, **43**, 3666–3761.
25. M. C. DeRosa, D. J. Hodgson, G. D. Enright, B. Dawson, C. E. B. Evans and R. J. Crutchley, *J. Am. Chem. Soc.*, 2004, **126**, 7619–7626.
26. S. Lamansky, P. Djurovich, D. Murphy, F. Abdel-Razzaq, R. Kwong, I. Tsyba, M. Bortz, B. Mui, R. Bau and M. E. Thompson, *Inorg. Chem.*, 2001, **40**, 1704–1711.
27. S. Lamansky, P. Djurovich, D. Murphy, F. Abdel-Razzaq, H.-E. Lee, C. Adachi, P. E. Burrows, S. R. Forrest and M. E. Thompson, *J. Am. Chem. Soc.*, 2001, **123**, 4304–4312.

28. S. Zhang, M. Hosaka, T. Yoshihara, K. Negishi, Y. Iida, S. Tobita and T. Takeuchi, *Cancer Res.*, 2010, **70**, 4490–4498.
29. X. Zheng, X. Wang, H. Mao, W. Wu, B. Liu and X. Jiang, *Nat. Commun.*, 2015, **6**, 5834.
30. X. Zheng, H. Tang, C. Xie, J. Zhang, W. Wu and X. Jiang, *Angew. Chem., Int. Ed.*, 2015, **54**, 8094–8099.
31. X. Zheng, H. Mao, D. Huo, W. Wu, B. Liu and X. Jiang, *Nat. Biomed. Eng.*, 2017, **1**, 57.
32. Z. Zhu, Y. Li, X. Li, R. Li, Z. Jia, B. Liu, W. Guo, W. Wu and X. Jiang, *J. Controlled Release*, 2010, **142**, 438–446.
33. A. N. Lukyanov and V. P. Torchilin, *Adv. Drug Delivery Rev.*, 2004, **56**, 1273–1289.
34. Z. Zhu, C. Xie, Q. Liu, X. Zhen, X. Zheng, W. Wu, R. Li, Y. Ding, X. Jiang and B. Liu, *Biomaterials*, 2011, **32**, 9525–9535.
35. T.-C. Cheng, S. R. Roffler, S.-C. Tzou, K.-H. Chuang, Y.-C. Su, C.-H. Chuang, C.-H. Kao, C.-S. Chen, I.-H. Harn, K.-Y. Liu, T.-L. Cheng and Y.-L. Leu, *J. Am. Chem. Soc.*, 2012, **134**, 3103–3110.
36. T. W. Jacobs, A. M. Gown, H. Yaziji, M. J. Barnes and S. J. Schnitt, *Am. J. Clin. Pathol.*, 2000, **113**, 251–258.

CHAPTER 13

# *Probing Tissue Oxygenation by Delayed Fluorescence of Protoporphyrin IX*

RINSE UBBINK AND EGBERT G. MIK*

Erasmus MC – University Medical Center Rotterdam, Department of Anesthesiology, 's-Gravendijkwal 230, 3015 CE, Rotterdam, The Netherlands
*E-mail: e.mik@erasmusmc.nl

## 13.1 Introduction

The adequate supply of oxygen to organs and tissues is of pivotal importance to sustain mammalian life. Aerobic metabolism is maintained through inhalation of air in the lungs and subsequent transport of the absorbed oxygen to tissues *via* the circulating blood. Flow of hemoglobin-bound oxygen through the macro- and microcirculation and diffusion of molecular oxygen into the tissue cells brings oxygen to its ultimate destination, the mitochondria. In the mitochondria oxygen is used in oxidative phosphorylation in order to efficiently produce adenosine triphosphate (ATP) that acts as the energy source for many cellular processes. Oxygen also plays a role in many other biochemical processes and mammalian tissue contains a large number of oxygen-consuming enzymes,[1] for example for reactive oxygen species generation in signal transduction.[2,3]

Accurate measurements of oxygen at various levels, *e.g.* in blood, organ, tissue and (intra)cellular compartments, is of importance in both research

Detection Science Series No. 11
Quenched-phosphorescence Detection of Molecular Oxygen: Applications in Life Sciences
Edited by Dmitri B. Papkovsky and Ruslan I. Dmitriev

Published by the Royal Society of Chemistry, www.rsc.org

and clinical situations. Over the last decades many techniques have been developed for oxygen measurements *in vivo*.[4,5] These techniques gain insight into the mechanisms of oxygen delivery but to some extent are useful also for diagnosis and treatment in patients at the bedside. In the late 1980s the phosphorescence quenching for measurement of oxygen in biological samples was introduced.[6] The possibility to use this technique *in vivo* in experimental animal studies provided researchers with a powerful and scalable tool for quantitative oxygen measurements.

Although phosphorescence quenching of injectable phosphorescent dyes allowed measurement of oxygen at the microvascular and interstitial levels,[7] the ultimate goal of measuring intracellular oxygen at the level of the mitochondria remained elusive. Furthermore, because it is depended on injection of metalloporphyrin derivatives the technique in its original form has not been used in clinical applications. In 2006 we introduced quenching of delayed fluorescence of protoporphyrin IX as a means for measuring oxygen tension ($PO_2$) in mitochondria.[8] This technique uses the optical properties of 5-aminolevulinic acid (ALA)-induced mitochondrial PpIX.

Oxygen-dependent quenching of delayed fluorescence of PpIX can be used in cultured cells, tissues, isolated organs and *in vivo*. Since the technique does not rely on the injection of potentially toxic metalloporphyrin complexes it is also applicable in humans. In this chapter, we will discuss the background of the technology, its technical implementation in a laboratory setting, and the implementation of the technique in a clinical measuring device. Furthermore, we will discuss its use for measuring cellular respiration *in vivo* in addition to mitochondrial oxygen tension (mito$PO_2$).

## 13.2 Background

### 13.2.1 Oxygen

Molecular oxygen is a diatomic gas that constitutes approximately 21% of the volume of air. The electron configuration of the oxygen molecule has two unpaired electrons with the same spin in degenerate orbitals. Therefore, oxygen is paramagnetic and the ground state of oxygen is a triplet state, which is very unusual compared to many other biochemically relevant molecules. The ground state being a triplet state has the important implication that molecular oxygen does not react directly with many other molecules, in contrast to the highly reactive oxygen radical singlet oxygen.

The amount of oxygen dissolved in a liquid depends on the solubility coefficient of oxygen in the solvent and the partial oxygen pressure ($PO_2$) above the solution. Since the solubility coefficients in biological compartments are not well known, it is common to measure and report oxygen levels as $PO_2$ value in kPa, mmHg, or Torr (1 kPa = 7.5 mmHg = 7.5 Torr).

### 13.2.2 Porphyrins

Porphyrins belong to the most abundant molecules on earth and are essential for aerobic life because of their key role in processes related to oxygen production, oxygen transport, and oxygen use.[9] Porphyrins are cyclic macromolecules composed of four modified pyrrole subunits. Porphyrins easily form complexes with metal ions like iron and magnesium and the metal ion defines much of the specific photochemical properties of a porphyrin.

Porphyrins are very good absorbers of light in the visible range. The absorption of a photon leads to photoexcitation of the porphyrin-metal complex. Photoexcitation can lead to population of an excited triplet state. Oxygen, having a triplet state as ground state, is a very effective quencher of this excited triplet state. In the process of quenching, energy is transferred to oxygen. This energy transfer leads to the formation of singlet oxygen, the basic mechanism in porphyrin-based photodynamic therapy.[10] Since the amount of oxygen is inversely related to the triplet state lifetime, the quenching properties can also be used to quantitatively measure the amount of oxygen.

For use in biological samples and *in vivo* it is mandatory that the triplet state lifetime can be easily measured, for example by phosphorescence. The method of measuring $PO_2$ in biological systems by means of oxygen-dependent quenching of phosphorescence was introduced by Vanderkooi *et al.*[6] at the end of the 1980s. Complexes of porphyrins with certain heavy metals show high phosphorescence yield and are very efficient oxygen probes. Palladium-meso-tetra(4-carboxyphenyl)porphine (Pd-porphyrin), bound to albumin before injection, has become a standard phosphorescent dye for microvascular $PO_2$ measurements *in vivo*.[11,12]

### 13.2.3 Protoporphyrin IX

Oxygen-dependent quenching of phosphorescence has proven itself to be a very useful technology for biological oxygen measurements. The technique is very scalable and applications range from single cell microscopy[13] to intravital microscopy[14,15] to fiber-based measurements on organs like the heart[16] and kidney.[17] In our laboratory we use phosphorescence quenching to measure microvascular oxygen tension in various animal models for studying pathophysiological mechanisms and experimental therapies in perioperative and intensive care.[18–21] In order to improve performance of the oxygen-sensitive phosphors, for example to reduce pH-dependency or improve the two-photon absorption cross-section, new porphyrin-based oxygen-sensitive dyes have been developed over recent years.[22–24]

The direct and quantitative measurement of microvascular $PO_2$ clearly has potential clinical applications. However, due to the need for injection of the phosphors and concerns about potential toxicity of the metalloporphyrin complexes the technique did not make the clinical step. This latter

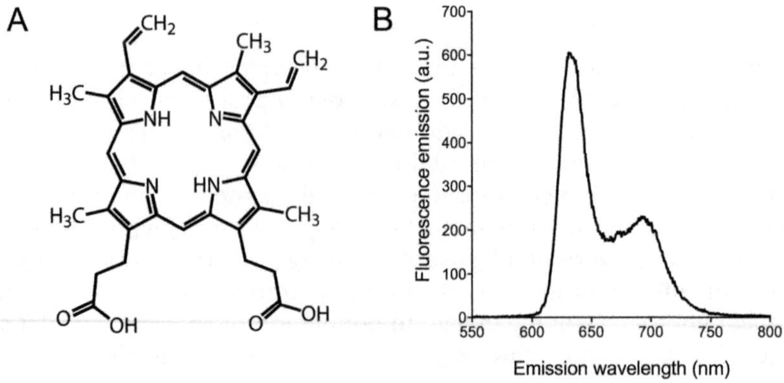


**Figure 13.1** A. Structure of protoporphyrin IX. B. Fluorescence emission spectrum of protoporphyrin IX, excited at 405 nm. The two-peaked emission spectrum is typical for this porphyrin.

limitation could potentially be overcome by using an endogenously synthesized porphyrin. But this porphyrin should meet the following requirements: it should have oxygen-sensitive optical properties, which are easy to measure, and which can be calibrated so the amount of oxygen can be measured as an absolute value. For this reason, we started the search for an endogenous porphyrin as oxygen-sensitive dye several years ago and found protoporphyrin IX (PpIX) to have the right properties. The structure formula of PpIX and its fluorescence emission spectrum are shown in Figure 13.1.

PpIX is endogenously present and is synthesized inside the mitochondria.[25] PpIX is the final precursor of heme in the heme biosynthetic pathway, and the conversion of PpIX to heme is a rate-limiting step. Administration of the porphyrin precursor 5-aminolevulinic acid (ALA) to cells, tissues, and animals substantially enhances the PpIX concentration.[26] Furthermore, PpIX possesses a triplet state that reacts strongly with oxygen.[27] While PpIX shows no detectable phosphorescence[28] it does emit a weak oxygen-sensitive delayed fluorescence,[8] with the same emission spectrum as the prompt fluorescence and of which the lifetime relates quantitatively to the oxygen tension. Since ALA-enhanced PpIX is located in the mitochondria, PpIX acts as an intramitochondrial oxygen-sensitive dye.

### 13.2.4 Delayed Fluorescence Quenching

Figure 13.2 shows the Jablonski diagram describing interactions between protoporphyrin IX and oxygen. Excitation of PpIX to the first singlet state ($S_1$) can lead to population of the triplet state ($T_1$) by a process called intersystem crossing. During intersystem crossing the electron is relaxed to the triplet state by changing its spin orientation without emission of a photon. The probability that a porphyrin molecule in the $T_1$ state relaxes to the $S_0$

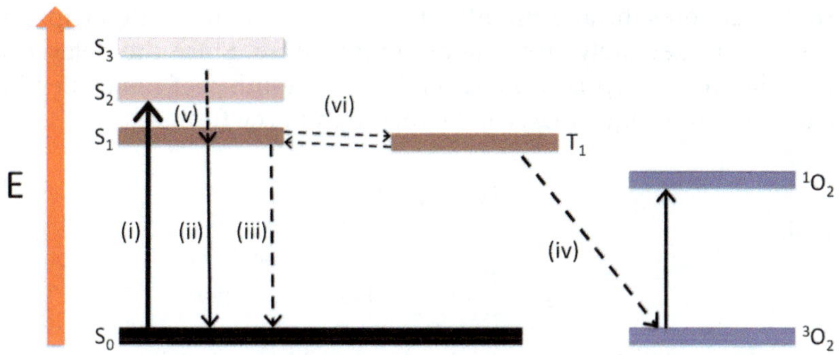


**Figure 13.2** Jablonski diagram of steady and transition states of protoporphyrin IX and molecular oxygen. Absorption (i), and (delayed) fluorescence (ii), radiationless transitions (iii), energy transfer (iv), internal conversion (v), bidirectional intersystem crossing (vi), and $S_0$, $S_1$, $S_2$, and $S_3$ denote the ground state and the first three excited singlet states of protoporphyrin IX, respectively. $T_1$ denotes the first excited triplet state. $^3O_2$ and $^1O_2$ denote the triplet ground state of oxygen and excited singlet oxygen, respectively.

state by spontaneous relaxation is denoted by rate constant $k_s$. Relaxation to $S_0$ can also occur by "triple-state quenching", a process in which a colliding oxygen molecule absorbs the energy from the porphyrin.

Triplet-state quenching results in relaxation of the porphyrin without emission of a photon and causes the triplet-state lifetime to be dependent on the collision frequency. The collision frequency is determined by the amount of oxygen and the chance that a single oxygen molecule causes a quenching event. The latter is defined by the Smoluochowski equation:

$$k_q = 4\pi N\gamma(D_o + D_f)(R_o + R_f) \tag{13.1}$$

in which $k_q$ is known as the quenching constant, $N$ is Avogadro's number, $\gamma$ the quenching efficiency, $D_o$ and $D_f$ are the diffusion coefficients of oxygen and the porphyrin, respectively, and $R_o$ and $R_f$ are the quenching radius of oxygen and the porphyrin, respectively.

The rate of relaxation of $T_1$ to $S_0$ is thus determined by both the transition probability by spontaneous relaxation ($k_s$) and the quenching probability ($k_q$). The decay rate of PpIX molecules in the excited triplet state after excitation with a pulse of light is given by the following differential equation:

$$\frac{d[T_1](t)}{dt} = -k_s[T_1](t) - k_q PO_2[T_1] \tag{13.2}$$

where $[T_1]$ denotes the amount of PpIX molecules in the excited triplet state, and $PO_2$ is the oxygen tension in the surrounding medium. Under ideal circumstances, *i.e.*, when the excitation pulse duration is much shorter than the triplet decay time, the solution of this differential equation yields:

$$[T_1](t) = [T_1]_0 e^{-(k_s + k_q PO_2)t} \tag{13.3}$$

where $[T_1]_0$ denotes the amount of PpIX molecules in the excited triplet state at $t = 0$, *i.e.* immediately after the excitation pulse. Since the delayed fluorescence intensity ($I(t)$) is proportional to the number of molecules in the excited triplet state eqn (13.3) can be rewritten in the form:

$$I(t) = [I]_0 e^{\left(-\frac{t}{\tau}\right)} \tag{13.4}$$

in which

$$\frac{1}{\tau} = \frac{1}{\tau_0} + k_q PO_2 \tag{13.5}$$

This last equation is known as the Stern–Volmer relationship, in which $\tau$ is the delayed fluorescence lifetime and $\tau_0 = 1/k_s$ is the time constant of the decay of the delayed fluorescence in the absence of oxygen, *i.e.*, the decay time of spontaneous relaxation. Since the measured $PO_2$ in the case of aminolevulinic acid-induced mitochondrial PpIX is the oxygen tension in the mitochondria (mito$PO_2$) the latter can be calculated from the delayed fluorescence lifetime as follows:

$$\text{mitoPO}_2 = \frac{\frac{1}{\tau} - \frac{1}{\tau_0}}{k_q} \tag{13.6}$$

Methods to retrieve the delayed fluorescence lifetime(s) in case of non-homogenous mito$PO_2$ (the situation that is normally present due to respiring mitochondria causing oxygen gradients) are explained later in this chapter.

### 13.2.5 Quenching Constants

To be able to calculate mito$PO_2$ from the measured delayed fluorescence lifetime one has to know the calibration constants, *i.e.* the quenching constant ($k_q$) and the lifetime in the absence of oxygen ($\tau_0$), from *in situ* mitochondrial protoporphyrin IX. Mito$PO_2$ reflects the local balance between oxygen supply and consumption.[29] Due to mitochondrial respiration and the presence of a diffusion gradient from a higher extracellular to lower intracellular oxygen tension one can only perform calibrations in non-respiring cells and tissues. Only under such circumstances one can assume the mito$PO_2$ to equal an applied extracellular $PO_2$.

We have performed calibration experiments in which we clamped the extracellular $PO_2$ to a fixed and known value in either cell culture medium or perfusate while mitochondrial respiration was blocked by cyanide and rotenone in various models. Our first calibrations were performed in cultured cells[8] and these were quickly followed by calibrations in isolated rat hepatocytes, cardiomyocytes, isolated perfused livers and hearts.[30,31] Recently we also performed calibrations in rat skin *in vivo*.[32] At a temperature of 37 °C, the values for the quenching constants have been found to be approximately $k_q = 830$ mmHg $s^{-1}$ and $\tau_0 = 0.8$ milliseconds. No important differences have been found between the various tissues.

## 13.3 Measuring Mitochondrial $PO_2$

### 13.3.1 Aminolevulinic Acid

Protoporphyrin IX is endogenously present in mammalian tissues but due to a negative feedback mechanism on its synthesis the normal levels are very low. In practice this means that delayed fluorescence of PpIX can only be detected after enhancement of its levels by administration of the porphyrin precursor 5-aminolevulinic acid (ALA). ALA is the first compound in the porphyrin synthesis pathway and is produced by the enzyme ALA synthase from glycine and succinyl-CoA. For research and clinical purposes ALA is available in various forms, for example as an easily water-soluble powder for cell culture, injection and oral administration, or as plasters for administration of ALA to the skin.

In the cytosol ALA is converted to coproporphyrinogen III (CPIII) *via* two intermediate steps. CPIII is transported into the mitochondria and converted to protoporphyrinogen IX and subsequently protoporphyrin IX. The conversion of PpIX to heme, by incorporation of an iron atom in the porphyrin structure, is a rate-limiting step. Therefore, administration of ALA leads to accumulation of PpIX in the mitochondria (Figure 13.3) and effectively induces an oxygen-measuring probe in these organelles.

ALA itself is very hydrophilic and dissolves easily in water, making it easy to administer to biological samples such as cell cultures. Furthermore, ALA can be dissolved in a buffer and injected in the blood stream of an experimental animal to allow $mitoPO_2$ measurements in various organs. Alternatively, local

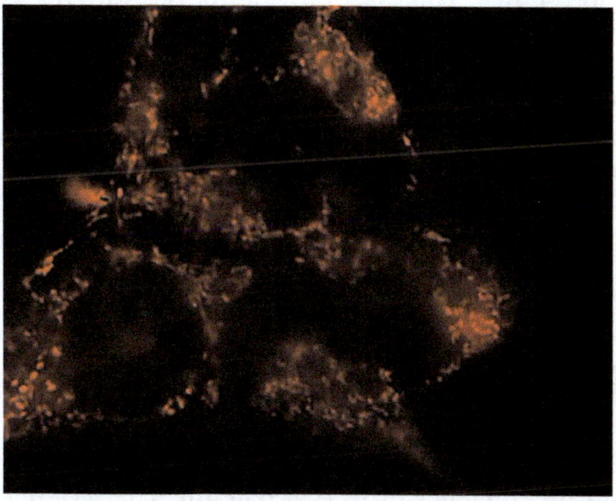

**Figure 13.3** Fluorescence microscopy picture of cultured neuroblastoma cells 4 hours after addition of 5-aminolevulinic acid (ALA) to the culture medium, using a Cy3 bandpass filter. The red spots are the mitochondria that became fluorescent due to ALA-induced accumulation of protoporphyrin IX. In previous work we showed that the red fluorescence co-localizes with the mitochondria-specific probe MitoTracker Green.[8]

application of ALA to tissues in the form of an ALA cream or ALA plaster is also possible. We used the latter in our first measurements of mitochondrial oxygen tension in humans[33] and for our first implementation of mitochondrial oxygen measurements in a clinical device.[34] Depending on the application, type of cells or tissues and ALA administration method the optimal timing for delayed fluorescence measurements after ALA administration is typically 2–4 hours.

## 13.3.2 Laboratory Setup

The basic principle of the delayed fluorescence measurements is straightforward and very similar to phosphorescence lifetime measurements. A complicating factor is that the relatively weak delayed fluorescence signal is easily overwhelmed by intense prompt fluorescence. Because prompt and delayed fluorescence share the same emission spectrum optical filtering is not possible. Therefore, the prompt fluorescence potentially leads to distortion of the signal through a temporary saturation of the sensitive photodetector and electronics. Pulsed excitation with a q-switched laser in combination with time-gated detection is used to overcome such problems.

### *13.3.2.1 Excitation Source*

Various and extensive descriptions of setups that have been used for mito$PO_2$ measurements have been published previously.[8,31,35,36] Currently, we use a compact computer-controlled tunable laser (Opolette 355-I, Opotek, Carlsbad, CA) as excitation source. This laser provides pulses with a specified duration of 4–10 nanoseconds and typically 2–4 mJ per pulse over the tunable range of 410–670 nm. The Opolette is coupled into a multimode fiber optic with a core diameter of 1000 μm by a Fiber Delivery System (Opotek), which acts as interconnect between excitation light source and the actual measuring fiber (varies with specific task). For example, for measuring mito$PO_2$ at the surface of organs, we use a bifurcated reflection probe (FCR-7IR400-2-ME, Avantes b.v., Eerbeek, the Netherlands). Excitation light levels at the output of the excitation branch are typically in the order of 50–200 μJ per pulse.

### *13.3.2.2 Gated Detector*

The PpIX signal is detected by a gated microchannel plate photomultiplier tube (MCP-PMT R5916U series, Hamamatsu Photonics, Hamamatsu, Japan), having a quantum efficiency of 24% at 650 nm. The MCP-PMT is mounted on a gated socket assembly (E3059-501, Hamamatsu Photonics) and cooled to −30 °C by a thermoelectric cooler (C10373, Hamamatsu Photonics). The MCP-PMT is operated at a voltage in the range of 2300–3000 V by a regulated high-voltage DC power supply (C4848-02, Hamamatsu Photonics). The emission branch of the reflection probe is coupled to the detector by in-house built optics. The MCP-PMT is configured for the "normally on" mode, and the PpIX emission light is filtered by a combination of a 590 nm longpass filter (OG590, Newport, Irvine, CA) and a broadband (675 ± 25 nm) bandpass filter (Omega Optical, Brattleboro, VT).

#### *13.3.2.3 Data-acquisition System*

The output current of the photomultiplier is voltage-converted by an amplifier with an input impedance of 50 Ω, 500 times voltage amplification, and a bandwidth around 50 MHz. Data acquisition is performed by a PC-based data-acquisition system containing a 10 megasamples per second (MS $s^{-1}$) simultaneous sampling data acquisition board (NI–PCI-6115, National Instruments, Austin, TX). The data acquisition runs at a rate of 10 MS $s^{-1}$, and typically, a number of laser pulses (repetition rate 20 Hz) are averaged before analysis to increase signal-to-noise ratio. Control of the setup and analysis of the data are performed with software written in LabView (National Instruments, Austin, TX, USA). Examples of typical delayed fluorescence signals obtained from *in vivo* experiments are shown in Figure 13.4.

#### *13.3.2.4 Signal Analysis*

The delayed fluorescence lifetime in case of a homogenous oxygen concentration and in the absence of self-quenching (*i.e.* at sufficiently low PpIX concentration) will decay monoexponentially according to eqn (13.4). However, the oxygen distribution in cells and tissues is in general not homogenous, due to the presence of oxygen gradients. In such cases, the delayed fluorescence signal can be described by an integral over an exponential kernel:

$$y(t) = \int_0^t \exp(-\lambda t) f(\lambda) \mathrm{d}\lambda \tag{13.7}$$

where $f(\lambda)$ denotes the spectrum of reciprocal lifetimes that should be determined from the finite data set $y(t)$. The underlying distribution of oxygen tensions can be recovered by assuming that the delayed fluorescence signal can be described by a sum of rectangular distributions with adequately small chosen width ($2\delta$), resulting in the following fit equation:[37]

$$Y^*(t) = Y(t)\exp(k_0 t)\frac{k_q \delta t}{\sinh(k_q \delta t)} = \sum w_i \exp(-k_q Q_i t) \tag{13.8}$$

where $Y(t)$ is the normalized phosphorescence data, $k_0$ is the first-order rate constant for delayed fluorescence decay in the absence of oxygen, $k_q$ is the quenching constant, and $w_i$ is the weight factor for the according bin with central $PO_2$ $Q_i$ and width $2\delta$ ($w_i >= 0$ and $\Sigma w_i = 1$). We have used this approach for the recovery of mito$PO_2$[30,31] histograms.

Deconvolution of detailed lifetime distributions in the time domain is relatively slow and especially for fast real-time signal analysis it is more convenient to determine mean mito$PO_2$ and an estimate of its variance directly

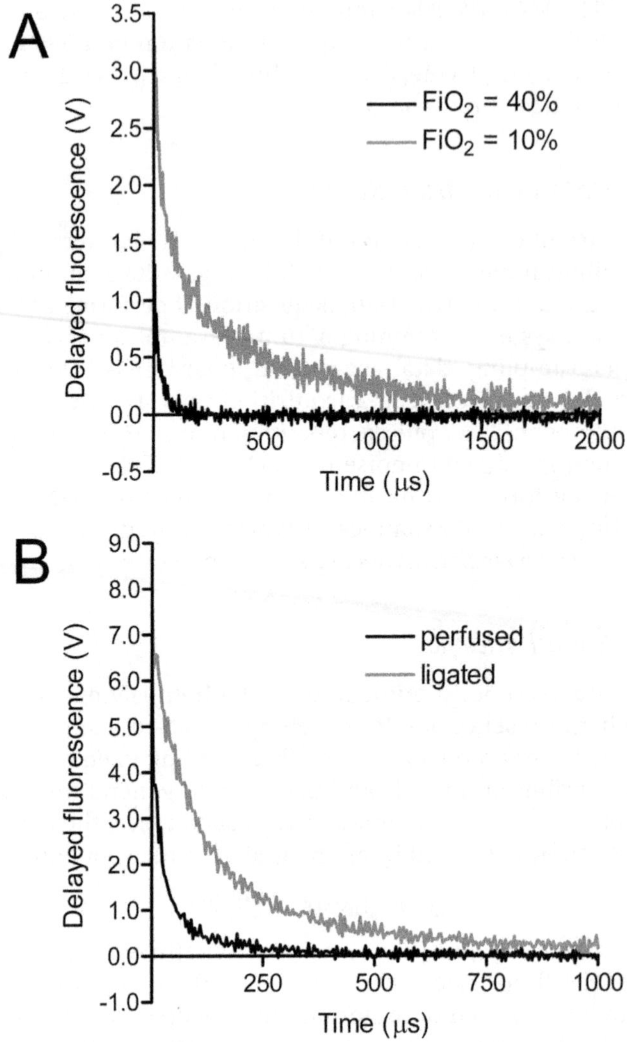


**Figure 13.4** Examples of delayed fluorescence signals obtained *in vivo* in an anesthetized and mechanically ventilated rat after intravenous administration of 5-aminolevulinic acid. A. Protoporphyrin IX delayed fluorescence measured in skin from the hind limb before and after ligation of the limb and cessation of blood flow. B. Protoporphyrin IX delayed fluorescence measured in the rectus abdominis muscle during mechanical ventilation with two different inhaled oxygen concentrations. Reprinted from *Biophysical Journal*, **95**, E. G. Mik, T. Johannes, C. J. Zuurbier, A. Heinen, J. H. Houben-Weerts, G. M. Balestra, J. Stap, J. F. Beek and C. Ince, *In vivo* Mitochondrial Oxygen Tension Measured by a Delayed Fluorescence Lifetime Technique, 3977–3990,[31] Copyright 2008, with permission from Elsevier.

from the photometric signal. This can be achieved by fitting distributions of quencher concentration to the delayed luminescence data.[15] The fitting function for a simple rectangular distribution with a mean $PO_2$ $Q_m$ and a $PO_2$ range from $Q_m - \delta$ till Qm + $\delta$ is:

$$Y_R(t) = \exp\left(-\left(k_0 + k_q Q_m\right)t\right)\frac{\sinh(k_q \delta t)}{k_q \delta t} \tag{13.9}$$

where $Y_R(t)$ is the normalized delayed fluorescence data, $k_0$ is the first-order rate constant for delayed fluorescence decay in the absence of oxygen, $k_q$ is the quenching constant, and $\delta$ is half the width of the rectangular distribution. In terms of quenching constants and the Stern–Volmer relationship, eqn (13.9) can be rewritten as:

$$Y_R(t) = \exp\left(-\left(\frac{1}{\tau_0} + k_q \langle PO_2 \rangle\right)t\right)\frac{\sinh\left(k_q \delta t\right)}{k_q \delta t} \tag{13.10}$$

where $\langle PO_2 \rangle$ is the mean $PO_2$ within the sample volume and $\tau_0$ the lifetime in the absence of oxygen. The second moment of the assumed rectangular (or uniform) distribution, the variance ($\sigma^2$), can be calculated from $\delta$ as:

$$\sigma^2 = \frac{\delta^2}{3} \tag{13.11}$$

Fitting eqn (13.10) to the photometric signal directly provides mean $mitoPO_2$ and an estimation of the heterogeneity of $mitoPO_2$ within the sample volume. A rectangular distribution is only a rough first approximation to many heterogeneous systems. Although the retrieval of mean $mitoPO_2$ has been proven reliable and robust, changes in variance should be interpreted with care. Variance should not be used for interpretation if it is uncertain whether a simple rectangular distribution is a feasible model for fitting of the data.

### *13.3.2.5 Multi Compartment Measurements*

For *in vivo* oxygen measurements so-called near-infrared phosphors have been developed which emit phosphorescence in the near-infrared region of the optical spectrum, while being exited with visible light.[38] Since the emission spectrum (with a peak typically above 800 nm) does not overlap with the emission from PpIX, an additional detector allows a near-infrared oxygen-sensing dye to be used simultaneously with PpIX delayed fluorescence measurements. Using Oxyphor G2 (Oxygen Enterprises Ltd., Philadelphia, PA, USA), a Pd-meso- tetra-(4-carboxyphenyl)-tetrabenzoporphyrin, we have been able to simultaneously measure $mitoPO_2$ and extracellular $PO_2$ in suspensions of isolated cells.[8] Also, Oxyphor G2 can be injected in the blood stream of experimental animals for microvascular oxygen measurements. This principle (see Figure 13.5) allowed us to measure simultaneously mitochondrial and microvascular oxygen tension in various animal models.[32,35,39]

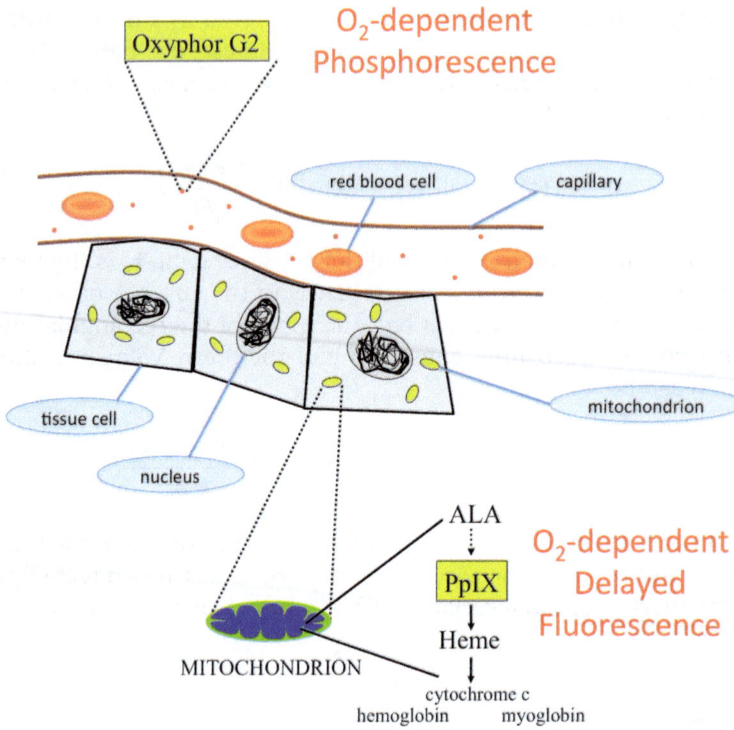


**Figure 13.5** Scheme of the measuring concept for simultaneous microvascular and mitochondrial oxygen measurements. Oxyphor G2 (a near-infrared emitting oxygen-sensing phosphorescent probe) is directly injected into the bloodstream and is used as microvascular oxygen probe by means of oxygen-dependent quenching of phosphorescence. PpIX is induced in the mitochondria by administration of its precursor ALA and is used as mitochondrial oxygen probe by means of oxygen-dependent quenching of delayed fluorescence. Reprinted with permission from Bodmer *et al.*[35] Copyright © 2012 WILEY-VCH Verlag GmbH & Co. KGaA, Weinheim.

## 13.4 *In vivo* Cellular Respirometry

Mitochondrial oxygen tension reflects the balance between local oxygen supply and oxygen consumption. Therefore, assumed measurements are performed *in vivo*, a change in mito$PO_2$ can be caused by an extracellular event (*e.g.* a change in blood flow or hematocrit), by an intracellular event (*e.g.* mitochondrial damage or metabolic adaptation) or by a combination of both. Under certain complex pathophysiological circumstances, for example during sepsis, it is often unclear whether the metabolic derailment is caused by insufficient oxygen supply (due to circulatory failure and shock) or cellular dysfunction.[40] Therefore, methods that probe tissue oxygenation and are able to distinguish between deviations caused by supply or demand are most welcome in both the research and clinical setting alike.

In addition to mito$PO_2$ the PpIX delayed fluorescence technique can be used to gain insight into local mitochondrial oxygen consumption (mito$VO_2$) by directly assessing the oxygen disappearance rate (ODR) under the measuring probe while occluding the microcirculation by applying pressure with the probe. The oxygen disappearance curve, *i.e.* the kinetics of mito$PO_2$ after cessation of local blood flow, can be analyzed by an adapted Michaelis–Menten approach. In contrast to *in vitro* respirometry in oxygen-closed reaction vessels, oxygen can diffuse back into the measuring volume. This factor has to be taken in account when performing analysis *in vivo*. We have previously shown that in skin the following formula can be used if auto-consumption of oxygen by the measuring technique is negligible:[41]

$$dP_n/d_n = -(V_0 \cdot P_n)/(P_{50} + P_n) + Z(P_0 - P_n) \tag{13.12}$$

In this equation, $P_n$ is the measured $PO_2$ after excitation flash number $n$, $P_0$ is the mean $PO_2$ before stop-flow, $Z$ is the inflow coefficient of oxygen, $P_{50}$ is the $PO_2$ at which cellular oxygen consumption is reduced to 1/2 $V_0$, and $dP_n/d_n$ is the rate of oxygen disappearance.

An organ in which both mito$PO_2$ and the mitochondrial ODR can be easily measured is the skin. Using this approach, we recently demonstrated that endotoxemia causes changed cellular respiration despite tissue oxygenation being restored to normal values by resuscitation.[42,43]

## 13.5 Clinical Implementation

### 13.5.1 The Skin as a "Canary" of the Body

For delayed fluorescence measurements the human skin is an easily accessible and non-invasive location allowing assessment of mitochondrial oxygenation and oxygen consumption in a clinically relevant organ. The fact that ALA can be applied topically by means of a cream or plaster eases the adoption of the technique because of the lack of need for systemic administration of the compound and the associated concerns about generalized (photo) toxicity. Furthermore, optical probes can be attached to the skin without being invasive in contrast to, for example, measurements in muscle or internal organs that would require some kind of percutaneous procedure.

Measurements in the skin are clinically meaningful because of the fact that, like the gut, the skin acts as a "canary" of the body. This means that, for example, in certain pathophysiological circumstances like hemodilution[44] cutaneous mito$PO_2$ changes foretell changes in systemic parameters and other organs. In addition, changes in cutaneous oxygen disappearance rate correlate with ODR changes in other organs and tissues.[43] Clinical studies to further evaluate the "canary" hypothesis are needed[45] and several clinical studies in various institutions are currently underway.

### 13.5.2 Priming of the Skin

Preparation of the skin for mitochondrial oxygen measurements is needed in order to induce enough mitochondrial PpIX for delayed fluorescence detection. We use a self-adhesive patch containing 8 mg ALA (Alacare®; Photonamic GmbH & Co, Wedel, Germany) for induction of PpIX. Our preferred place to measure is the skin on the anterior chest wall above or just next to the sternum. Some skin preparation enhances ALA penetration and shortens the time until adequate signal is obtained.[33] To this end, hair (if present) is shaved and the skin is rubbed with a fine abrasive pad, *e.g.* as present on a standard electrocardiogram (ECG) sticker, to remove the top parts of the stratum corneum. After approximately 4 hours an adequate delayed fluorescence signal can be expected at the site of the patch. To prevent photobleaching of the PpIX the skin is protected from ambient light after removal of the ALA patch.

### 13.5.3 The COMET Measuring System

The oxygen-dependent quenching of delayed fluorescence of PpIX technique for measuring mito$PO_2$ in cells, tissues and *in vivo* has shown to be robust and applicable in the laboratory setting. The next step was to bring this technology to the patient's bedside. New requirements needed to be fulfilled including ease of transport, applicability to a patient in bed, protection against spilled water, laser safety and protection against ambient light entering the optical system, since the photo multiplier tube will be damaged immediately if it is exposed to direct sunlight or intense surgical lighting in the operating room.

After successful fulfillment of all requirements a new monitor was engineered called "COMET", an acronym for Cellular Oxygen METabolism. The COMET measuring system enables physicians to measure oxygen tension and oxygen consumption at the subcellular level in the mitochondria. Since April 2016 the COMET is a CE-marked device made by Photonics Healthcare, Utrecht, The Netherlands. The laser used for this measuring system is a delicate device and especially the size was a real challenge. The laboratory setup is fixed and fills a wall and the first attempt to build a transportable device for clinical use was still refrigerator sized.[43] Although it allowed the first study in humans,[33] this first prototype device was not useable in the clinical setting.

The COMET without the cradle is 22 × 33 × 29 cm in size, weighs around 10 kg and without restrictions can be used in a clinical environment (see Figure 13.6). The use of an ALA-containing self-adhesive patch and a specially developed optical Skin Sensor makes it applicable at the bedside. The light source in the COMET is a 515 nm pulsed laser with pulse duration of 60 ns and a 10 Hz repetition rate. The fluorescent signal is projected on a gated red-sensitive photomultiplier tube. A 12 inch TFT-LCD screen enables users to interact and start a single, multiple or dynamic measurements. Apart from the main switch to turn on the device, the COMET has no physical buttons.

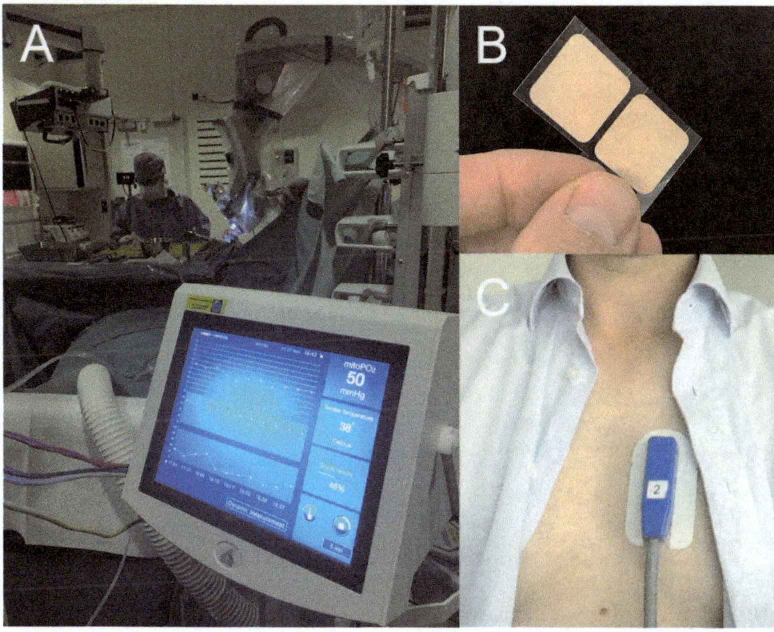


**Figure 13.6** The COMET measuring system, the first clinical device for mitochondrial oxygen measurements based on delayed fluorescence of protoporphyrin IX. A. Picture of a COMET monitor during measurements for a clinical study in the operating room. B. Impression of the used ALA-containing plasters for priming of the skin. C. The Skin Sensor positioned on the chest of a healthy volunteer.

If a USB storage device is inserted in the USB-port on the rear panel the data is exported in a comma separated file format. An extensive description of the COMET device can be found elsewhere.[34]

### 13.5.4 Example Measurements

To demonstrate the usability of the COMET monitor we provide two examples from actual measurements with the device. The first example is a measurement of cutaneous mito$PO_2$ in laboratory experiment in a pig. The animal was anesthetized, mechanically ventilated under hemodynamic monitoring and control. After reaching a stable baseline situation, a protocol of progressive continuous normovolemic hemodilution was started. This means that from an arterial access blood was continuously removed at a constant rate, while at the same rate intravenous fluid was simultaneously administered *via* venous access. The effect of this intervention is that the hematocrit value continually decreased, while the circulation and blood pressure remained largely intact. The effect on cutaneous mito$PO_2$ is shown in Figure 13.7.

At time point A the hemodilution protocol starts. Initially compensatory mechanisms, *i.e.* increase in cardiac output and blood flow to the tissues,

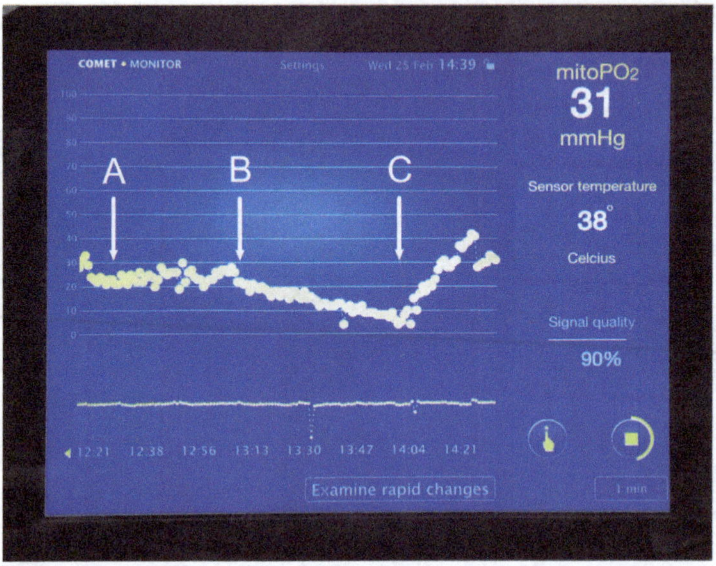


**Figure 13.7** An interval measurement of cutaneous mitochondrial oxygen tension ($mitoPO_2$) performed in a pig during progressive hemodilution and subsequent blood transfusion. In this example, which is a photograph of the screen of a COMET device, $mitoPO_2$ is measured at an interval of 1 minute. At time point A the hemodilution starts, at time point B a critical hematocrit is reached and $mitoPO_2$ becomes dependent on the hemoglobin level, at time point C the animal receives a blood transfusion.

keep up with the decrease in hematocrit and $mitoPO_2$ remains stable. At time point B a critical hematocrit is reached in which the compensatory mechanisms start to fail. From here $mitoPO_2$ becomes dependent on the hematocrit and starts to decline with ongoing hemodilution. At time point C a blood transfusion is given which restores the hematocrit above the critical value and $mitoPO_2$ restores to values above baseline.

The second example shows a dynamic $mitoPO_2$ measurement in a healthy volunteer (Figure 13.8). At time point A the Skin Sensor is positioned without causing pressure on the skin. At time point B a steady signal was obtained and a number of baseline measurements were done. At time point C pressure was applied on the Skin Sensor and the local microcirculation in the skin under the sensor was occluded. Because oxygen supply was stopped while cellular oxygen consumption went on, $mitoPO_2$ immediately decreased. The rate of this decrease could be analyzed according to eqn (13.12) in order to quantify the ODR. At time point D the pressure was released. In a healthy microcirculation the blood flow will immediately restore and show a temporarily physiological overshoot. Such overcompensation caused $mitoPO_2$ to temporarily rise above baseline values (time point E).

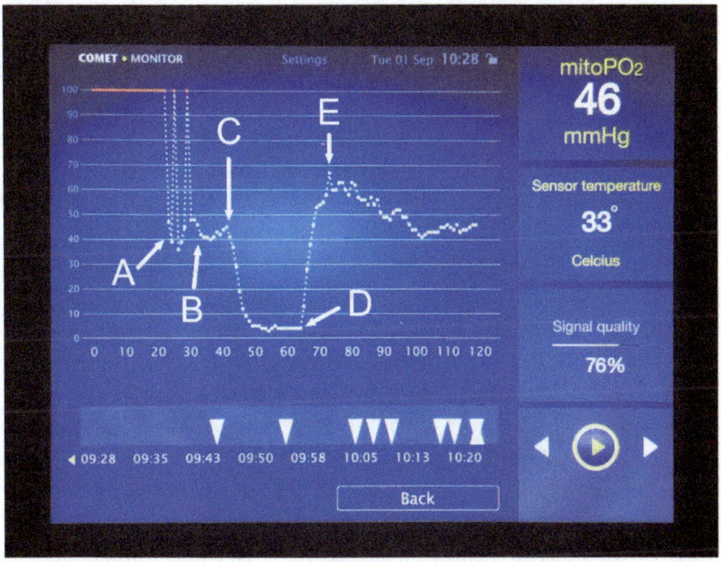


**Figure 13.8** A dynamic measurement of cutaneous mitochondrial oxygen tension (mito$PO_2$) performed in a healthy volunteer. In this example, which is a photograph of the screen of a COMET device, mito$PO_2$ is measured at an interval of 1 second. At time point A the measuring probe is positioned, at time point B a baseline is measured, at time point C pressure is applied on the measuring probe to occlude the local microvessels and cease oxygen supply to the skin, at time point D the pressure is released and at time point E one sees the physiological hyperemic reaction resulting in a temporary overshoot of mito$PO_2$ compared to baseline. In one screenshot one gets information about tissue oxygenation (baseline level), cellular oxygen consumption (during application of pressure) and microvascular function (*i.e.* the normal physiological hyperemic reaction).

## 13.6 Conclusion

Oxygen-dependent quenching of delayed fluorescence of ALA-enhanced mitochondrial PpIX can be used to measure tissue oxygenation on the mitochondrial level. The technique is scalable from isolated cells to *in vivo* measurements and is safe and robust enough for bedside measurements in clinical applications.

## Conflict of Interest Statement

Dr E. G. Mik is founder and shareholder and R. Ubbink is shareholder of Photonics Healthcare B. V., Utrecht, The Netherlands. Photonics Healthcare develops and markets the COMET measuring system. Photonics Healthcare holds the exclusive licenses to several patents regarding this technology, filed and owned by the Academic Medical Center in Amsterdam and the Erasmus Medical Center in Rotterdam, The Netherlands.

## References

1. J. M. Vanderkooi, M. Erecinska and I. A. Silver, *Am. J. Physiol.*, 1991, **260**, C1131–C1150.
2. W. Droge, *Physiol. Rev.*, 2002, **82**, 47–95.
3. Q. Xi, S. Y. Cheranov and J. H. Jaggar, *Circ. Res.*, 2005, **97**, 354–362.
4. R. Springett and H. M. Swartz, *Antioxid. Redox Signaling*, 2007, **9**, 1295–1301.
5. H. M. Swartz and J. F. Dunn, *Adv. Exp. Med. Biol.*, 2003, **530**, 1–12.
6. J. M. Vanderkooi, G. Maniara, T. J. Green and D. F. Wilson, *J. Biol. Chem.*, 1987, **262**, 5476–5482.
7. D. F. Wilson, O. S. Finikova, A. Y. Lebedev, S. Apreleva, A. Pastuszko, W. M. Lee and S. A. Vinogradov, *Adv. Exp. Med. Biol.*, 2011, **701**, 53–59.
8. E. G. Mik, J. Stap, M. Sinaasappel, J. F. Beek, J. A. Aten, T. G. van Leeuwen and C. Ince, *Nat. Methods*, 2006, **3**, 939–945.
9. G. Layer, J. Reichelt, D. Jahn and D. W. Heinz, *Protein Sci.*, 2010, **19**, 1137–1161.
10. A. E. O'Connor, W. M. Gallagher and A. T. Byrne, *Photochem. Photobiol.*, 2009, **85**, 1053–1074.
11. L. W. Lo, C. J. Koch and D. F. Wilson, *Anal. Biochem.*, 1996, **236**, 153–160.
12. M. Sinaasappel and C. Ince, *J. Appl. Physiol.*, 1996, **81**, 2297–2303.
13. M. C. Hogan, *J. Appl. Physiol.*, 2001, **90**, 1871–1876.
14. M. Sinaasappel, C. Donkersloot, J. van Bommel and C. Ince, *Am. J. Physiol.*, 1999, **276**, G1515–G1520.
15. L. Zheng, A. S. Golub and R. N. Pittman, *Am. J. Physiol.*, 1996, **271**, H365–H372.
16. C. J. Zuurbier, M. van Iterson and C. Ince, *Cardiovasc. Res.*, 1999, **44**, 488–497.
17. T. Johannes, E. G. Mik and C. Ince, *J. Appl. Physiol.*, 2006, **100**, 1301–1310.
18. T. Johannes, C. Ince, K. Klingel, K. E. Unertl and E. G. Mik, *Crit. Care Med.*, 2009, **37**, 1423–1432.
19. T. Johannes, E. G. Mik and C. Ince, *Shock*, 2009, **31**, 97–103.
20. T. Johannes, E. G. Mik, K. Klingel, H. J. Dieterich, K. E. Unertl and C. Ince, *Shock*, 2009, **31**, 521–528.
21. T. Johannes, E. G. Mik, K. Klingel, P. T. Goedhart, C. Zanke, B. Nohe, H. J. Dieterich, K. E. Unertl and C. Ince, *Clin. Exp. Pharmacol. Physiol.*, 2009, **36**, 870–879.
22. T. V. Esipova, A. Karagodov, J. Miller, D. F. Wilson, T. M. Busch and S. A. Vinogradov, *Anal. Chem.*, 2011, **83**, 8756–8765.
23. A. Fercher, S. M. Borisov, A. V. Zhdanov, I. Klimant and D. B. Papkovsky, *ACS Nano*, 2011, **5**, 5499–5508.
24. R. N. Gillanders, O. V. Arzhakova, A. Hempel, A. Dolgova, J. P. Kerry, L. M. Yarysheva, N. F. Bakeev, A. L. Volynskii and D. B. Papkovsky, *Anal. Chem.*, 2010, **82**, 466–468.
25. R. Poulson, *J. Biol. Chem.*, 1976, **251**, 3730–3733.
26. H. Fukuda, A. Casas and A. Batlle, *Int. J. Biochem. Cell Biol.*, 2005, **37**, 272–276.

27. J. Dalton, C. A. McAuliffe and D. H. Slater, *Nature*, 1972, **235**, 388.
28. S. J. Chantrell, C. A. McAuliffe, R. W. Munn and A. C. Pratt, *J. Chem. Soc., Faraday Trans. 1*, 1977, **60**, 858–865.
29. E. G. Mik, *Anesth. Analg.*, 2013, **117**, 834–846.
30. E. G. Mik, C. Ince, O. Eerbeek, A. Heinen, J. Stap, B. Hooibrink, C. A. Schumacher, G. M. Balestra, T. Johannes, J. F. Beek, A. F. Nieuwenhuis, P. van Horssen, J. A. Spaan and C. J. Zuurbier, *J. Mol. Cell. Cardiol.*, 2009, **46**, 943–951.
31. E. G. Mik, T. Johannes, C. J. Zuurbier, A. Heinen, J. H. Houben-Weerts, G. M. Balestra, J. Stap, J. F. Beek and C. Ince, *Biophys. J.*, 2008, **95**, 3977–3990.
32. F. A. Harms, S. I. Bodmer, N. J. Raat, R. J. Stolker and E. G. Mik, *Opt. Lett.*, 2012, **37**, 2625–2627.
33. F. A. Harms, R. J. Stolker and E. G. Mik, *PLoS One*, 2016, **11**, e0159544.
34. R. Ubbink, M. A. Bettink, R. Janse, F. A. Harms, T. Johannes, F. M. Munker and E. G. Mik, *J. Clin. Monit. Comput.*, 2017, **31**, 1143–1150.
35. S. I. Bodmer, G. M. Balestra, F. A. Harms, T. Johannes, N. J. Raat, R. J. Stolker and E. G. Mik, *J. Biophotonics*, 2012, **5**, 140–151.
36. F. A. Harms, W. M. de Boon, G. M. Balestra, S. I. Bodmer, T. Johannes, R. J. Stolker and E. G. Mik, *J. Biophotonics*, 2011, **4**, 731–739.
37. A. S. Golub, A. S. Popel, L. Zheng and R. N. Pittman, *Biophys. J.*, 1997, **73**, 452–465.
38. I. Dunphy, S. A. Vinogradov and D. F. Wilson, *Anal. Biochem.*, 2002, **310**, 191–198.
39. G. M. Balestra, M. C. Aalders, P. A. Specht, C. Ince and E. G. Mik, *J. Biophotonics*, 2015, **8**, 615–628.
40. C. Ince and E. G. Mik, *J. Appl. Physiol.*, 2016, **120**, 226–235.
41. F. A. Harms, W. J. Voorbeijtel, S. I. Bodmer, N. J. Raat and E. G. Mik, *Mitochondrion*, 2013, **13**, 507–514.
42. F. A. Harms, S. I. Bodmer, N. J. Raat and E. G. Mik, *Crit. Care*, 2015, **19**, 343.
43. F. A. Harms, S. I. Bodmer, N. J. Raat and E. G. Mik, *J. Clin. Monit. Comput.*, 2015, **29**, 509–519.
44. L. H. Romers, C. Bakker, N. Dollee, S. E. Hoeks, A. Lima, N. J. Raat, T. Johannes, R. J. Stolker and E. G. Mik, *Anesthesiology*, 2016, **125**, 124–132.
45. E. O. O'Brien and U. Schmidt, *Anesthesiology*, 2016, **125**, 20–21.

CHAPTER 14

# Microfluidic Systems and Optical Oxygen Sensors: A Perfect Match for Advancing Bioprocessing and Microbiology

BIRGIT UNGERBÖCK AND TORSTEN MAYR*

Graz University of Technology, Institute of Analytical Chemistry and Food Chemistry, Stremayrgasse 9, 8010, Graz, Austria
*E-mail: torsten.mayr@tugraz.at

## 14.1 Introduction

Nowadays, a general trend towards miniaturization can be observed, with novel ideas affecting a variety of different disciplines. This trend towards miniaturization also includes the miniaturization of laboratory operations onto singular devices, which can be summarized under the concept of the so-called "Lab-on-a-Chip" (LoaC). In this concept sequences of laboratory operations are performed automatically and carried out on a small scale - realized by microfluidic technology, which is defined by the small volume of fluid that is processed by these systems (commonly μL to fL).

The versatile field of microfluidics has faced an enormous rise during the last two decades since its origin as a miniaturized gas chromatograph at Stanford University[1] and the more general idea of the miniaturized total

Detection Science Series No. 11
Quenched-phosphorescence Detection of Molecular Oxygen: Applications in Life Sciences
Edited by Dmitri B. Papkovsky and Ruslan I. Dmitriev

Published by the Royal Society of Chemistry, www.rsc.org

analysis system μTAS,[2] with integration of all necessary analytical steps onto one single device. Since then, microfluidic technology has been used to achieve improvements in many research fields, such as point-of-care testing, diagnostics, industrial and environmental monitoring, analytics, microreaction technology and pharmaceutical and life sciences. Today, there is no doubt that miniaturized devices exhibit tremendous potential – not only in research, but also regarding commercial applications. Throughout the last few years, the amount of microfluidics-based products has been constantly growing and the global microfluidics market is projected to face a further enormous rise within the next years.[3]

The success of microfluidics can mainly be attributed to the advantages of miniaturized devices such as minimized reagent and sample consumption, increased automation possibility, high controllability and reproducibility and reduced manufacturing costs due to mass-production, *e.g.* from inexpensive polymers. These advantages lead to applications, which have been hard to realize before: microfluidic devices allow the study and analysis of complex cellular processes, they enable new automated and affordable applications in environmental monitoring and food analysis and facilitate the fast and cost-efficient development of new diagnostic abilities or of new (bio)-processes in the chemical and pharmaceutical industry.

Already at an early stage of development of μTAS the idea was born to use microfluidic platforms for chemical sensing and it was predicted that the μTAS progress would be linked to the use of a variety of sensors.[2] Such developments are still important in order to move the field from the so-called "Chip-in-a-Lab", to the real "Lab-on-a-Chip". However, the integration of sensors into microfluidic devices is challenging due to miniaturization-specific reasons: first, the small number of molecules within the reduced analysis volume results in the need for highly sensitive sensing devices and the constraint that the analyte concentration may not be changed during measurement. Second, the sensing device must be scalable to even smaller dimensions, which is not possible for all existing sensing technologies. And third, the costs of sensor integration play an important role as many microfluidic devices are thought to serve as disposable tools.

Optical luminescent sensors provide a range of interesting features to handle the above-mentioned challenges: they are highly sensitive, generally do not consume the analyte under investigation, are easy to miniaturize and relatively low-cost. A variety of different parameters can be measured by luminescent chemical sensors, ranging from ions over neutral molecules and gases to physical parameters.

Herein, oxygen represents one of the most important analytes in a range of applications such as bioprocess monitoring, cell culture applications and tissue engineering due to its significant importance for living organisms and biological systems. For example, oxygen is a key substrate in aerobic bioprocesses as an important nutrient for microorganisms or as a substrate for enzymatic conversions. Moreover, it is involved in a range of important

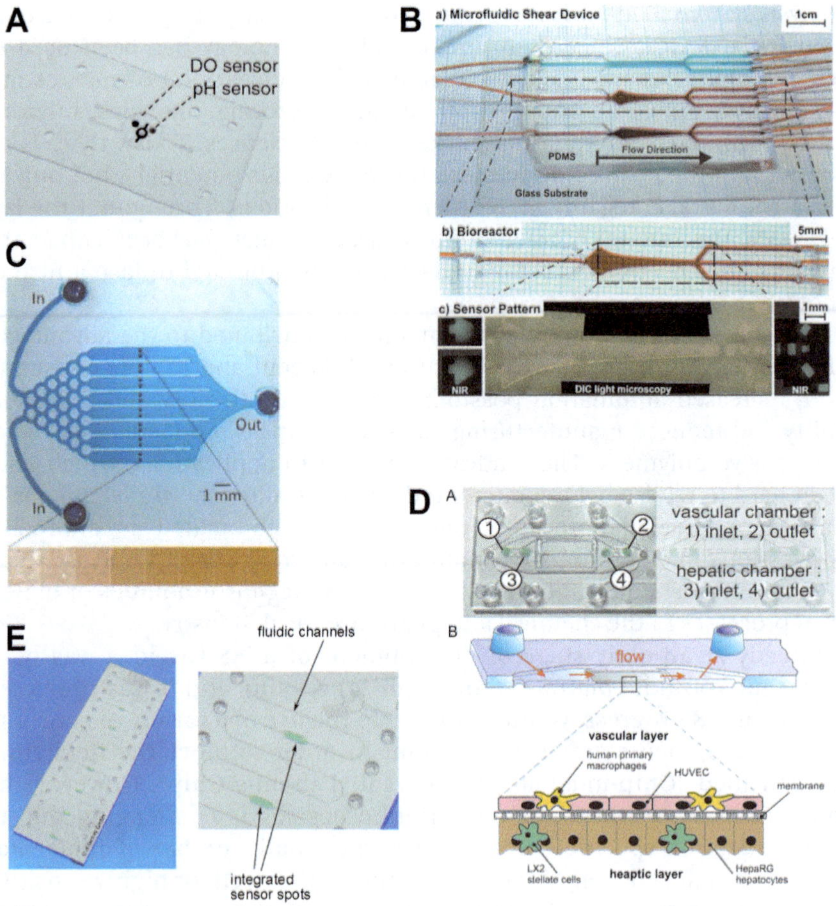


**Figure 14.1** Examples of microfluidic systems with integrated oxygen sensors: (A) commercially available optical oxygen sensor spots glued into a microbioreactor used for fermentation. Reproduced with permission from Zhang *et al.*, *Biotechnology & Bioengineering*, A well-mixed, polymer-based microbioreactor with integrated optical measurements,[9] John Wiley and Sons, Copyright © 2005 Wiley Periodicals, Inc. (B) (TOP) PDMS-based microfluidic device with integrated oxygen sensors and external interfacing. The channels were filled with red and blue aqueous dye solution to aid visualization. (MIDDLE) Close-up showing the central bioreactor device providing linearly increasing shear stress. (BOTTOM) Differential interference contrast (DIC) light microscopy and NIR fluorescence images of the sensor patterns integrated into the bioreactor and inlet/outlet channels. Reproduced from ref. 10 with permission from The Royal Society of Chemistry. (C) Microfluidic $O_2$ gradient generator using mixing chambers to generate distinct oxygen profiles. Oxygen gradient generated by the devices was imaged by the fluorescence quenching of oxygen sensors (orange fluorescence imaged below the micrographs). Reproduced from ref. 11 with permission from The Royal Society of Chemistry. (D) Oxygen sensors integrated in to a Liver-on-Chip. Sensor spots positioned at

cellular processes such as energy-generation in the form of ATP, enzymatic reactions and generation of metabolic substrates and therefore serves as an indicator for the presence, viability and/or metabolic cell status.

That special biological role makes oxygen an analyte of major importance, which should be commonly monitored to ensure optimal bioprocess performance, to create physiological conditions for cells or tissues or to investigate the metabolic status of cells or tissues. In this regard, bioprocesses, cell culture or tissue engineering, which are carried out in microfluidic devices, are no exception. Morever: due to their high surface-to-volume ratio and the variety of different materials of which they are made, oxygen levels in microfluidic devices are hard to assess. A microfluidic channel is often seen as a closed system, but can in fact also be highly permeable to oxygen – depending on microfluidic chip material and thickness. A careful choice of chip material and design as well as oxygen monitoring are therefore necessary in order to determine or assess the cell's environmental conditions in a microfluidic channel.

The importance to monitor oxygen in microfluidic applications has already been recognized by various scientists and in recent years several publications reported on the successful integration of luminescent oxygen sensors into miniaturized devices. Some examples of microbioreactors and microfluidic systems with integrated oxygen sensor are shown in Figure 14.1. Several reviews have been published on this subject including the review by Grist *et al.*,[4] who surveyed previously proposed optical oxygen sensors and discussed their applicability in microfluidic cell culture, a review by Pfeiffer and Nagl,[5] focusing on integration methods for luminescent oxygen, pH and temperature sensors into microfluidic platforms, or a review on integration and application of optical chemical sensors in microbioreactors by Gruber *et al.*[6] Other reviews have focused on different main subjects such as oxygen imaging in microfluidic devices[7] or oxygen transfer characteristics of microbioreactors,[8] but additionally provide an overview over state of the art on-line measurement techniques in microbioreactors.

The integration of luminescent oxygen sensors into miniaturized devices remains a challenging task, because a smart combination of sophisticated sensing chemistry, sensor format and the optical measurement system is required to fulfil application-specific sensing requirements while considering compatibility issues with chip material and fabrication method of the microfluidic devices. The present book chapter provides a discussion of

---

the inlets (1, 3) and the outlets (2, 4) of the upper and lower channel systems, respectively. Reproduced from *Biomaterials*, **71**, K. Rennert, S. Steinborn, M. Gröger, B. Ungerböck, A.-M. Jank, J. Ehgartner, S. Nietzsche, J. Dinger, M. Kiehntopf, H. Funke, F. T. Peters, A. Lupp, C. Gärtner, T. Mayr, M. Bauer, O. Huber and A. S. Mosig, A microfluidically perfused three dimensional human liver model, 119–131,[12] Copyright 2015, with permission from Elsevier. (E) Silicon-glass microreactor with integrated oxygen sensors for online monitoring of biocatalytic conversions – reproduced with permission from Micronit GmbH.

challenges, which affect sensor integration in microfluidics, and provides a critical overview over frequently used sensor formats and read-out methods. Furthermore, this book chapter provides a detailed review of different methods of sensor layer integration and discusses which issues herein need to be considered with respect to microfluidic chip fabrication. Finally, applications of oxygen sensors in microfluidic devices are categorized according to the respective application purpose, taking representative work in this area as examples, in order to point out importance and future directions of oxygen sensing in microfluidic channels.

## 14.2 Challenges

Generally, optical luminescent sensors are promising tools for integration into microfluidic environments due to their sensitivity, reduced size and low price. However, it is important to consider various miniaturization-specific demands, such as the need for high brightness and photostability of the indicators, compatibility of the oxygen sensors with the sample under investigation (*e.g.* cells), chip material and chip fabrication process and also particularly strict demands regarding prize and user-friendliness.

A huge variety of luminescent oxygen indicators are nowadays available with a range of different features such as varying brightness, spectral properties, luminescence lifetime and dissolving behaviour in aqueous or organic media.[13] Generally, luminescent indicators and detection system have to be tailored to the final application and to be compatible with each other (discussed in book chapters 1,3,4 and 5). Here, criteria for the choice of luminescent indicators are discussed with regard to their integration into microfluidics and especially promising features of luminescent indicators are highlighted.

Owing to low channel depths of microfluidic devices bright indicators can be beneficial in order to increase the amount of detectable light or to reduce the amount of sensing material through the application of thin-film optodes or lower concentration of nanoparticles. Oxygen indicators, which are known for their high brightness, comprise Ir(III) coumarin complexes, Pt(II) porphyrins and NIR emitting Pt(II) benzoporphyrins.[14–16] Moreover, the brightness of moderately luminescent indicators can be increased by application of light harvesting.[17] This method is based on the application of two luminescent dyes, one serving as antenna dye by absorbing excitation light energy, which is further partially transferred to an indicator dye *via* FRET.

The use of indicators with longwave shifted excitation and emission can be beneficial for measurements of cells and tissues, in order to reduce biological autofluorescence upon UV irradiation. Additionally, light of short wavelengths - and especially UV light - is known for having a perturbing impact on cell growth and viability. The use of long wavelengths also plays an important role for potential applications in tissue monitoring, because longwave shifted light is less affected by absorption and scattering of cell material and thus the penetration depth of excitation light is increased.

Promising oxygen indicators for such applications are NIR emitting probes such as Pt(II) and Pd(II) complexes with π-extended benzo-naphtoporphyrins[18] as well as azabenzoporphyrins.[19] The need for spectral compatibility of luminescent indicators with available optical imaging equipment may, however, contraindicate the use of NIR dyes, because CCD chips of microscope cameras as well as some photodetectors tend to exhibit a lower sensitivity in this spectral region.

Apart from optical features of the used indicators it might be necessary to also consider the physical or covalent immobilization of the indicator in the host matrix. In recent years, studies revealed that the sensors with covalent linkage features improve sensor performance and are not affected by dye leaching or migration into the chip material compared to sensors, which use physical entrapment methods.[20] Nevertheless, physical entrapment is most probably going to succeed in many cases comparing the effort for applying these different immobilization strategies.

Due to the fact that many microfluidic applications deal with living organisms, the interaction of sensor material and cells plays an important role for considering the ideal sensing setup. Adverse effects may even be pronounced in microfluidic applications, because the very low volumes within microfluidic channels lead to close distance of sensor and cells. In this context, toxicity of luminescent indicators plays a secondary role, because the dye is usually excluded from direct contact with the investigated biological material by immobilization in a host matrix (see Section 14.3). However, toxic interaction between sensor and cells can be caused by production of singlet oxygen, which by definition happens during the dynamic quenching process of oxygen sensors. *In situ* generated singlet oxygen can theoretically diffuse through the matrix material and negatively affect cell growth before it returns to its ground state.

It should be noted here that singlet oxygen generation can also lead to other adverse effects. For example, singlet oxygen can react with luminescent indicators or the sensor matrix, increasing the photobleaching rate or a general change in polymer permeability, respectively.[21] This can further lead to depletion of oxygen, resulting in decreased oxygen levels within the microfluidic environment. Potential ways to reduce toxic or oxygen-depleting effects of singlet oxygen are either to decrease excitation light intensities or the dye loading - leading to a reduced production of singlet oxygen - or to increase diffusion path lengths between the sensing layer and biological material to increase the probability that singlet oxygen is deactivated before contact with cells.

A further challenge in microfluidic oxygen sensing is the choice of a robust detection method. Oxygen can be quantified by measuring luminescence intensity or luminescence lifetime of the indicator. The simplest method for sensing oxygen is the measurement of luminescence intensity from an instrumental point of view. However, even at high sensor layer homogeneity the use of that relatively simple method is limited due to high scattering possibility at microfluidic channel walls. Different detection methods are described and the choice of a suitable detection method is discussed in Section 14.5.

Another restriction for the integration of luminescent oxygen sensors in microfluidics is the demand for compatibility of the sensor with microfluidic chip material and microfluidic chip fabrication. Adhesion of the sensor layer to and background fluorescence from the chip material plays a major role, but also subsequent steps in chip fabrication (*e.g.* bonding) have to be considered. Many processes involve the use of high temperatures, plasma treatment or the use of solvents: interaction of such processes with the sensor material may compromise the sensor performance. Hence, the stability of integrated sensors towards chip fabrication steps should be tested at an early stage of development.

## 14.3 Sensor Formats

Optical oxygen sensors can be fabricated yielding different desired formats, only limited by the processing possibilities of the matrix material, which typically hosts the indicator dye. This makes them a very versatile platform to be integrated into microfluidics. Basically, three different sensor formats have been applied in microfluidic devices: dissolved indicators, sensor layers and micro/nanoparticles. A schematic overview of these different sensor formats is presented in Figure 14.2. All these formats are described and discussed regarding their applicability in microfluidics throughout the next section.

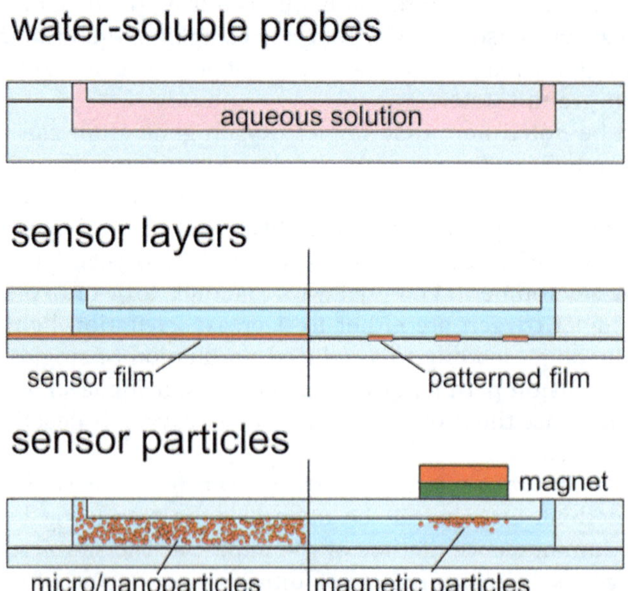


**Figure 14.2** Schematic overview over different sensor formats for the application in microfluidic devices.

### 14.3.1 Water-soluble/Macromolecular Probes

Water-soluble probes are the simplest possibility to integrate a sensing functionality into a microfluidic device, although their use implies several disadvantages. A drastic limitation of sensing performance is the possible interference with their environment due to the lack of an encapsulation matrix. Moreover, these probes enable adjustment of sensor sensitivity and selectivity only by selection of different dyes, which could otherwise be achieved by fine-tuning with a suitable matrix. A way to overcome these severe drawbacks is the covalent linkage to other molecules, wherein the dendritic encapsulation of the luminophore into a cage of branched molecules (dendrimers) is one of the most promising techniques.[22]

### 14.3.2 Sensor Layers

Sensor layers represent the most commonly used format of optical sensors in microfluidics due to the fact that their use entails several advantageous features: their stationary nature is convenient in application, the sensor material is not wasted during the measuring process and the risk of sensor–sample interaction is minimized. Moreover, they allow for single-point measurements or for imaging applications in a user-friendly way, which is relatively unsusceptible to errors because they can be fabricated with high homogeneity. Challenges, which come along with the fix integration of sensing layers, are stricter requirements regarding photostability and a leaching-free entrapment method of the indicator in the matrix material to warrant the long-term stability of a sensor chip and to prohibit possible toxic interference of dyes and cells.

It is important that the fabrication of sensor layers and the layers themselves are compatible with microfluidic chip material regarding adhesion and with the applied microfabrication techniques – for example bonding procedures, which involve the use of organic solvents or elevated temperatures. Several techniques to integrate sensor layers into ready-made and self-made devices exhibiting different degree of complexity have been developed. Generally, they can be deposited from a so-called "sensor cocktail" (matrix material, indicator and additives dissolved in a suitable solvent) onto a substrate by spin- and knife-coating, screen printing, photo-polymerization, spray-coating and ink-jet printing. Subsequently the applied wet film is allowed to dry, polymerize or cure. A more detailed overview on sensor layer fabrication methods, applied either as patterned or un-patterned films, is presented in Section 14.4.

### 14.3.3 Micro/Nanoparticles

Sensor beads represent a sensing platform comprising the advantages of water-soluble probes and sensor layers. On the one hand they can be used as flexible dissolved indicators regarding their integration into different systems, while on the other hand they exhibit the advantages of indicator encapsulation into a suitable matrix. Therefore, sensor particles represent a simple and versatile tool also for their application in microfluidic environments. They can simply be added to the fluid, which is pumped through microfluidic channels and therefore additional effort of chip fabrication development is avoided. Moreover, photobleaching behaviour plays a minor role due to the fact, that the sensing element is not stationary, but permanently exchanged.[23] However, new challenges arise concerning the brightness of the applied sensor particles. Reduced channel depths lead to insufficient emission intensities, because the small volume limits the amount of sensor material. Beside the use of ultrabright indicators,[24] a method to circumvent this disadvantage is the use of magnetic sensor particles.[25] The combination of both magnetic and sensing properties allows one to combine advantageous features of nanosensor particles and sensor layers: they are easy to integrate into any microfluidic device, while their magnetic separation from the surrounding solution allows the generation of *in situ* sensor spots with increased brightness compared to nanoparticle dispersions.

Aside from applying luminescent nanosensor particles dispersed in the fluid, sensor particles have also been applied to generate stationary sensor layers by physical adsorption of oxygen-sensitive nanoparticles onto the glass surface of microchannels[26] or by swelling particles in organic solvent prior to applying them as stationary sensor spot.[21]

In conclusion, it is important to note that the sensor format should be chosen according to the respective measurements requirements and correlated to the detection method, *e.g.* the requirement for single-point measurements or laterally resolved information. The "optimal" sensor setup also strongly depends on considerations regarding the applied microfluidic device and the final application – meaning issues such as the device material, or if the goal is a commercial application requiring a sensor solution, which is compatible with an industrial fabrication procedure.

## 14.4 Sensor Layer Integration and Fabrication

The sensor layer can be deposited by various techniques, including direct staining of chip material, spin- and knife-coating, screen-printing, photopolymerization, spray-coating and inkjet printing. We introduce these methods very briefly in the following section.

A straightforward method is *direct staining* of the chip material with indicator dyes. With this method parts of the microfluidic device are transformed into the sensor itself. Various groups reported on staining PDMS material as an oxygen permeable membrane and sensing layer in a multilayer microfluidic device.[27–30] A requirement to apply this method is the suitable permeability of the hosting material for oxygen and good solubility of the indicator therein.

*Spin coating* uses rotational force and is a versatile and established technique in industrial processes and laboratories. Spin coating can produce layers in homogenous thickness down to a few hundred nanometers.[31,32] Structuring of the layer requires a post-coating process such as laser ablation as shown be Grist *et al.* (Figure 14.3A).[33]

*Blade or knife-coating* is a simple and well-known technique for producing optical sensors layers and also employed for micro-bioreactors. The sensor formulation is spread on a planar substrate using a knife. The gap between the surface and the knife controls the thickness of the resulting film. Sensor spots were cut-out from foils with knife-coated sensor layers and glued into the miniaturized chambers.[34]

*Spray coating* employs airbrushes to apply a sensor formulation as a homogenous layer onto a substrate. The method is suitable to coat larger areas homogenously. The smallest area reported is 2 mm × 2 mm.[35] Structured layers can be achieved by applying stencils. With this method sensor spots with a diameter of 100 μm were produced (see Figure 14.3D).[36]

*Screen-printing* is well-established for preparing electrochemical sensors in high volumes.[37] In this technique, the sensor formulation is applied through a coated mesh with a squeegee blade to a substrate, except in the areas that are impermeable for the sensor formulation by a blocking stencil. A drawback of this method is that the sensor film can only be applied to a flat surface and not in microfluidic channels or grooves. The technique was demonstrated in a microfluidic chip by Mayr *et al.* shown in Figure 14.3C.[38]

*Inkjet printing* is well-known from graphical printing and is a promising technique for integration of sensor layers in microfluidic device and mass-production, because it enables coating of large areas as well as free-form/maskless structuring.[39] Herein, we consider micro-dispensing, where the deposition of droplets is guided by a piezoelectrically, liquid or pressure driven tappets from a reservoir through a nozzle, as the most suitable technique since it can handle viscous sensor formulation or sensing materials containing particles.[40] Examples of structured layers using this technique are depicted in Figure 14.3B.

*Photo-polymerization* enables the preparation of small sensor structures at a high dimensional accuracy. In the simplest form, a photo-active matrix is cured during to the illumination with UV-light focused with a microscope objective.[41] Sensor spots prepared with this technique are shown in Figure 14.3F. Structured sensor layers with high precision in a size down to 50 μm were demonstrated by using photolithographic masks in combination with photoresists and plasma etching.[32] Etzkorn *et al.* deposited the mixture, the polymeric sensor matrix and a photoresist on a substrate. After exposure to UV-light the sensor pattern and application of a developer a sensitive ring shaped layer with a diameter of 100 μm was obtained.[42] Photo-resists were also used to produce PDMS-stamps for applying sensor layers by micro-contact printing.[10]

The choice of the deposition technique depends on different factors such as structure size, availability and cost of sensing material, viscosity of the sensor formulation and compatibility of the chip material with solvents of the sensor formulation. Although various techniques have been successfully demonstrated many of them are limited regarding compatibility with mass production. Herein, we consider inkjet printing as the most promising technique for sensor integration, since it can be integrated easily into the chip fabrication process.

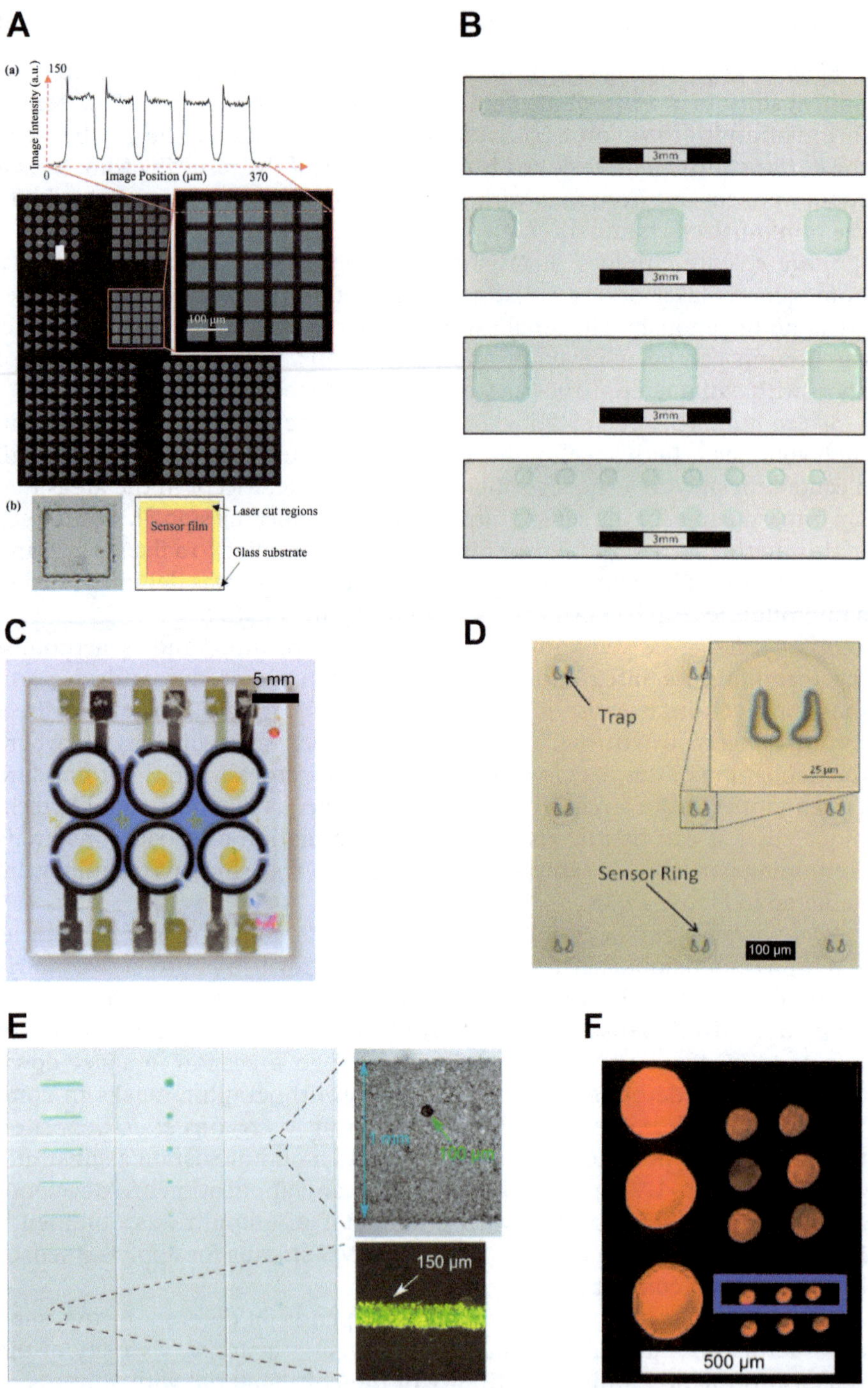


**Figure 14.3** Examples of deposition techniques. (A) Spin-coated sensor layer structured by laser ablation – (a) Phosphorescence intensity images of patterned sensor films. An intensity cross-section of the array of squares is shown. (b) Brightfield image of one of the squares in the patterned array. Reproduced from ref. 33 with permission from the Royal Society of Chemistry. (B) Inkjet-printed sensor layers and spots. Reproduced

## 14.5 Detection Principles

The analytical information carried in the integrated sensor can be read-out by various methods from outside through the transparent chip material. This contactless detection principle enables the separation of the optical read-out system from the sensor material. This is of particular interest for the use of disposable microdevices, because the incorporation of relatively expensive optical components can be omitted. The quenching of the phosphorescence of indicator dyes by molecular oxygen is quantified by different detection principles based on phosphorescence intensity and lifetime. This section summarizes the detection principle applied to oxygen sensing in microfluidic devices.

Detection of *phosphorescence intensity* (Figure 14.4A) is the most straightforward way to quantify oxygen, since it requires simple instrumentation.[10,41,45,46] However, luminescence intensity measurements are prone to many interfering effects including variations in the optical path, fluctuation in the light source and detector, variations of the concentration of the indicator dye, photobleaching and light scattering. Therefore, the use of ratiometric methods or luminescence lifetime is recommended to overcome the drawbacks of intensity-based read-out.

*Wavelength ratiometric* methods rely on the detection of two bands in the emission spectra (Figure 14.4B). Either the indicator dye itself exhibits two

with permission from ref. 43. (C) Screen-printed sensor spots. Image of a polymer sensor chip with six ring-shaped organic photodiodes and sensor spots (orange). Reprinted from *Analytical and Bioanalytical Chemistry*, Filter-free integrated sensor array based on luminescence and absorbance measurements using ring-shaped organic photodiodes, **404**, 2012, 2841–2849,[44] T. Abel, M. Sagmeister, B. Lamprecht, E. Kraker, S. Köstler, B. Ungerböck and T. Mayr. Copyright © Springer-Verlag 2012, with permission from Springer. (D) Sensor rings structured by microfabrication using photoresists. Microscopic image of a *n* 3 × 3 array of single-cell self-assembly traps encircled by an oxygen sensor embedded in SU8 rings. The inset is a magnified image of a trap and sensor ring (reproduced from ref. 42 with permission from IOP Publishing). (E) Structured sensor layers prepared by airbrush spraying in combination with stencils on glass. Photograph of sensing lines or spots range from 100 µm to 1000 µm in width or diameter and microscopic bright-field images of a 150 µm wide sensor line and 100 µm in diameter sensor spot integrated into a 1 mm wide microfluidic channel. Reproduced from *Sensors and Actuators B: Chemical*, 2016, **228**, 748–757,[35] https://doi.org/10.1016/j.snb.2016.01.050, © 2016 The Authors. Published by Elsevier B.V. under the terms of the CC BY 4.0 licence, https://creativecommons.org/licenses/by/4.0/. (F) Sensor spots prepared by photo-polymerisation on a cover glass. Sensor spots were polymerized using a 20×, 40× or 60× microscope objectives and light of high power UV-LED. Reproduced from *Microchimica Acta*, In-line monitoring of pH and oxygen during enzymatic reactions in off-the-shelf all-glass microreactors using integrated luminescent microsensors, **184**, 2016, 621–626,[41] S. A. Pfeiffer, S. M. Borisov and S. Nagl. Copyright © Springer-Verlag Wien 2016, with permission from Springer.

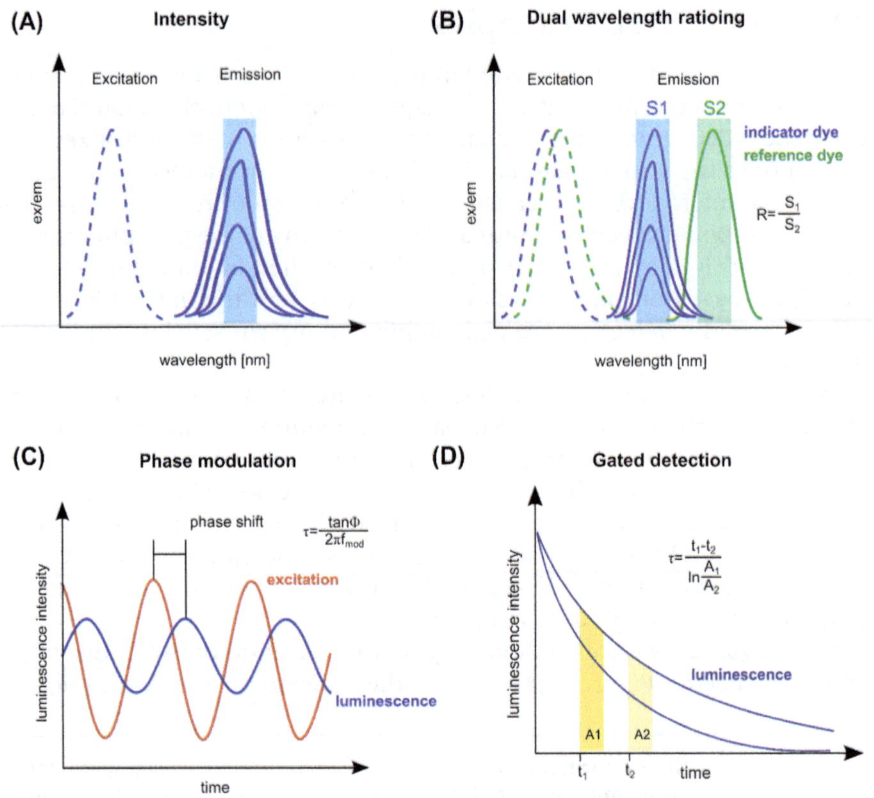


**Figure 14.4** Schematic representation of detection principles applied for oxygen sensing in microfluidic devives. (A) Luminescence intensity measurements of a band of wavelengths. (B) Dual wavelength ratioing, (C) Lifetime determination by phase modulation and (D) Lifetime determination by gated detection. Reproduced from *Trends in Biotechnology*, **31**, J. M. Bolivar, T. Consolati, T. Mayr and B. Nidetzky, Shine a light on immobilized enzymes: real-time sensing in solid supported biocatalysts, 194–203,[53] Copyright 2013, with permission from Elsevier.

emission bands, with only one of them being sensitive to oxygen or an inert reference dye is added, providing the reference emission.[31,47] The ratio of the two emission signals is related to the analyte concentrations. Thus, the use of the ratiometric methods improves oxygen measurements by increasing the reliability with respect to light source variations, scattering, detection efficiency and sensing layer thickness. However, signal variations due photobleaching or leaching of the luminophore, ambient light or wavelength-dependent scattering, still compromise the accuracy of the oxygen measurement.

*Phase modulation* is a method to determine luminescence lifetime in the frequency domain (Figure 14.4C). The indicator dyes are excited with sinusoidally modulated light. The detectors register the emitted light with a certain delay in comparison to the excitation. This delay is measured as a phase shift by a lock-in amplifier and used to calculate the lifetime.[34,42,48,49] Instrumentation is small and inexpensive for lifetimes in the microsecond range.[35]

*Time-gated detection* is a frequently used method to determine the lifetime in the time domain (Figure 14.4D). A phosphorescent dye is excited by a pulse and the decay is recorded in successive time-gates. The lifetime can be calculated and related to the analyte concentration.[31,50–52]

Further considerations for choosing the read-out method relate to the question if single-point or spatially resolved information on oxygen concentration is needed. On the one hand, microscopy setups in combination with digital cameras are popular for quantification of phosphorescence of the integrated sensors. This configuration yields in spatial information, *e.g.* to visualize gradients. On the other hand, single-point measurements employing optical fibers, which guide excitation and emission light between a miniaturized read-out instrument and the integrated sensors require less bulky instrumentation and provide the highest precision.

## 14.6 Applications

Due to the significant importance of oxygen for living organisms and biological systems, applications for microfluidic oxygen sensing are diverse and several measurement setups have already been successfully realized in the fields of bioprocess monitoring, microfluidic cell culture and tissue engineering. Generally, these applications may suit two main purposes: applications, where oxygen is measured to ensure stable reaction or cell culture conditions (oxygen serving as a key substrate in aerobic bioprocesses or an important nutrient for cells) and applications, where oxygen is the main analyte - indicating the reaction progress or the viability or metabolic status of living systems. Furthermore, applications of microfluidic oxygen sensing can be categorized according to the application field (Figure 14.5). The following section provides an overview over the main applications in (1) microbioreactors and (2) microfluidic cell culture tissue monitoring and utilizes examples to describe the different fields of applications in detail.

*Microbioreactors* with integrated oxygen sensors are well suited to accelerate bioprocess development. Such reactors can be made disposable by using polymer chip material, thus reducing assembly, cleaning, and sterilization efforts. At the same time, microbioreactors require fewer resources and produce less waste, which is of interest for bioassays in general, and specifically for investigations such as continuous culture assays and studies of expensive enzymes.

Applications of oxygen sensors in microbioreactors can be treated under three sub-headings. First, they have been used to acquire oxygen data of high-throughput fermentations. While single disposable microbioreactors with integrated sensors greatly simplify the effort per experiment, parallel operation of multiple microbioreactors has the potential to make the technology viable for high-throughput data acquisition.[9,34,54–56] Even higher throughput can be achieved by using droplet microbioreactors, which uses a two-phase flow to transport tiny droplets, each representing a single bioreactor.[57]

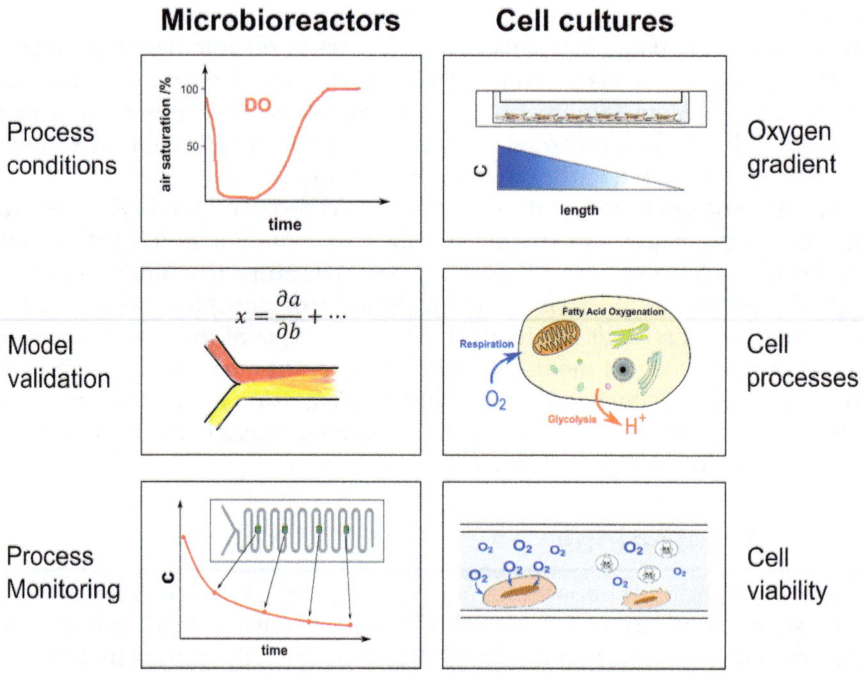


**Figure 14.5** Overview of application fields of microreactors and microfluidic systems utilizing integrated oxygen measurements.

Second, oxygen sensors have been applied in microbioreactors to validate models, which describe the bioprocesses or diffusion processes therein.[29] For example, experimental results can be compared to results derived from mathematical models with varying complexity to assess the required model complexity for achieving precise predictions and to depict the governing transport characteristics at the microscale.[24]

Third, microfluidic oxygen sensing can be used to observe the reaction progress in the case that oxygen is consumed or produced during product formation along a microfluidic channel of a flow-through microreactor by either single point measurements at different distances from the microreactor inlet[35] or by oxygen imaging of the whole microreactor.[58] Such microreactor systems with integrated oxygen sensors thus provide useful tools for determining kinetic parameters at the microscale.

Another main application of oxygen sensors in microfluidic environments is their use in microfluidic cell culture and tissue monitoring. The application of microfluidics to cell-based research has faced an enormous rise during the last decade. Microfluidic structures have been used for almost all processes in cell research: cell sampling, cell trapping and sorting, cell treatment and cell analysis.[59,60] Microfluidic environments are ideal for working with cells due to the comparable scale of microfluidic structures and cells themselves. Additionally, microstructures present in microfluidic devices can be used to provide a 3D cell culture environment, which is closer to *in vivo* conditions than traditional 2D cell culture environments. Moreover, microfluidic technology allows the generation of predictable and reproducible chemical gradients, which affect cell growth, migration, and differentiation within tissues.[28,61] Finally, the reduced sample requirement makes microscale devices ideally suited for studies of samples with small cell numbers, such as primary cancer cells obtained from needle biopsies, or stem cells.

Applications of oxygen sensors in microfluidic cell culture and tissue monitoring can also be categorized in three main applications. First, oxygen sensors can be used to develop *in vivo*-like microfluidic environments regarding oxygen concentration or gradients.[10,27,28,46,62–64] Such devices can then be used as new *in vitro* research tools to study human organ-like structures at the cellular level under conditions close to the physiological situation – a relatively new application field, called Organ-on-a-Chip.[12] Second, such sensors can be used to measure oxygen to draw conclusions regarding complex cellular processes. For example, on-line oxygen sensing in tissue can serve to measure mitochondrial function, which plays a critical role in the development of disease and chemical toxicity or for the regulation of stem cell pluripotency.[65] Oxygen sensing has also been applied to measure the oxygen consumption rates of single cells, providing a deeper understanding of cellular metabolism.[66–68] Finally, microfluidic devices equipped with oxygen sensors can be used in biosensing applications, with oxygen indicating an organism's reaction, for example death or metabolic changes in presence of toxic substances. For example, microdevices with integrated oxygen sensors can be applied as a pesticide detection platform for *in situ* analysis of pesticide concentration detected *via* metabolism/photosynthesis of algae in tap water.[69]

It should be noted here that in all these applications it might be important not only to investigate oxygen concentration as point measurements along a microfluidic channel or in a microfluidic chamber, but also to study the distribution of oxygen over the device dimensions (2D or even 3D).

To conclude, oxygen sensors have proven to be valuable tools in microbioreactors as well as in microfluidic cell culture and tissue engineering applications. Especially applications in microfluidic cell culture showed that there is a lack of knowledge on how oxygen is distributed, delivered or consumed within microfluidic channels. Integration of luminescent oxygen sensors therefore can serve to increase researcher's awareness towards the important role of oxygen for biological applications.

## 14.7 Conclusion

The combination of optical oxygen sensor and microfluidic devices is straightforward and provides attractive tools for biology, biotechnology and life sciences. Monitoring of oxygen as a key variable in miniaturized bioprocesses and cell cultures advances significantly applications in these fields.

Many approaches to realize microfluidic devices with integrated oxygen sensor have been successfully demonstrated. Sophisticated sensor materials have been developed to overcome challenges including brightness and stability of sensor materials, interference with biological samples and toxicity of materials to cells or tissues.

The advancement of sensor materials yielded the possibility to choose between different sensor formats according to the requirements of the specific application. The most suitable format can be chosen with consideration of the detection method, device geometry, chip material and bonding assembly methods. Integrated sensor layers are convenient for end-users and various techniques are available to achieve microbioreactors and cell culture devices featuring online oxygen monitoring.

Luminescent sensors offer the possibility to perform single point or 2D measurements yielding different levels of analytical information. The applied detection principles demand certain properties of the sensing material or *vice versa*. Most commonly optical microscopy or detection with optical fibres is used. These set-ups are cumbersome and inconvenient for running several microdevices in parallel. Therefore, most experiments are performed with single microdevices with bulky measurement periphery. To move from this "Chip-in-a-Lab" to the realisation of "Lab-on-a-Chip", miniaturized optoelectronic systems with multi-channel read-out capabilities are needed. Alternatively, the microchip designs are adapted to the layout of microtiter plates fitting to available high-throughput readers.

The contactless read-out of the integrated sensors from outside make the microfluidic chips attractive for the application as disposables because the detection system is separated. This is important for mass production of microfluidic chips because complicated integration processes can be omitted and production costs kept low. Up-to-now microfluidics with integrated sensors are custom-made single pieces prepared in research laboratories. Moving the technology to mass production will make it available for the non-sensor specialist and allow it to become a standard tool. Development and integration of complementary parameters such as pH, glucose and lactate will help to exploit the whole potential of microfluidics in applications such as bioprocessing, cell culturing and tissue engineering. We also expect that integrated oxygen sensor will play a key role for monitoring and analyzing of cell and tissue functionalities in the currently emerging Organ-on-Chip technology. This micro-physiological models are expected to have a significant impact on drug development and personalized treatment of diseases.

## References

1. S. C. Terry, J. H. Jerman and J. B. Angell, *IEEE Trans. Electron Devices*, 1979, **26**, 1880–1886.
2. A. Manz, N. Graber and H. M. Widmer, *Sens. Actuators, B*, 1990, **1**, 244–248.
3. L. R. Volpatti and A. K. Yetisen, *Trends Biotechnol.*, 2014, **32**, 347–350.
4. S. M. Grist, L. Chrostowski and K. C. Cheung, *Sensors*, 2010, **10**, 9286–9316.

5. S. A. Pfeiffer and S. Nagl, *Methods Appl. Fluoresc.*, 2015, **3**, 034003.
6. P. Gruber, M. P. C. Marques, N. Szita and T. Mayr, *Lab Chip*, 2017, **17**, 2693–2712.
7. S. Sun, B. Ungerböck and T. Mayr, *Methods Appl. Fluoresc.*, 2015, **3**, 034002.
8. *Oxygen Transfer Characteristics of Miniaturized Bioreactor Systems - Kirk - 2013-Biotechnology and Bioengineering*, Wiley Online Library, http://onlinelibrary.wiley.com/doi/10.1002/bit.24824/full, accessed 13 October 2017.
9. Z. Zhang, N. Szita, P. Boccazzi, A. J. Sinskey and K. F. Jensen, *Biotechnol. Bioeng.*, 2006, **93**, 286–296.
10. V. Nock, R. J. Blaikie and T. David, *Lab Chip*, 2008, **8**, 1300–1307.
11. J. F. Lo, E. Sinkala and D. T. Eddington, *Lab Chip*, 2010, **10**, 2394.
12. K. Rennert, S. Steinborn, M. Gröger, B. Ungerböck, A.-M. Jank, J. Ehgartner, S. Nietzsche, J. Dinger, M. Kiehntopf, H. Funke, F. T. Peters, A. Lupp, C. Gärtner, T. Mayr, M. Bauer, O. Huber and A. S. Mosig, *Biomaterials*, 2015, **71**, 119–131.
13. M. Quaranta, S. M. Borisov and I. Klimant, *Bioanal. Rev.*, 2012, **4**, 115–157.
14. S. A. Vinogradov and D. F. Wilson, *J. Chem. Soc. Perkin Trans. 2*, 1995, 103.
15. S. M. Borisov and I. Klimant, *Anal. Chem.*, 2007, **79**, 7501–7509.
16. S. M. Borisov, G. Nuss and I. Klimant, *Anal. Chem.*, 2008, **80**, 9435–9442.
17. T. Mayr, S. M. Borisov, T. Abel, B. Enko, K. Waich, G. Mistlberger and I. Klimant, *Anal. Chem.*, 2009, **81**, 6541–6545.
18. F. Niedermair, S. M. Borisov, G. Zenkl, O. T. Hofmann, H. Weber, R. Saf and I. Klimant, *Inorg. Chem.*, 2010, **49**, 9333–9342.
19. S. M. Borisov, G. Zenkl and I. Klimant, *ACS Appl. Mater. Interfaces*, 2010, **2**, 366–374.
20. L. H. Hutter, B. J. Müller, K. Koren, S. M. Borisov and I. Klimant, *J. Mater. Chem. C*, 2014, **2**, 7589–7598.
21. B. Ungerböck, S. Fellinger, P. Sulzer, T. Abel and T. Mayr, *Analyst*, 2014, **10**, 2551–2559.
22. S. A. Vinogradov, L.-W. Lo and D. F. Wilson, *Chem.–Eur. J.*, 1999, **5**, 1338–1347.
23. D. Aigner, B. Ungerböck, T. Mayr, R. Saf, I. Klimant and S. M. Borisov, *J. Mater. Chem. C*, 2013, **1**, 5685–5693.
24. B. Ungerböck, A. Pohar, T. Mayr and I. Plazl, *Microfluid. Nanofluid.*, 2014, **14**, 565–574.
25. G. Mistlberger and I. Klimant, *Bioanal. Rev.*, 2010, **2**, 61–101.
26. L. C. Lasave, S. M. Borisov, J. Ehgartner and T. Mayr, *RSC Adv.*, 2015, **5**, 70808–70816.
27. M. Skolimowski, M. W. Nielsen, J. Emnéus, S. Molin, R. Taboryski, C. Sternberg, M. Dufva and O. Geschke, *Lab Chip*, 2010, **10**, 2162–2169.
28. L. Wang, W. Liu, Y. Wang, J. Wang, Q. Tu, R. Liu and J. Wang, *Lab Chip*, 2013, **13**, 695.
29. P. C. Thomas, S. R. Raghavan and S. P. Forry, *Anal. Chem.*, 2011, **83**, 8821–8824.

30. J. W. Grate, C. Zhang, M. Wilkins, M. G. Warner, N. C. Anheier, J. Suter, R. Kelly and M. Oostrom, in *Micro- and Nanotechnology Sensors, Systems, and Applications V,* International Society for Optics and Photonics, 2013, vol. 8725, p. 872522.
31. B. Ungerböck, V. Charwat, P. Ertl and T. Mayr, *Lab Chip*, 2013, **13**, 1593–1601.
32. H. Zhu, X. Zhou, F. Su, Y. Tian, S. Ashili, M. R. Holl and D. R. Meldrum, *Sens. Actuators, B*, 2012, **173**, 817–823.
33. S. M. Grist, N. Oyunerdene, J. Flueckiger, J. Kim, P. C. Wong, L. Chrostowski and K. C. Cheung, *Analyst*, 2014, **139**, 5718–5727.
34. A. Zanzotto, N. Szita, P. Boccazzi, P. Lessard, A. J. Sinskey and K. F. Jensen, *Biotechnol. Bioeng.*, 2004, **87**, 243–254.
35. J. Ehgartner, P. Sulzer, T. Burger, A. Kasjanow, D. Bouwes, U. Krühne, I. Klimant and T. Mayr, *Sens. Actuators, B*, 2016, **228**, 748–757.
36. J. Ehgartner, P. Sulzer, T. Burger, A. Kasjanow, D. Bouwes, U. Krühne, I. Klimant and T. Mayr, *Sens. Actuators, B*, 2016, **228**, 748–757.
37. J. P. Metters, R. O. Kadara and C. E. Banks, *Analyst*, 2011, **136**, 1067.
38. T. Mayr, T. Abel, B. Ungerböck, M. Sagmeister, V. Charwat, P. Ertl, E. Kraker, S. Köstler, A. Tschepp and B. Lamprecht, *Proc. SPIE 8479, Organic Semiconductors in Sensors and Bioelectronics V*, 2012, p. 84790H.
39. J. Li, F. Rossignol and J. Macdonald, *Lab Chip*, 2015, **15**, 2538–2558.
40. I. B. Tahirbegi, J. Ehgartner, P. Sulzer, S. Zieger, A. Kasjanow, M. Paradiso, M. Strobl, D. Bouwes and T. Mayr, *Biosens. Bioelectron.*, 2017, **88**, 188–195.
41. S. A. Pfeiffer, S. M. Borisov and S. Nagl, *Microchim. Acta*, 2017, **184**, 621–626.
42. J. R. Etzkorn, W.-C. Wu, Z. Tian, P. Kim, S.-H. Jang, D. R. Meldrum, A. K.-Y. Jen and B. A. Parviz, *J. Micromech. Microeng.*, 2010, **20**, 095017.
43. P. Sulzer, *Masterthesis*, Graz University of Technology, 2016.
44. T. Abel, M. Sagmeister, B. Lamprecht, E. Kraker, S. Köstler, B. Ungerböck and T. Mayr, *Anal. Bioanal. Chem.*, 2012, **404**, 2841–2849.
45. L. Gitlin, C. Hoera, R. J. Meier, S. Nagl and D. Belder, *Lab Chip*, 2013, **13**, 4134.
46. Z. Lin, T. Cherng-Wen, P. Roy and D. Trau, *Lab Chip*, 2009, **9**, 257–262.
47. L. Wang, M. A. Acosta, J. B. Leach and R. L. Carrier, *Lab Chip*, 2013, **13**, 1586–1592.
48. M. Horka, S. Sun, A. Ruszczak, P. Garstecki and T. Mayr, *Anal. Chem.*, 2016, **88**, 12006–12012.
49. S.-H. Huang, Y.-H. Hsu, C.-W. Wu and C.-J. Wu, *Biomicrofluidics*, 2012, **6**, 044118.
50. M. Viefhues, S. Sun, D. Valikhani, B. Nidetzky, E. X. Vrouwe, T. Mayr and J. M. Bolivar, *J. Micromech. Microeng.*, 2017, **27**, 065012.
51. D. Sud, G. Mehta, K. Mehta, J. Linderman, S. Takayama and M.-A. Mycek, *J. Biomed. Opt.*, 2006, **11**, 050504.
52. G. Mehta, K. Mehta, D. Sud, J. Song, T. Bersano-Begey, N. Futai, Y. S. Heo, M.-A. Mycek, J. Linderman and S. Takayama, *Biomed. Microdevices*, 2007, **9**, 123–134.

53. J. M. Bolivar, T. Consolati, T. Mayr and B. Nidetzky, *Trends Biotechnol.*, 2013, **31**, 194–203.
54. N. Szita, P. Boccazzi, Z. Zhang, P. Boyle, A. J. Sinskey and K. F. Jensen, *Lab Chip*, 2005, **5**, 819–826.
55. A. Super, N. Jaccard, M. P. Cardoso Marques, R. J. Macown, L. D. Griffin, F. S. Veraitch and N. Szita, *Biotechnol. J.*, 2016, **11**, 1179–1189.
56. S. Grist, J. Schmok, M.-C. Liu, L. Chrostowski and K. Cheung, *Sensors*, 2015, **15**, 20030–20052.
57. M. Horka, S. Sun, A. Ruszczak, P. Garstecki and T. Mayr, *Anal. Chem.*, 2016, **88**, 12006–12012.
58. L. Gitlin, C. Hoera, R. J. Meier, S. Nagl and D. Belder, *Lab Chip*, 2013, **13**, 4134–4141.
59. H. Andersson and A. Van den Berg, *Sens. Actuators, B*, 2003, **92**, 315–325.
60. J. El Ali, P. K. Sorger and K. F. Jensen, *Nature*, 2006, **442**, 403–411.
61. J. Park, T. Bansal, M. Pinelis and M. M. Maharbiz, *Lab Chip*, 2006, **6**, 611.
62. C. J. Ochs, J. Kasuya, A. Pavesi and R. D. Kamm, *Lab Chip*, 2014, **14**, 459–462.
63. R. H. W. Lam, M.-C. Kim and T. Thorsen, *Anal. Chem.*, 2009, **81**, 5918–5924.
64. P. C. Thomas, M. Halter, A. Tona, S. R. Raghavan, A. L. Plant and S. P. Forry, *Anal. Chem.*, 2009, **81**, 9239–9246.
65. D. Bavli, S. Prill, E. Ezra, G. Levy, M. Cohen, M. Vinken, J. Vanfleteren, M. Jaeger and Y. Nahmias, *Proc. Natl. Acad. Sci. U. S. A.*, 2016, **113**, E2231–E2240.
66. J. R. Etzkorn, W.-C. Wu, Z. Tian, P. Kim, S.-H. Jang, D. R. Meldrum, A. K.-Y. Jen and B. A. Parviz, *J. Micromech. Microeng.*, 2010, **20**, 095017.
67. T. W. Molter, S. C. McQuaide, M. T. Suchorolski, T. J. Strovas, L. W. Burgess, D. R. Meldrum and M. E. Lidstrom, *Sens. Actuators, B*, 2009, **135**, 678–686.
68. H. Zhu, Y. Tian, S. Bhushan, F. Su and D. R. Meldrum, *IEEE Sens. J.*, 2012, **12**, 1668–1672.
69. I. B. Tahirbegi, J. Ehgartner, P. Sulzer, S. Zieger, A. Kasjanow, M. Paradiso, M. Strobl, D. Bouwes and T. Mayr, *Biosens. Bioelectron.*, 2017, **88**, 188–195.

CHAPTER 15

# $pO_2$ Measurements in Biological Tissues by Luminescence Lifetime Spectroscopy: Strategies to Exploit or Minimize Phototoxic Effects in Tumors

VERONIKA HUNTOSOVA*[a] AND GEORGES WAGNIERES[b]

[a]Center for Interdisciplinary Biosciences, Technology and Innovation Park, P. J. Safarik University in Kosice, Jesenna 5, 041 54 Kosice, Slovakia; [b]Laboratory for Functional and Metabolic Imaging (LIFMET-BMO), Institute of Physics, Swiss Federal Institute of Technology in Lausanne (EPFL), Station 6, Batiment de Chimie, CH-1015 Lausanne, Switzerland
*E-mail: veronika.huntosova@upjs.sk

## 15.1 Introduction

Monitoring the tissue oxygenation is of high interest for the treatment of different disorders and diseases by radiotherapy and PDT. Besides mitochondrial dysfunction,[1] tissue hypoxia[2,3] and metabolic disorders,[4] modifications of the oxygen consumption rate (OCR) plays a significant role in the development of many conditions, including cancer.[5,6] Existing methods to assess tissue oxygenation are characterized by different features and correspond to technical challenges specific to each biomedical application.[7] Besides

Detection Science Series No. 11
Quenched-phosphorescence Detection of Molecular Oxygen: Applications in Life Sciences
Edited by Dmitri B. Papkovsky and Ruslan I. Dmitriev

Published by the Royal Society of Chemistry, www.rsc.org

oxygen-sensitive polarographic microelectodes,[8] blood oxygenation level-dependent (BOLD) MRI contrast[9] and microspectrophotometry of hemoglobin,[10] phosphorescence quenching-based measurements are preferred for a large number of applications. Their ability to estimate the $pO_2$ minimally invasively in interstitial fluids,[11,12] in mitochondria[13,14] or in other cellular/tissular compartments[15–17] makes phosphorescence quenching-based approaches appropriate for numerous biological applications.

Rumsey *et al.* were the first to report, in 1988, measurements of the oxygen distribution in perfused rat liver based on the detection of the Pd-coproporphyrin phosphorescence intensity.[18] Phosphorescence of these porphyrins enabled measurement of the $pO_2$ in biological samples with a high spatial and temporal resolution (~10 µm, <0.1 s, respectively).[19] Since then, great progress has been made in this field. Modifications of oxygen sensitive molecules, their incorporation in nanocarriers, as well as the development and recent implementation of detection techniques, such as time-correlated single photon counting (TCSPC), enabled the determination of the oxygenation of cell/tissue with improved sensitivities and precisions.[20–26] Other progress is associated with the generation of $pO_2$ maps, a concept that corresponds to a strong demand expressed by the scientific community of this field, in particular to image the tissue oxygenation in the intra- and extra-vascular spaces.[27,28] This enables, in particular, the study of the correlation between the perfusion of tissues and their oxygenation. Interestingly, this approach also enables the study of dynamic effects after perturbations of the metabolism (change of oxygen consumption) and/or oxygen delivery. Detailed studies in this field have been reported by the Pittman and Golub group.[29] They used the quenching of the Pd-porphyrin phosphorescence to monitor the interstitial oxygen tension and oxygen consumption in skeletal muscle during contraction.[30,31] These groups have introduced the concept of "$VO_2$ analysis" (maximal oxygen consumption) to better-describe oxygen gradients and transports near arterioles in rat mesentery.[32,33] $VO_2$ analysis of oxygen requires the calculations of the OCR through respiration in a specific region, and oxygen diffusion from surrounding tissue into this region. Oxygen transport in the microcirculation and dedicated methods to measure the $pO_2$ was recently reviewed by Pittman in an intricate work.[29] The phosphorescence quenching method was also applied by Wilson's group in the neonatal porcine model to measure oxygenation of peripheral organs during cardiopulmonary bypass.[34] They were able to image the $pO_2$ in a beating heart and estimated the degree of oxygen deprivation during ischemia.[35] Hypoxia or insufficient tissue oxygenation caused by thrombosis or after injury are certainly not the only conditions requiring the use of $pO_2$ measurement methods. Indeed, hypoxia resulting from an imbalance between oxygen supply and consumption is frequently encountered in advanced tumors, mostly due to an inadequate perfusion. Numerous clinical studies underlined the impact of hypoxia on the aggressiveness of solid tumors and its correlation with a patient's prognosis and survival.[36,37] Importantly, most malignant tumors are heterogeneous in all respects, including regarding their oxygen supply and consumption. Visualization of tumor oxygenation and perfusion in the body can be performed with different techniques, including: dynamic contrast-enhanced MRI, positron emission, computed tomography, Doppler ultrasound, and electron paramagnetic resonance.[9,38–41] This list can be completed with phosphorescence quenching approaches to investigate superficial tissues of the tumor, or the bulk of such lesion when interstitial optical fibers are used to excite and collect the luminescence of a $pO_2$ molecular probe administered systemically.[16,41]

Excepting the use of Clark's electrodes, one alternative method to measure oxygen is to attach a matrix containing oxygen sensitive molecules at the tip of a probe that can be inserted into the tissue. This is the case for the commercially available OxyLite™ system (Oxford Optronics, UK), which is based on the use of a Ru-coordination dye that is subject to the quenching of its phosphorescence by oxygen. One advantage of this system is its high sensitivity and accuracy, in particular at very low $pO_2$.[42] Another interesting approach to measure the $pO_2$ in the bloodstream while avoiding an *i.v.* administration of oxygen sensitive molecules is based on the attachment of such molecules to supporting surfaces. Schafer *et al.* applied a Pt-porphyrin coating to a mammary window chamber model of breast cancer and detected physiologically induced oxygen changes in three dimensions in the tumor with precisions comparable to that of the OxyLab fiber optic based oxygen probe.[43]

As mentioned above, the excited molecule used to detect the oxygenation by phosphorescence-based methods frequently produce singlet oxygen. In addition, they can produce other reactive oxygen species and photoproducts that often lead to photodamage. Moreover, the photoproduct's luminescence may spectrally and temporally overlap with the oxygen-sensitive molecule. This may result in biased measurements of the tissue oxygenation. Therefore, ideal oxygen sensitive molecule does not exist. However, new developments and the design of optimized molecules give a promising vision in this field, as described below.

This chapter presents an overview of the methods used for tumor oxygenation measurements with luminescent oxygen sensitive molecules, with a special emphasis given to the monitoring of oxygen consumption during PDT. Since these molecules have a certain degree of phototoxicity, which is sometimes deliberately exploited, we describe applications of different oxygen sensitive molecules, combined or not, with PS used in PDT. Subcellular localization, water-solubility and photostability of these molecules are compared with their phototoxicity.

## 15.2 Assessment of the Tumor Oxygenation with the Phosphorescence Quenching-based Approach

Most advanced cancers are heterogeneous, in particular regarding their perfusion, cell proliferation rate, and metabolic activities. These parameters not only have an impact on the oxygenation of the lesions, but also on the biodistribution and uptake of exogenous agents administered intravenously or topically.[16,41,44–48] This is the case for oxygen sensitive molecules used to determine the level of oxygen with the phosphorescence quenching method.

In one pilot study, Cerniglia *et al.* used a Xenon flash lamp and a gated video camera to image the Pd-porphyrin phosphorescence in rat tumors.[46] They demonstrated oxygen partial pressures of 1–8 mmHg in tumors, whereas this value was 28–40 mmHg in healthy tissue. Vinogradov and Wilson used a frequency domain fiber-based spectrophosphometer to detect the emission of Pd-porphyrin administered intravenously to measure tumor oxygenation changes in rats breathing air or a 10% oxygen : 90% nitrogen mixture.[47] Similarly, a transparent dorsal window chamber in mice allowed Helmlinger *et al.* to measure the $pO_2$ distribution in LS174T human colon adenocarcinoma after *i.v.* administration of Pd-porphyrin.[45] The oxygen level of normal subcutaneous interstitial tissue, was about 30 mmHg in this model, whereas the oxygenation of tumors decreased below 10 mmHg. Ziemer *et al.* demonstrated with the same oxygen probe that different types of tumor had different oxygenation.[40] While murine melanoma presented oxygen concentrations almost identical to those of normal muscle tissue (~40 mmHg), the oxygenation of renal adenocarcinoma and Lewis lung carcinoma in mice significantly decreased to 24 and 1.8 mmHg, respectively. Developments in the field of oxygen imaging took profit of the availability of new oxygen sensitive molecules. This is the case for Dunphy *et al.* who used two types of Pd-porphyrins in rats bearing subcutaneous glioma tumors.[48] They reported a $pO_2$ of 45 mmHg in normal tissue surrounding the tumor, whereas this value decreased to 10 mmHg in the tumor. More recently, the phosphorescence spectroscopy of Ir-coordination compounds was used to monitor tumor hypoxia in a mouse model of colon adenocarcinoma, before and after treatment with cisplatin.[41] Another class of stable phosphorescent oxygen sensitive molecules involves Ru-coordination complexes. The phosphorescence lifetime of such a complex administered *i.v.* was used to discriminate between tumor and normal tissue in the CAM model.[16]

## 15.3 Photodynamic Therapy

Due to its minimally invasive character and partial selectivity, photodynamic therapy (PDT) is an interesting approach for treating early stage superficial cancers in hollow organs and other regions of the body accessible to light application.[49–52] PDT is also considered as a treatment of interest for, among others, actinic keratosis (in dermatology), and certain forms of age-related macular degeneration (in ophthalmology). To obtain the desired cytotoxic and healing effects, PDT relies on the presence in the targeted tissue of a photosensitizer (PS), molecular oxygen ($O_2$), and administration of light at wavelengths absorbed by the PS. A local application of light and the sometimes selective tumor localization of the PS ensures, in many situations, selective destruction of the tumor cells while preserving their surrounding

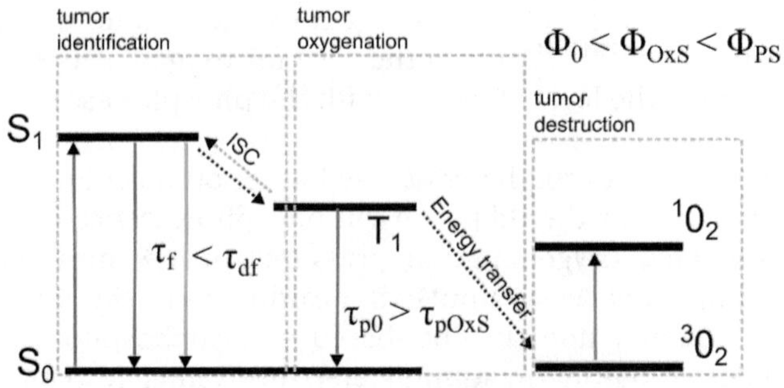


**Figure 15.1** Simplified Jablonski's diagram presenting the physico-chemical processes involved in tumor photodiagnosis (tumor identification), oxygenation measurement, and photodynamic therapy (destruction). Blue arrows represent: (i) photosensitizer excitation and emission for tumor detection (typical fluorescence lifetime: 1–10 ns), and (ii) oxygen-sensitive molecule (OxS) emission for monitoring tumor oxygenation (typical lifetime of the triplet state or the delayed fluorescence (green arrow): 1–1000 μs). Red arrow represents resonant energy transfer-mediated excitation of molecular oxygen, which can induce the photodestruction of biological tissues/cells by the formation of singlet oxygen. Dotted arrows represent the ISC – intersystem crossing. $S_0$-PS ground state, $S_1$-PS singlet state, $T_1$-PS triplet state, $\tau_f$-PS fluorescence lifetime, $\tau_{df}$-PS delayed fluorescence lifetime, $\tau_{p0}$-PS triplet state lifetime in the absence of oxygen, $\tau_{pOxS}$-triplet state lifetime in the presence of oxygen, $^3O_2$-triplet state of molecular oxygen, $^1O_2$-singlet state of molecular oxygen. The excitation of the PS in its $S_1$ state can also be due to triplet–triplet interactions of two PS molecules and singlet oxygen feedback delayed fluorescence (SOFDF), two mecanisms that are not indicated in this diagram.

healthy tissue. The phototoxic reactions of interest for PDT all have in common that they require $O_2$ to induce cell death.[53] The mechanisms leading to cell death are generally separated into two categories, Type I and Type II photodynamic reactions.[54–56] In Type I reactions, the excited PS interacts with a biological molecule and transfers a proton or an electron to it. This species may then react further with oxygen present locally and produce reactive oxygen species (ROS). In Type II reactions, the excited PS transfers its energy to $O_2$ to form the extremely cytotoxic singlet oxygen molecule (see Figure 15.1). It is likely that Type I and Type II reactions coexist in many cases, and the ratio of one to the other depends, among others, not only on the PS but also on the micro-environment of the cell.[57] In both cases, the resulting species can readily oxidize neighboring biological molecules and damage the cell beyond repair, ultimately provoking cell death. The extreme reactivity and short half-life of ROS and singlet oxygen also mean that they will not diffuse very far before encountering a molecule that they will react with.[54] It is accepted that PDT-induced cell death can also result from damage to its surrounding vessels.[58] These vascular damages induced by PDT often result in hypoxia and nutrient deprivation.

In some cases, it was observed that light treatments with high fluence rates resulted in less tumor damage.[59] This suggests that those high fluence rates depleted locally the oxygen level needed for effective PDT. However, fractionation of irradiation can increase the efficacy of PDT, in part due to the temporary recovery of the vascular system and tissue re-oxygenation during the dark intervals.[60]

Although most PSs have a high singlet oxygen production quantum yield,[57] which explains their high phototoxicity,[61] they also frequently have relatively high fluorescence or phosphorescence quantum yields.[62]

These features can be exploited from many respects: (i) their fluorescence is used, in certain cases, to detect, localize and/or delineate certain cancerous lesions,[63,64] (ii) the monitoring of the light dose to be applied in PDT can be based on the PS fluorescence brightness since it is, in a first approximation, proportional to the PS concentration. Therefore, a low light dose is applied to tissues presenting a bright fluorescence, and *vice versa*,[65,66] and, importantly in the context of this chapter, their phosphorescence and/or delayed fluorescence intensity and lifetime frequently depend on the $pO_2$. Therefore, the PS itself can be used to measure this parameter and, if needed, to adapt the light dose used for PDT accordingly.[67]

The spectral properties of a tumor-selective PS and those of an oxygen-sensitive molecular probe can be selected in such a way that they do not interfere when they are co-administered to monitor the tissue oxygenation shortly prior or during PDT. This approach opens new possibilities to: (i) identify/detect tumor cells, (ii) determine their oxygenation level, and, (iii) monitor the light dose used for PDT. In such a situation, the use of a poorly phototoxic oxygen sensitive molecular probe is preferred. Time-resolved measurements of the PS and oxygen probe luminescence performed before PDT provides interesting information about the aggregation state (*i.e.*, phototoxicity of the PS) and the oxygenation of the tissue/cells.[68–70] An illustrative example of this approach is presented in Figure 15.2. The localization and aggregation state of a photosensitive molecule (hypericin) are determined in BT474 mammosphere by fluorescence intensity and lifetime measurements, respectively. The poorly phototoxic oxygen sensitive Ru-coordination compound ($[Ru(Phen)_3]^{2+}$) was administered to the same sample to image mammosphere oxygenation. Classical confocal microscopy images presenting the co-localization of both molecules are also presented in Figure 15.2 with a better resolution than the lifetime images. Such lifetime images of both molecules can give useful information about the localization and concentration of the most biologically active form of hypericin (the monomer is the most phototoxic,[71,72] with fluorescence a lifetime of 5 ns – blue color) and the level of molecular oxygen by $[Ru(Phen)_3]^{2+}$ (a low level of oxygen corresponds to lifetimes >700 ns – green/blue colors).[16]

Although most applications of PDT involve the administration of photosensitizers synthesized exogenously, *i.e.* outside of the patient's body, several PDTs are based on the use of a biochemical precursor of the PS: topical or systemic administration of δ-aminolevulinic acid (ALA) or one of its derivatives, leads, after a suitable time ranging between one and ten hours, to

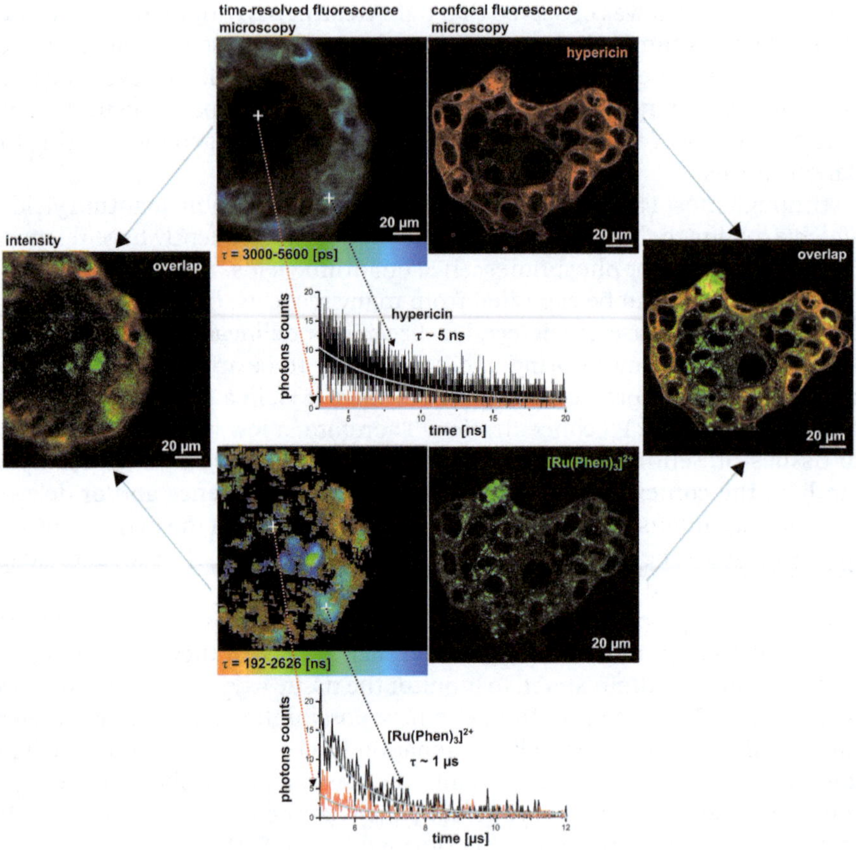


**Figure 15.2** Imaging of the photosensitizer hypericin and the oxygen sensitive molecules $[Ru(Phen)_3]^{2+}$ in living BT474 mammospheres (76 μm in diameter) by time-resolved fluorescence/phosphorescence microscopy (left part) and confocal fluorescence microscopy (right part). Hypericin (500 nM, 1 h, red color) luminescence (lifetime) is presented in the upper row, whereas the $[Ru(Phen)_3]^{2+}$ (200 μM, 24 h, green color) luminescence (lifetime) is presented in the lower row. Middle row presents overlaps of the hypericin and $[Ru(Phen)_3]^{2+}$ luminescence intensity images. Lifetime images show: (i) the hypericin fluorescence lifetimes after an excitation pulse at 555 nm, and (ii) $[Ru(Phen)_3]^{2+}$ phosphorescence lifetimes detected after an excitation pulse at 470 nm. Both emissions were detected with a hybrid TCSPC detector above 600 nm (similarly as described in ref. 101 and 102). Two different regions of interest were selected: (i) in the middle of a mammosphere (red line in decays graphs) and, (ii) close to the surface of a mammosphere (black line in decays graphs). Note, that the time-resolved images were detected with worse confocality (40x water immersion objective), excitation with a "white" laser (SuperK EXTREME, 80 MHz, 5 ps pulse, the scan rate is defined by changing the line time and the zoom state (zoom: 2.21x): hypericin 2.065 ms line time, 0.529 s/frame in FLIM mode and $[Ru(Phen)_3]^{2+}$ 10.23 ms line time, 2.62 s/frame in PLIM mode) and resolution than the confocal fluorescence images (63x oil

biosynthesis and accumulation of photoactivable porphyrins (PAPs), including protoporphyrin IX (PpIX), in the cells of the target tissue.[73] δ-ALA and PpIX are naturally occurring intermediates in the heme biosynthesis,[73] the penultimate step of which is the synthesis of PpIX in the mitochondria. The heme biosynthesis is controlled by feedback mechanisms preventing accumulation of PpIX in a functional cell,[54,55] but these are bypassed when an excess of precursors is administered exogenously. Thus, PPIX temporarily accumulates, preferentially in hypermetabolic or neoplastic cells.[54,55,73,74] Note that ALA + derivative-based PDT procedures are now approved for the treatment of premalignant and malignant conditions in dermatology.[49] The formulation of an ALA derivative (Hexvix®) is also approved for the detection of superficial bladder cancer by fluorescence cystoscopy.[75] As we will see in more detail below, PpIX is interesting in many respects: (i) it is produced in the mitochondria, an organelle that plays a key role in the metabolism, (ii) ALA is a clinically approved precursor of PpIX for several diagnostic and therapeutic applications, (iii) the PpIX DF lifetime strongly depends on the $pO_2$, which enables to measuring $pO_2$ values well below 5 mmHg, (iv) the PpIX DF is sufficiently bright to enable its detection in biological tissues using imaging techniques.

## 15.4 Oxygen Consumption During Photodynamic Therapy

The photochemical reactions resulting from the excitation of molecular oxygen by the PS induce oxygen consumption in the treated tissue. This consumption significantly influences the dosimetry and the potency of PDT. A summary of the oxygenation detection methods and oxygen sensitive molecules used to measure the consumption of oxygen in tissues treated by PDT is presented in Table 15.1. Clark-type polarographic oxygen microelectrodes,[76–80] electrochemical microsensor systems,[81] hypoxia antibody-based markers and radioactivity assays,[82] photoacoustic lifetime imaging,[83] as well as fiber-based phosphorescence lifetime[27,84] and delayed fluorescence lifetime[67] measurements and imaging systems are used to assess the $pO_2$ in biological tissues during PDT.

The first direct detection of oxygen consumption in a tumor during Photofrin® mediated PDT was performed with Clark-types of microelectrodes.

---

immersion objective, excitation with solid-state CW lasers at 405 and 555 nm for $[Ru(Phen)_3]^{2+}$ and hypericin, respectively). While the fluorescence lifetime of hypericin provides information about its aggregation state which influences its phototoxicity (the most phototoxic monomeric state corresponds to long lifetimes), $[Ru(Phen)_3]^{2+}$ luminescence lifetime gives information about the oxygenation state of the system (longer lifetimes refer to less oxygenated areas). The intensity of both molecules indicates their intracellular localization in mammospheres. Altogether, this information can serve as input parameters for local dosimetry.

**Table 15.1** List of the detection methods, oxygen-sensitive molecules and photosensitizers used to monitor the level of oxygen in various samples. Changes of this level measured before and after PDT are reported for selected cases. (ALA – δ-Aminolevulinic acid, PpIX – protoporphyrin IX, PdTCPP – palladium *meso*-tetra(4-carboxyphenyl)porphyrin, EF3 – 2-(2-nitroimidazol-1[H]-yl)-*N*-(3,3,3-trifluoropropyl)acetamide, Photofrin – porfimer sodium, $AlPcS_4$ – Al(III) phthalocyanine chloride tetrasulfonic acid, mTHPC – *meta*-tetra(hydroxyphenyl)chlorin, PAD-S31 – 13,17-bis (1-carboxypropion) carbamoylethyl-3-ethenyl-8-ethoxyiminoethylidene-7-hydroxy-2,7,12,18-tetramethyl-porphyrin sodium).

| Method for $O_2$ detection | $O_2$ Sensitive molecule | Photosensitizer | Specimen | $O_2$ Before PDT | $O_2$ After PDT | Ref. |
|---|---|---|---|---|---|---|
| Delayed fluorescence lifetime measurement | ALA–PpIX | ALA–PpIX | Chicken embryo chorioallantoic membrane | 12% $pO_2$ | 0% $pO_2$ | 67 |
| Phosphorescence lifetime measurement | PdTCPP | ALA–PpIX | Wistar rat liver | – | – | 84 |
| PLIM – phosphorescence lifetime imaging | PdTCPP | PdTCPP | Young piglet eyes | 20 mmHg | 2 mmHg | 27 |
| PALI – photoacoustic lifetime imaging | Methylene blue | Methylene blue | Nude mice bearing LNCaP tumors | – | – | 83 |
| Hypoxia marker (antibody staining)/radioactivity assay | EF3 | Photofrin | RIF tumor on the shoulders of mice | 3–5 mmHg | During PDT 1–2 mmHg/after PDT 3–5 mmHg | 82 |
| Needle style oxygen electrode/ prompt photosensitizer fluorescence | $AlPcS_4$ | $AlPcS_4$ | In PBS solution | – | – | 86 |
| Clark-type oxygen microelectrode (Eppendorf) | – | PdTCPP | EMT6/Ro spheroids | – | – | 85 |
| | – | Photofrin | White rabbit ears, VX-2 skin carcinoma | – | – | 76 |
| | – | mTHPC | AML-5 leukemia cells | – | – | 77 |
| | – | Photofrin | RIF tumor in mice | 10 mmHg | 2 mmHg | 78 |
| | – | ALA–PpIX | Wistar rat colon | – | – | 79 |
| | – | PAD-S31 | Renca, mouse renal carcinoma cells | – | – | 80 |
| Oxygen needle sensor | Needle sensor | Photofrin | Aqueous bacterial suspension (Staphylococcus aureus) | – | – | 87 |
| Electrochemical microsensor system | Chronoamperometric oxygen sensor | PpIX | T-47D, breast cancer cells | – | – | 81 |

Tromberg *et al.* attached transcutaneous electrodes on rabbit ears, with and without tumor, and monitored $pO_2$ differences during irradiations.[76] They were able to evidence oxygen consumption and reoxygenation of the irradiated tissues. On the other hand, Sitnik *et al.* used polarographic needle probes to measure the $pO_2$ within RIF tumor grown in mice.[78] Applications of Photofrin® mediated PDT of tumors illuminated with different fluence rates led to significant decrease of $pO_2$ (representatively from 10 mmHg to 2 mmHg). Encouraged by those first studies Clark-type electrodes have also been employed during *meta*-tetra(hydroxyphenyl)chlorin (mTHPC),[77] 13,17-bis (1-carboxypropion) carbamoylethyl-3-ethenyl-8-ethoxyiminoethylidene-7-hydroxy-2,7,12,18-tetramethyl-porphyrin sodium (PAD-S31),[80] palladium *meso*-tetra(4-carboxyphenyl)porphyrin (PdTCPP),[85] δ-Aminolevulinic acid (ALA)–protoporphyrin IX (PpIX)[79] and Al(III) phthalocyanine chloride tetrasulfonic acid ($AlPcS_4$)[86] mediated PDT. Needle-based probes are not only used to measure the $pO_2$ in tissues, but also in solutions, including to measure the oxygen consumption in bacterial suspension[87] treated with Photofrin® or $AlPcS_4$.[86] Busch *et al.* observed a correlation between hypoxia marker intensity decreases and Eppendorf $pO_2$ measurements (ranging between 1–5 mmHg) during Photofrin® mediated PDT.[82]

The first efforts to monitor the tissue oxygenation during PDT using the phosphorescence quenching method were performed in 1998 by McIlroy *et al.*[84] These tissue oxygenation studies were based on the use of PdTCPP bound to bovine serum albumin (BSA) before, during and after treatment in rat liver by ALA–PpIX mediated PDT. This group established a method to measure the $pO_2$ within and at the periphery of necrotic regions induced by PDT. Phosphorescence lifetimes of PdTCPP were detected between 700 and 800 nm, with an excitation at 527 nm (pulse duration: 1 ns; 10 μJ per pulse; repetition rate: 10 Hz). Phosphorescence measurements were performed during PpIX irradiation at 635 nm, where PdTCPP poorly absorbs. The decays were fitted after a 20 μs delay relative to the laser pulse to avoid contributions from the prompt fluorescence. For high PDT light doses, $pO_2$ decreased from 22 mmHg to 2 mmHg.

Lately, the photosensitizing properties of oxygen molecular probes have become more and more acknowledged, making them good candidates to perform PDT. Mitra and Foster denoted that 514 nm CW argon ion laser light photobleached PdTCPP in solutions containing BSA and in cell spheroids, an effect that was accompanied by oxygen consumption.[85] Stepinac *et al.* applied time and frequency domain PLIM (phosphorescence lifetime imaging) after intravenous administration of PdTCPP to determine the oxygen consumption in piglet's eyes vasculature treated by PDT at 532 nm (500 Hz; mean irradiance: 5 mW $cm^{-2}$).[27] The $pO_2$ decreased from 25 mmHg to 1 mmHg during the 8 minutes following the beginning of the irradiations, and the PdTCPP phosphorescence lifetime increased from 100 to 600 μs.

Shao *et al.* combined PDT with a photoacoustic lifetime imaging system to treat tumor xenografts in mice with methylene blue, which was used as PS and oxygen marker.[83] In this study, PDT was induced with light at 660 nm emitted by a laser diode, whereas photoacoustic probing was performed with a pulsed laser emitting at 810 nm for $pO_2$ mapping. This approach, based on the combination of two techniques, is a new way to individualize the light dose applied for PDT in clinical practice.

As mentioned above, PpIX is preferentially produced in mitochondria. This accumulation is particularly selective in certain cancers, a property used in fluorescence imaging to identify cancer cells, and tumor boundaries.[88,89] On the other hand, Mik's group proposed to exploit the quenching of the PpIX delayed fluorescence to assess the metabolism of cutaneous tissues in the clinic.[14] This group reported values of the PpIX delayed fluorescence lifetime ranging between 10 and 1000 µs for physiological and "low" $pO_2$, respectively. Similarly, Piffaretti *et al.* measured the PpIX delayed fluorescence lifetime to monitor the oxygen consumption during PDT in the chick's chorioallantoic membrane (CAM) vasculature.[67] They developed a fiber-based technique to excite (405 nm, 0.5 ns pulse duration, maximal light dose 100 mJ $cm^{-2}$) and detect the PpIX delayed fluorescence while PDT was induced with a 405 nm CW laser (light dose: up to 2.25 J $cm^{-2}$). A good correlation between PDT-induced vascular damage, oxygen decrease, and the light dose was found. As a consequence, this group demonstrated that the PDT efficacy can be predicted, individualized and optimized in real-time with one molecule used for both PDT and $pO_2$ measurement.

## 15.5 Strategies to Minimize Tissue Damage While Measuring the $pO_2$

As mentioned above, oxygen is consumed and highly reactive ROS (namely singlet oxygen and free radicals) are formed by most oxygen molecular probes excited in their triplet state in physiological environments. Depending on their actual localization, the interaction of these ROS with cellular substrates may lead to apoptosis, necrosis or other types of cell demise.[57,61] PSs localizing in mitochondria tend to induced apoptosis, whereas those localizing in the lysosomes, endosomes and/or peroxisomes preferentially generate necrosis.[57,61] Although these photo-effects are required for photosensitizer-mediated PDT, they must be minimized during measurements of the $pO_2$. It should be noted that these photo-effects can result in the degradation of the $pO_2$ molecular probe itself (photobleaching) and lead to the generation of its photoproducts. This can significantly affect measurements of the $pO_2$ since some of these photoproducts often present a luminescence spectroscopy that is similar to that of the $pO_2$ probe.

### 15.5.1 Impact of the Biodistribution and Subcellular Localization of Oxygen-sensitive Molecules on Their Phototoxicity

Lifetimes of ROS formed during PDT are very short (~µs), and their diffusion in tissue is limited.[90] Therefore, one strategy to reduce the phototoxic effects of oxygen-sensitive molecules during $pO_2$ measurements consists to control their localization. It should be noted that this strategy must take into account the type of compartment where the $pO_2$ is measured. These compartments frequently involve: the intravascular, extravascular, interstitial, extracellular and intracellular spaces.[14,23,44,91] Recently, Dmitriev and Papkovsky published a report on the optimal spectral properties, cell specificities and intracellular localization of oxygen-sensitive molecules in different cell and tissue models.[92] The hydrophobic properties of such molecules define their localization and clearance rates. In general, hydrophilic oxygen sensitive molecules are preferred *in vivo*. However, their penetration into cells is often reduced.[44] Examples of intra/extra-cellular localizations of three selected oxygen sensitive molecules presenting different oxygen sensitivity ranges (within 1 μs or 1 ms) are presented in Figure 15.3. PdTCPP and $[Ru(Phen)_3]^{2+}$ are water-soluble molecules that localize in the extracellular space. However, after long (24 h) incubations, they can be found inside cells, mostly in endocytotic vesicles.[44] On the other hand, the hydrophobic PpIX, dissolved in dimethyl sulfoxide (below 1% in aqueous solution), was not detected in the extracellular space,

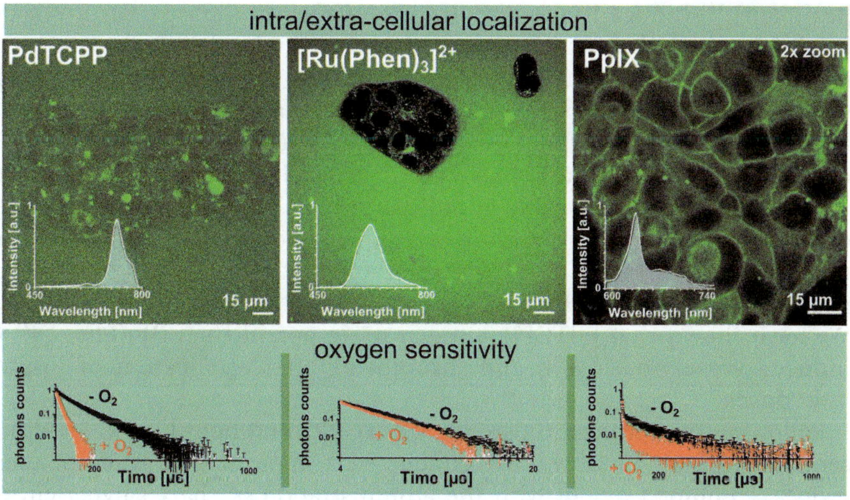


**Figure 15.3** Fluorescence confocal images showing the intracellular/extracellular localization (green) of 100 μM PdTCPP (incubation time: 24 h, $\lambda_{exc}$ = 488 nm, $\lambda_{em}$ > 600 nm), 200 μM $[Ru(Phen)_3]^{2+}$ (incubation time: 24 h, $\lambda_{exc}$ = 405 nm, $\lambda_{em}$ > 600 nm) and 100 μM PpIX (incubation time: 24 h, $\lambda_{exc}$ = 405 nm, $\lambda_{em}$ > 600 nm) in BT 474 mammospheres. Representative emission spectra of fully dissolved forms are presented in the inserts (lower left corners in each image). Representative phosphorescence (solutions of PdTCPP and $[Ru(Phen)_3]^{2+}$) and delayed fluorescence (PpIX built up in the chicken embryo chorioallantoic membrane after ALA topical application) decays in air saturated (red) and hypoxic conditions (black) are presented in the lower line. More information about the spectral properties and detection techniques are described in ref. 16, 44, 94, 100 and 103. Intracellular/extracellular localization, water-solubility and oxygen sensitivity ranges are the most important indicators to select relevant oxygen sensitive molecules for intravascular/extravascular and hypoxic tissue oxygen measurements.

even 24 h after incubation. However, its bright fluorescence intensity was observed in the cellular membranes. This Figure 15.3 illustrates that changing the lipophilic properties of oxygen sensitive molecules enables to target different tissue compartments. Another example of different localizations of oxygen sensitive molecule and PS is presented in Figure 15.2.

In all cases, it is important that oxygen consumption and singlet oxygen production by oxygen sensitive molecule are as limited as possible. This statement is at the origin of the numerous studies aiming at identifying oxygen-sensitive molecules presenting minimal singlet oxygen production yields.

### 15.5.2 Correlation Between the Production of ROS and the Phototoxicity While Measuring the $pO_2$

Although the phototoxicity of oxygen sensitive molecules used for phosphorescence lifetime-based measurements is usually minimal, their excitation with light may induce photodamage. Recently, we have shown that $[Ru(Phen)_3]^{2+}$ induce limited phototoxic effects on the vasculature of the CAM.[16] The threshold limit was identified at a relatively high concentration (10 mg $kg^{-1}$ of body weight) and light dose (10 J $cm^{-2}$). $[Ru(Phen)_3]^{2+}$ was shown to localize in endosomal vesicles, a property that is probably in relation with its poor phototoxicity.[44] Although its singlet oxygen quantum yield is not particularly low,[93] its hydrophilic character makes it biocompatible, and weakly phototoxic.[16,94] In addition, $[Ru(Phen)_3]^{2+}$ is quite photostable[95] and, in comparison with porphyrins, its photoproducts are rarely detected.[85,96–100] The phototoxicity of $[Ru(Phen)_3]^{2+}$ was compared with that of PdTCPP, a $pO_2$ probe that was previously used for PDT.[27] We have shown that PdTCPP is much more potent than $[Ru(Phen)_3]^{2+}$ to induced vessels occlusion and thrombosis after application of limited light doses.[94]

Following the localization study of PdTCPP, $[Ru(Phen)_3]^{2+}$ and PpIX presented in Figure 15.3, we have conducted a phototoxicity study of these probes, combined or not with the administration of the PS hypericin on U87 MG cells monolayers. Flow cytometric analysis of such treated cells is presented in Figure 15.4A and B. It should be noted that all three oxygen sensitive molecules mentioned above present a red emission above 600 nm and a relatively poor absorption around 590 nm, a spectroscopy which corresponds to that of propidium iodide. For this reason, we have performed flow cytometry measurements with and without propidium iodide staining (red and black points, respectively in Figure 15.4A) to indicate the presence of necrotic cells. Apoptotic cells were revealed staining them with Annexin V/FITC that fluoresces in the green. This analysis showed that the combination of PpIX with hypericin generates the most important photodamage, whereas the combination of this PS with $[Ru(Phen)_3]^{2+}$ induces minimal damage (Figure 15.4B). MTT assay based on the assessment of formazan production in mitochondria confirmed these different cell responses (Figure 15.4C).

The potency of PpIX could be due to its localization in the mitochondrial and cellular membranes. It should also be noted, that the second PpIX absorption Q-band is very close to 590 nm.

Ceroni *et al.* studied the phototoxicity of Pd-compounds used as oxygen sensors and compared them with Photofrin.[96] They were not able to detect any phototoxic effects on RIF cells at Pd-compounds concentrations ranging between 2.5 and 20 μM, when the samples were illuminated with CW light at 630 nm and with a dose of 2 J $cm^{-2}$. However, as reported by Stepinac *et al.*, photoactivated PdTCPP caused redistribution of occludin in retinal vessels and a breakdown of the blood-retinal barrier.[27]

These results suggest that the phototoxicity of $pO_2$ sensitive molecule could importantly influence dosimetry and identification of PS phototoxicity when applying both methods, $pO_2$ measurements, and PDT, simultaneously. To precisely determine PS phototoxicity in such cases, at the cellular level, the combination treatment should be performed. However, many $pO_2$ sensitive molecules are in the same spectral regions as PS or apoptotic/necrotic (PI or Annexin V/FITC) fluorescent markers used in flow-cytometry. The solution to this problem will be to estimate correction by performing the measurements in the absence of apoptotic/necrotic fluorescent markers, as it was demonstrated in Figure 15.4. By this approach, one can much better define which $pO_2$ probe is minimally invasive and suitable for oxygenation or oxygen consumption detection during PDT.

### 15.5.3 Alteration of $pO_2$ Measurements by the Luminescence of Oxygen Molecular Probes Photoproducts

As mentioned above, photoproducts of oxygen molecular probes produced during oxygenation measurements can introduce artifacts since the luminescence spectroscopy of some of them is similar to that of the probe itself. This is the case for PpIX produced endogenously. We have also recently shown that this effect, which takes place in solutions and in the CAM, decreases the PpIX oxygen sensitivity.[100] We demonstrated in this study that the generation of photoproducts can be avoided if light doses at 405 nm lower than several hundreds of mJ $cm^{-2}$ are used for $pO_2$ measurements. In addition, we have shown that bias induced by the PpIX photoproducts luminescence can be avoided if the detection of PpIX DF lifetime is performed between 620 and 640 nm, a spectral domain where the PpIX photoproducts do not luminesce. It should be noted that these photoproducts may also contribute to the formation of singlet oxygen and, consequently, are sometimes phototoxic.

Although the artifacts induced by the photobleaching of molecular probes as well as the bias induced by the luminescence of their photoproducts are very important issues, they are virtually unexplored. It is likely that works dedicated to these issues will strongly increase the reliability, accuracy and use of this $pO_2$ measurement technique.

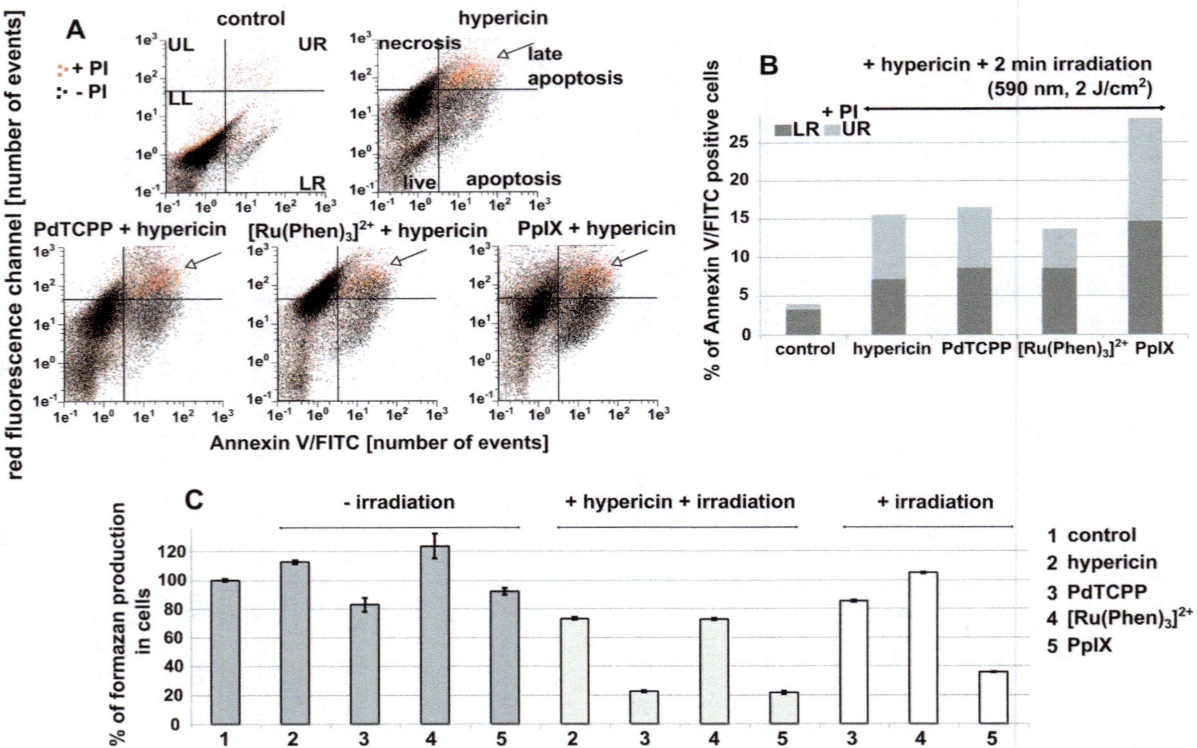


**Figure 15.4** (A) Flow cytometry analysis of damaged U87 MG (human glioma) cells ($2 \times 10^4$ cells) performed 3 h after irradiation (2 J cm$^{-2}$ at 590 nm; irradiance: 17 mW cm$^{-2}$; the whole culture was illuminated). Cells were incubated during 24 h with 100 µM PdTCPP, 200 µM $[Ru(Phen)_3]^{2+}$, 100 µM PpIX, and 1 h with 0.5 µM hypericin. A combination of PdTCPP and hypericin, $[Ru(Phen)_3]^{2+}$

## 15.6 Conclusion

Measuring the tissue oxygenation using the phosphorescence quenching-based approach presents many advantages as compared with other techniques. The minimally invasive aspect of this approach combined with the fact that the $pO_2$ is measured directly and in real time with an outstanding sensitivity and spatial resolution probably explains why it is already successfully used in the clinic, in addition to its numerous non-clinical applications. Nevertheless, care must be taken to avoid artifacts or unwanted phototoxic effects if such $pO_2$ measurements are not performed in optimal conditions, both in terms of oxygen-sensitive molecular probe and light dose. Strategies to prevent such artefacts include the use of: (i) molecular probes localizing in tissue/cell compartments where the production of ROS has acceptable consequences, (ii) spectral conditions which are such that the luminescence of photoproducts does not interfere with that if the molecular probe, and (iii) excitation light dose of the molecular probe that are low enough to avoid the generation of photoproducts or phototoxic effects. This last condition is frequently obtained using high sensitivity detectors and optimal light collection efficiency. The availability of last generation technologies enabling one to produce mappings of the tissue oxygenation with two-photon excitation of molecular probes will further increase the potential of this approach. Finally, the development of new oxygen-sensitive molecules and formulations minimizing the production of ROS while improving their specificities for tissue/cells compartments of interest will broaden the application fields of this $pO_2$ measurement approach.

---

and hypericin, as well as PpIX and hypericin were tested as a model of PSs co-administration with oxygen sensitive molecules. Cells were washed before irradiation. Propidium iodide was used as an indicator of necrotic cells and Annexin V/FITC as an indicator of apoptotic cells. Flow-cytometry analysis of cells populations is separated into four quadrants: LL- live cells, LR- early apoptotic cells, UR- late apoptotic cells and UL- necrotic cells. Black cell populations (−PI) in the figures are labeled only with Annexin V/FITC and red cell populations represent the same cells labeled with Annexin V/FITC and propidium iodide (+PI). Black arrows point to damaged cells, usually designed as late apoptotic cells. (B) Percentages of positive cells to Annexin V/FITC staining represented in A: dark gray – LR, light gray – UR, determined in cell populations labeled with propidium iodide (+PI). (C) MTT assay of U87 MG cells in the presence of hypericin, combined or not, with PdTCPP, RuPhen or PpIX in the dark (dark gray columns) and after irradiation (light gray columns in the presence of hypericin and white columns in the absence of hypericin). The assay was performed 5 h after irradiation. Phototoxicity of hypericin was reported in the following ref. 71, 104 and 105. These results demonstrate the synergistic phototoxic effects of PdTCPP, PpIX and hypericin (higher photodamages). A limited potency of hypericin was observed when co-administered with $[Ru(Phen)_3]^{2+}$. These findings suggest that the phototoxicity of oxygen sensitive molecule should also be taken into consideration when they are used to monitor the light dosimetry in PDT.

## Acknowledgement

This work was supported by the Slovak Research and Development Agency APVV-15-0485, and the Swiss National Science Foundation (Grant No. CR32I3_159746).

## References

1. M. T. Lin and M. F. Beal, *Nature*, 2006, **443**, 787–795.
2. M. Hockel and P. Vaupel, *J. Natl. Cancer Inst.*, 2001, **93**, 266–276.
3. K. L. Eales, K. E. R. Hollinshead and D. A. Tennant, *Oncogenesis*, 2016, **5**, e190.
4. T. N. Seyfried, R. E. Flores, A. M. Poff and D. P. D'Agostino, *Carcinogenesis*, 2014, **35**, 515–527.
5. S. Reuter, S. C. Gupta, M. M. Chaturvedi and B. B. Aggarwal, *Free Radicals Biol. Med.*, 2010, **49**, 1603–1616.
6. D. Dreher and A. F. Junod, *Eur. J. Cancer*, 1996, **32A**, 30–38.
7. R. N. Pittman, *Regulation of Tissue Oxygenation. Measurement of Oxygen*, Morgan & Claypool Life Sicences, San Rafael (CA), 2011, ch. 10.
8. D. G. Buerk, *Oxygen Sensing*, 2004, **381**, 665–690.
9. L. Gagnon, S. Sakadzic, F. Lesage, J. J. Musacchia, J. Lefebvre, Q. Q. Fang, M. A. Yucel, K. C. Evans, E. T. Mandeville, J. Cohen-Adad, J. R. Polimeni, M. A. Yaseen, E. H. Lo, D. N. Greve, R. B. Buxton, A. M. Dale, A. Devor and D. A. Boas, *J. Neurosci.*, 2015, **35**, 3663–3675.
10. R. N. Pittman and B. R. Duling, *J. Appl. Physiol.*, 1975, **38**, 315–320.
11. S. A. Vinogradov, M. A. Fernandez-Seara, B. W. Dupan and D. F. Wilson, *Comp. Biochem. Physiol., Part A: Mol. Integr. Physiol.*, 2002, **132**, 147–152.
12. J. M. Vanderkooi, G. Maniara, T. J. Green and D. F. Wilson, *J. Biol. Chem.*, 1987, **262**, 5476–5482.
13. E. G. Mik, T. Johannes, C. J. Zuurbier, A. Heinen, J. H. P. M. Houben-Weerts, G. M. Balestra, J. Stap, J. F. Beek and C. Ince, *Biophys. J.*, 2008, **95**, 3977–3990.
14. E. G. Mik, J. Stap, M. Sinaasappel, J. F. Beek, J. A. Aten, T. G. van Leeuwen and C. Ince, *Nat. Methods*, 2006, **3**, 939–945.
15. J. K. Asiedu, J. Ji, M. Nguyen, N. Rosenzweig and Z. Rosenzweig, *J. Biomed. Opt.*, 2001, **6**, 116–121.
16. V. Huntosova, S. Gay, P. Nowak-Sliwinska, S. K. Rajendran, M. Zellweger, H. van den Bergh and G. Wagnieres, *J. Biomed. Opt.*, 2014, **19**, 077004.
17. H. Kurokawa, H. Ito, M. Inoue, K. Tabata, Y. Sato, K. Yamagata, S. Kizaka-Kondoh, T. Kadonosono, S. Yano, M. Inoue and T. Kamachi, *Sci. Rep.*, 2015, **5**, 10657.
18. W. L. Rumsey, J. M. Vanderkooi and D. F. Wilson, *Science*, 1988, **241**, 1649–1651.
19. D. F. Wilson, J. M. Vanderkooi, T. J. Green, G. Maniara, S. P. DeFeo and D. C. Bloomgarden, *Adv. Exp. Med. Biol.*, 1987, **215**, 71–77.
20. X. D. Wang and O. S. Wolfbeis, *Chem. Soc. Rev.*, 2014, **43**, 3666–3761.

21. M. Quaranta, S. M. Borisov and I. Klimant, *Bioanal. Rev.*, 2012, **4**, 115–157.
22. J. Ji, N. Rosenzweig, I. Jones and Z. Rosenzweig, *Anal. Chem.*, 2001, **73**, 3521–3527.
23. R. I. Dmitriev and D. B. Papkovsky, *Cell. Mol. Life Sci.*, 2012, **69**, 2025–2039.
24. O. S. Finikova, A. Y. Lebedev, A. Aprelev, T. Troxler, F. Gao, C. Garnacho, S. Muro, R. M. Hochstrasser and S. A. Vinogradov, *ChemPhysChem*, 2008, **9**, 1673–1679.
25. S. Sakadzic, E. Roussakis, M. A. Yaseen, E. T. Mandeville, V. J. Srinivasan, K. Arai, S. Ruvinskaya, A. Devor, E. H. Lo, S. A. Vinogradov and D. A. Boas, *Nat. Methods*, 2010, **7**, 755–759.
26. C. F. Wu, B. Bull, K. Christensen and J. McNeill, *Angew. Chem., Int. Ed.*, 2009, **48**, 2741–2745.
27. T. K. Stepinac, S. K. Chamot, E. Rungger-Brandle, P. Ferrez, J. L. Munoz, H. van den Bergh, C. E. Riva, C. J. Pournaras and G. A. Wagnieres, *Invest. Ophthalmol. Visual Sci.*, 2005, **46**, 956–966.
28. W. Becker, V. Shcheslavskiy and A. Ruck, *Adv. Exp. Med. Biol.*, 2017, **1035**, 19–30.
29. R. N. Pittman, *Microcirculation*, 2013, **20**, 117–137.
30. L. M. Smith, A. S. Golub and R. N. Pittman, *Microcirculation*, 2002, **9**, 389–395.
31. W. H. Nugent, B. K. Song, R. N. Pittman and A. S. Golub, *Microvasc. Res.*, 2016, **105**, 15–22.
32. A. S. Golub, M. C. Barker and R. N. Pittman, *Am. J. Physiol.: Heart Circ. Physiol.*, 2007, **293**, H1097–H1106.
33. A. S. Golub and R. N. Pittman, *Am. J. Physiol.: Heart Circ. Physiol.*, 2008, **294**, H2905–H2916.
34. W. M. DeCampli, G. Schears, R. Myung, S. Schultz, J. Creed, A. Pastuszko and D. F. Wilson, *J. Thorac. Cardiovasc. Surg.*, 2003, **125**, 472–480.
35. W. L. Rumsey, M. Pawlowski, N. Lejavardi and D. F. Wilson, *Am. J. Physiol.*, 1994, **266**, H1676–H1680.
36. H. J. Feldmann, *Lung Cancer*, 2001, **33**, S77–S83.
37. J. C. Walsh, A. Lebedev, E. Aten, K. Madsen, L. Marciano and H. C. Kolb, *Antioxid. Redox Signaling*, 2014, **21**, 1516–1554.
38. B. F. Jordan and P. Sonveaux, *Front. Pharmacol.*, 2012, **3**, UNSP 94.
39. D. Jennings, N. Raghunand and R. J. Gillies, *Cancer Metastasis Rev.*, 2008, **27**, 589–613.
40. L. S. Ziemer, W. M. F. Lee, S. A. Vinogradov, C. Sehgal and D. F. Wilson, *J. Appl. Physiol.*, 2005, **98**, 1503–1510.
41. Y. Zeng, Y. Liu, J. Shang, J. Ma, R. Wang and L. Deng, *PLoS One*, 2015, **10**, e0126302.
42. B. X. Wen, M. Urano, J. L. Humm, V. E. Seshan, G. C. Li and C. C. Ling, *Radiat. Res.*, 2008, **169**, 67–75.
43. R. Schafer and A. F. Gmitro, *Biomed. Opt. Express*, 2015, **6**, 639–650.
44. V. Huntosova, K. Stroffekova, G. Wagnieres, M. Novotova, Z. Nichtova and P. Miskovsky, *Metallomics*, 2014, **6**, 2279–2289.

45. G. Helmlinger, F. Yuan, M. Dellian and R. K. Jain, *Nat. Med.*, 1997, **3**, 177–182.
46. G. J. Cerniglia, D. F. Wilson, M. Pawlowski, S. Vinogradov and J. Biaglow, *J. Appl. Physiol.*, 1997, **82**, 1939–1945.
47. D. F. Wilson, S. A. Vinogradov, B. W. Dugan, D. Biruski, L. Waldron and S. A. Evans, *Comp. Biochem. Physiol., Part A: Mol. Integr. Physiol.*, 2002, **132**, 153–159.
48. I. Dunphy, S. A. Vinogradov and D. F. Wilson, *Anal. Biochem.*, 2002, **310**, 191–198.
49. P. Agostinis, K. Berg, K. A. Cengel, T. H. Foster, A. W. Girotti, S. O. Gollnick, S. M. Hahn, M. R. Hamblin, A. Juzeniene, D. Kessel, M. Korbelik, J. Moan, P. Mroz, D. Nowis, J. Piette, B. C. Wilson and J. Golab, *Ca-Cancer J. Clin.*, 2011, **61**, 250–281.
50. Z. Huang, *Technol. Cancer Res. Treat.*, 2005, **4**, 283–293.
51. M. B. Ericson, A. M. Wennberg and O. Larko, *Ther. Clin. Risk Manage.*, 2008, **4**, 1–9.
52. A. Weiss, H. van den Bergh, A. W. Griffioen and P. Nowak-Sliwinska, *Biochim. Biophys. Acta, Rev. Cancer*, 2012, **1826**, 53–70.
53. C. Fritsch, P. Lehmann, W. Stahl, K. W. Schulte, E. Blohm, K. Lang, H. Sies and T. Ruzicka, *Br. J. Cancer*, 1999, **79**, 1603–1608.
54. S. Collaud, A. Juzeniene, J. Moan and N. Lange, *Curr. Med. Chem.: Anti-Cancer Agents*, 2004, **4**, 301–316.
55. Q. Peng, K. Berg, J. Moan, M. Kongshaug and J. M. Nesland, *Photochem. Photobiol.*, 1997, **65**, 235–251.
56. H. van den Bergh and J. P. Ballini, in *Lasers in Ophthalmology – Basic, Diagnostic and Surgical Aspects*, ed. F. Fankhauser and S. Kwasniewska, Kugler Publications, The Hague, 2003, pp. 183–195.
57. A. P. Castano, T. N. Demidova and M. R. Hamblin, *Photodiagn. Photodyn. Ther.*, 2004, **1**, 279–293.
58. N. L. Oleinick, A. L. Nieminen and S. M. Chiu, Cell killing by photodynamic therapy, in *Advances in Photodynamic Therapy: Basic, Translational and Clinical*, ed. M. R. Hamblin and P. Mroz, Artech House, Boston, MA, USA, 2008, pp. 115–133.
59. S. A. Blant, A. Woodtli, G. Wagnieres, C. Fontolliet, H. vandenBergh and P. Monnier, *Photochem. Photobiol.*, 1996, **64**, 963–968.
60. T. H. Foster, R. S. Murant, R. G. Bryant, R. S. Knox, S. L. Gibson and R. Hilf, *Radiat. Res.*, 1991, **126**, 296–303.
61. A. P. Castano, T. N. Demidova and M. R. Hamblin, *Photodiagn. Photodyn. Ther.*, 2005, **2**, 1–23.
62. L. B. Josefsen and R. W. Boyle, *Met.-Based Drugs*, 2008, **2008**, 276109.
63. K. El Hoshy, M. Bosseila, D. El Sharkawy and R. Sobhi, *Photodiagn. Photodyn. Ther.*, 2016, **14**, 4–8.
64. E. Belykh, N. L. Martirosyan, K. Yagmurlu, E. J. Miller, J. M. Eschbacher, M. Izadyyazdanabadi, L. A. Bardonova, V. A. Byvaltsev, P. Nakaji and M. C. Preul, *Front. Surg.*, 2016, **3**, UNSP 55.
65. J. Barge, T. Glanzmann, M. Zellweger, D. Salomon, H. van den Bergh and G. Wagnieres, *Photodiagn. Photodyn. Ther.*, 2013, **10**, 683–693.

66. T. Glanzmann, C. Hadjur, M. Zellweger, P. Grosjean, M. Forrer, J. P. Ballini, P. Monnier, H. van den Bergh, C. K. Lim and G. Wagnieres, *Photochem. Photobiol.*, 1998, **67**, 596–602.
67. F. Piffaretti, A. M. Novello, R. S. Kumar, E. Forte, C. Paulou, P. Nowak-Sliwinska, H. van den Bergh and G. Wagnieres, *J. Biomed. Opt.*, 2012, **17**, 115007.
68. M. Y. Berezin and S. Achilefu, *Chem. Rev.*, 2010, **110**, 2641–2684.
69. W. Becker, *J. Microsc.*, 2012, **247**, 119–136.
70. S. Yeh, M. Patterson, J. Hayward and Q. Fang, *Photonics*, 2014, **1**, 530–564.
71. V. Huntosova, Z. Nadova, L. Dzurova, V. Jakusova, F. Sureau and P. Miskovsky, *Photochem. Photobiol. Sci.*, 2012, **11**, 1428–1436.
72. V. Huntosova, L. Alvarez, L. Bryndzova, Z. Nadova, D. Jancura, L. Buriankova, S. Bonneau, D. Brault, P. Miskovsky and F. Sureau, *Int. J. Pharm.*, 2010, **389**, 32–40.
73. N. Fotinos, M. A. Campo, F. Popowycz, R. Gurny and N. Lange, *Photochem. Photobiol.*, 2006, **82**, 994–1015.
74. H. Fukuda, A. Casas and A. Batlle, *Int. J. Biochem. Cell Biol.*, 2005, **37**, 272–276.
75. G. Wagnieres, P. Jichlinski, N. Lange, P. Kucera and H. van den Bergh, in *Handbook of Photomedicine*, ed. M. R. Hamblin and Y.-Y. Huang, CRC Press (Taylor & Francis), 2014, pp. 411–426.
76. B. J. Tromberg, A. Orenstein, S. Kimel, S. J. Barker, J. Hyatt, J. S. Nelson and M. W. Berns, *Photochem. Photobiol.*, 1990, **52**, 375–385.
77. M. T. Jarvi, M. S. Patterson and B. C. Wilson, *Biophys. J.*, 2012, **102**, 661–671.
78. T. M. Sitnik, J. A. Hampton and B. W. Henderson, *Br. J. Cancer*, 1998, **77**, 1386–1394.
79. A. Curnow, J. C. Haller and S. G. Bown, *J. Photochem. Photobiol., B*, 2000, **58**, 149–155.
80. S. Kawauchi, Y. Morimoto, S. Sato, T. Arai, K. Seguchi, H. Asanuma and M. Kikuchi, *Lasers Med. Sci.*, 2004, **18**, 179–183.
81. J. Marzioch, J. Kieninger, J. A. Sandvik, E. O. Pettersen, Q. Peng and G. Urban, *Eurosensors*, 2015, **120**, 468–471.
82. T. M. Busch, S. M. Hahn, S. M. Evans and C. J. Koch, *Cancer Res.*, 2000, **60**, 2636–2642.
83. Q. Shao, M. A. Biel and S. Ashkenazi, *Optical Methods for Tumor Treatment and Detection: Mechanisms and Techniques in Photodynamic Therapy Xxiii*, 2014, vol. 8931.
84. B. W. McIlroy, A. Curnow, G. Buonaccorsi, M. A. Scott, S. G. Bown and A. J. MacRobert, *J. Photochem. Photobiol., B*, 1998, **43**, 47–55.
85. S. Mitra and T. H. Foster, *Biophys. J.*, 2000, **78**, 2597–2605.
86. M. A. Weston and M. S. Patterson, *Phys. Med. Biol.*, 2013, **58**, 7039–7059.
87. T. Maisch, J. Baier, B. Franz, M. Maier, M. Landthaler, R. M. Szeimies and W. Baumler, *Proc. Natl. Acad. Sci. U. S. A.*, 2007, **104**, 7223–7228.
88. C. T. Andrade, J. D. Vollet, A. G. Salvio, V. S. Bagnato and C. Kurachi, *Photodiagn. Photodyn. Ther.*, 2014, **11**, 409–415.

89. N. Dognitz, D. Salomon, M. Zellweger, J. P. Ballini, T. Gabrecht, N. Lange, H. van den Bergh and G. Wagnieres, *J. Photochem. Photobiol., B*, 2008, **93**, 140–148.
90. J. Park, J. Lee and C. Choi, *PLoS One*, 2011, **6**, e23211.
91. Y. Feng, J. H. Cheng, L. Zhou, X. G. Zhou and H. F. Xiang, *Analyst*, 2012, **137**, 4885–4901.
92. R. I. Dmitriev and D. B. Papkovsky, *Methods Appl. Fluoresc.*, 2015, **3**, 034001.
93. J. Varchola, V. Huntosova, D. Jancura, G. Wagnieres, P. Miskovsky and G. Bano, *Photochem. Photobiol. Sci.*, 2014, **13**, 1781–1787.
94. V. Huntosova, E. Gerelli, D. Horvath and G. Wagnieres, *J. Biophotonics*, 2017, **10**, 708–717.
95. Z. Rosenzweig and R. Kopelman, *Sens. Actuators, B*, 1996, **36**, 475–483.
96. P. Ceroni, A. Y. Lebedev, E. Marchi, M. Yuan, T. V. Esipova, G. Bergamini, D. F. Wilson, T. M. Busch and S. A. Vinogradov, *Photochem. Photobiol. Sci.*, 2011, **10**, 1056–1065.
97. Z. Jurasekova, V. Huntosova, P. Miskovsky and D. Jancura, *Eur. Biophys. J. Biophys. Lett.*, 2017, **46**, S330.
98. R. Dedic, A. Stibal, V. Vyklicky, M. Franek, A. Svoboda and J. Hala, *J. Innovative Opt. Health Sci.*, 2015, **8**, 1550037.
99. I. S. Vinklarek, M. Scholz, R. Dedic and J. Hala, *Photochem. Photobiol. Sci.*, 2017, **16**, 507–518.
100. V. Huntosova, E. Gerelli, M. Zellweger and G. Wagnieres, *J. Photochem. Photobiol., B*, 2016, **164**, 49–56.
101. M. Misuth, J. Joniova, D. Belej, S. Hrivnak, D. Horvath and V. Huntosova, *J. Biophotonics*, 2017, **10**, 423–432.
102. D. Belej, Z. Jurasekova, M. Nemergut, G. Wagnieres, D. J. Cura and V. Huntosova, *J. Inorg. Biochem.*, 2017, **174**, 37–44.
103. F. Piffaretti, K. Santhakumar, E. Forte, H. E. van den Bergh and G. A. Wagnieres, *J. Biomed. Opt.*, 2011, **16**, 069801.
104. V. Huntosova, M. Novotova, Z. Nichtova, L. Balogova, M. Maslanakova, D. Petrovajova and K. Stroffekova, *Toxicol. In Vitro*, 2017, **40**, 184–195.
105. M. Misuth, J. Joniova, D. Horvath, L. Dzurova, Z. Nichtova, M. Novotova, P. Miskovsky, K. Stroffekova and V. Huntosova, *Cell. Signalling*, 2017, **34**, 11–22.

CHAPTER 16

# In vivo *Brain Functional Imaging Using Oxygenation-related Optical Signal*

VASSILIY TSYTSAREV*[a,b] AND DMITRI B. PAPKOVSKY[c]

[a]Department of Anatomy and Neurobiology, University of Maryland School of Medicine, HSF-2, 20 Penn Street, Baltimore, MD 21201, USA; [b]Fischell Department of Bioengineering, University of Maryland, Jeong H. Kim Engineering Building, 8278 Paint Branch Dr, College Park, Baltimore, MD 20742, USA; [c]School of Biochemistry and Cell Biology, University College Cork, Cork, Ireland
*E-mail: tsytsarev@umaryland.edu

## 16.1 Introduction

In the last few decades brain optical imaging has developed considerably. These methodologies of brain research are based on the concept that some intrinsic optical properties of the neural tissue, or properties of exogenous materials introduced into it, are related (directly or indirectly) to certain metabolic parameters of the tissue, and responsive to and can report on their local changes within the brain. Different parts of the spectrum can be used to monitor these parameters and related physiological changes *in vivo*, usually in experiments with the exposed brain of live animals. At the same time, use of optical imaging methods has limitations related to the high levels of light scattering and absorption by neural (and other) tissues. In this chapter, we will

Detection Science Series No. 11
Quenched-phosphorescence Detection of Molecular Oxygen: Applications in Life Sciences
Edited by Dmitri B. Papkovsky and Ruslan I. Dmitriev

Published by the Royal Society of Chemistry, www.rsc.org

describe the range of available oxygen-based optical imaging approaches that are currently being applied in experimental neuroscience. Functional imaging of neural activity in the brain, the so-called functional brain mapping, remains one of the most important tasks for modern experimental neuroscience.

Due to the relatively high level of energy consumption and metabolism, brain parenchyma is one of the most metabolically active tissues in mammalian organisms. Brain activity and metabolism never stop, and they constantly consume oxygen to generate the energy required for normal functioning of millions of neurons and astrocytes. Every minute the brain of an adult human consumes approximately 3.5 ml of oxygen per 100 g of tissue.[1]

Normal neuronal activity is rapidly impaired if oxygen supply is reduced or interrupted even for a short time, since oxygen storage capacity in the brain parenchyma is very limited.[2] Not only neurons are sensitive to physiological changes in tissue oxygen levels, but astrocytes also respond to decreases in oxygen concentration with elevations in intracellular calcium.[2] Oxygen deficiency also inhibits mitochondrial respiration of astrocytes, leading to mitochondrial depolarization, enhanced production of free radicals and lipid peroxidation, activation of phospholipase C, and release of calcium ions from intracellular calcium stores.

The intensity of oxygen consumption is determined by many physiological parameters, it is also different in the asleep and awake brain, in normal and pathological conditions. Real-time live monitoring of oxygen-related metabolism in the different areas of the brain, including cortical and subcortical structures, have become possible after the advent of the functional brain mapping methods. Thus, the first techniques were based on the unidirectional clearance of the short-lived isotope, oxygen-15 and measurement of the steady-state balance between oxygen inflow and oxygen-15 decay.[3,4] A similar method, called autoradiography, is based on the transient analysis of oxygen consumption.[5–7]

Oxygen (micro)electrodes, which do not allow real time functional mapping, were also successfully employed to establish the spatiotemporal relationships between neuronal activity, cerebral blood volume, and local oxygenation during epilepsy.[8,9] It is clear that their application for oxygenation mapping in human medicine has many limitations, however some important physiological data has been obtained here. Thus, oxygen electrodes were employed to monitor local tissue oxygenation during epileptiform events in humans, revealing that intense seizures elicit a decrease in local oxygenation.[10]

Implantable oxygen microelectrodes being small in size and with the ability to record at a sub-second temporal resolution are attractive for determining the relationships between metabolism, neural activity and behavior: activity-dependent local changes in metabolism and oxygenation in the hippocampus were determined and test rats demonstrated significant increases in hippocampal oxygen consumption during maze testing.[11]

The spatial heterogeneity of brain tissue partial pressure of oxygen in the different parts of the cerebral cortex was also investigated using different methods. Thus, monitoring tissue oxygen with oxygen microelectrodes, dual responses (increase and decrease) in tissue oxygen in response to the functional stimulation at the somatosensory cortex was recorded.[12] Such dual

**Table 16.1** Main characteristics of the different functional brain imaging techniques based on the monitoring of tissue oxygenation.

| Imaging method | Spatial resolution | Temporal resolution | Invasivity |
|---|---|---|---|
| Functional magnetic resonance imaging (fMRI) | Up to 1 mm | Up to 0.1 s | Non-invasive |
| Functional near-infrared spectroscopy (fNRIS) | 2–5 cm, up to 1 cm in infants | Up to 1 ms, but usually 10–100 ms | Non-invasive |
| Optical imaging of intrinsic signal (IOS), so-called optical recording of intrinsic signals (ORIS) | Up to 50 μm | Up to 1 ms, but usually 10–100 ms | Invasive |
| Diffuse optic tomography (DOT) | 2–5 cm, better in infants | Technically high (up to 1 ms) but usually 10–100 ms | Non-invasive |
| Photoacoustic imaging (PAI) | Up to 1 μm in invasive animal experiments, probably up to 10 mm in transcranial imaging | Up to 100 ms, determined by many physiological and technical factors | Invasive; potential can be non-invasive (transcranial) in human studies |
| Phosphorescence chemical oxygen-sensitive probe | Up to 50 μm; in combination with 2-photon imaging can be improved up to 1 μm | Up to 5–10 ms | Invasive |

characters of the physiological response are very common, and they were also observed using fMRI, IOS, and oxygen-sensitive methods.[13–16]

Functional brain mapping techniques, based on the monitoring of local cerebral oxygenation, provide valuable tools not only for basic, but also for translational studies. The investigative potential of such techniques becomes greatly enhanced, if they are able to operate in an imaging format, with high spatial and temporal resolution. A number of optical imaging techniques can provide such capabilities. The main characteristics of these optical imaging techniques are summarized in Table 16.1, and their detailed description is given below.

## 16.2 Intrinsic Optical Signal (IOS)

One of the first brain optical imaging techniques – intrinsic optical signal imaging (IOS), measures changes in the optical properties of brain tissue. IOS permits visualization of neural activity with a spatial resolution about 50 microns or even better. The most valuable component of the optic signal comes from hemoglobin absorption, which is reflective to changes in local oxygenation and deoxygenation that relate to neural activity-dependent physiological process.[14,17–19]

The optical signal measured by IOS can be separated into two different time components. The first component reflects the increase in deoxyhemoglobin concentration resulting from elevated oxygen consumption of neurons activated in a particular cortical area. This process increases light absorption in the red and near-infrared region of spectrum. The second component reflects the increases in local cerebral blood-flow (CBF) and oxyhemoglobin concentration.

This response phenomenon was also confirmed by animal experiments with the 4-Aminopyridine (4-AP) model of epileptic seizures. As demonstrated with the rat 4-AP model, the fast decrease in oxygen concentration was seen to coincide with the epileptic seizure onset, followed by a gradual return to baseline.[8] IOS based on the mapping of local changes in hemoglobin oxygenation in the brain also demonstrated that epileptic seizures induce an immediate increase in deoxyhemoglobin and a decrease in tissue oxygen concentration.[8]

The important advantage of IOS over the isotopic-based methods is that it can be performed without any extrinsic chemical probes. Therefore, IOS is safe and can be used even on the human brain, when the latter gets exposed during neurosurgical operations. Nevertheless, IOS has serious limitations: the changes in optic properties, associated with transformation of oxyhemoglobin to deoxyhemoglobin and back, are rather small (<1%), and cannot be measured reliably without extensive signal averaging, noise removal and other complicated mathematical operations.[20–23] Another limitation of IOS is that it measures tissue oxygen indirectly (*via* the absorption of oxy- and deoxy-hemoglobin) and senses changes in bound rather than free/dissolved oxygen. As a result, data interpretation is more complex (changes in blood flow and hemoglobin saturation should be accounted for) and influenced by other factors. These limitations of IOS can be overcome by using the other oxygen imaging techniques.

## 16.3 Functional Magnetic Resonance Imaging (fMRI)

The very similar biological phenomenon can be observed and utilized in several other techniques, particularly functional magnetic resonance imaging (fMRI)[24–26] and near infrared functional spectroscopy (fNIRS).[27,28] Functional magnetic resonance imaging (fMRI) is currently regarded as the "gold standard" for visualizing local differences in oxygen consumption and cerebral blood flow. Such blood oxygenation level dependent (BOLD) imaging is based on the fundamental differences in the paramagnetic properties of oxy- and deoxy-hemoglobin.[13] Cerebral blood circulation is controlled by the local oxygen and carbon dioxide tensions in the neural tissue. When a specific area of the brain increases its activity in response to a motor, sensory or cognitive task, the depletion of oxygen (delivered by erythrocytes to local capillaries) leads to a significant initial drop in oxyhemoglobin and an increase in deoxyhemoglobin. As a result of this, cerebral blood flow (CBF) increases a few seconds later, which brings more oxyhemoglobin and removes the excess of deoxyhemoglobin. This process can be visualized by fMRI and used for functional brain mapping. Although BOLD–fMRI is widely available, is a non-invasive technique which does not have depth limitation, it also has some disadvantages.[27] Thus, fMRI sessions usually take a longer time than other brain imaging methods, due to the need for signal averaging. Also imaging artifacts frequently occur at tissue boundaries, since the quality of the MRI signal varies in different parts of the brain. fMRI is costly due to the need for liquid helium for setup functioning, and fMRI is difficult or even impossible to apply to infants or patients that cannot be immobilized for at least a few minutes.

## 16.4 Functional Near-Infrared Spectroscopy (fNIRS)

In optical imaging, UV and visible light poorly penetrates through biological tissue, but photons of the red and near infrared parts of the spectrum travel much further through tissue. Photons with wavelengths between 700 nm and 1100 nm can penetrate skin, muscle, and bone tissue and reach the cerebral cortex of the human brain. fNIRS is an optical imaging technique, which enables non-invasive, relatively portable and low-cost systems for monitoring oxygenation of different regions of the brain, without administering exogenous chemical probes or contrast agents.[29,30]

In fNIRS imaging, photons follow a "banana-shaped" path from the light emitter, through skin and skull, then passing through neighboring regions of cerebral cortex and reaching the detector. Temporal resolution of fNIRS is comparable with fMRI (Table 16.1) since it relies on the same metabolic process, however its spatial resolution is much less due to high light scattering and low numbers of light emitters and detectors. Nevertheless, NIRS has one main advantage: the subject can be monitored not just when s/he is placed in a chamber or chair and immobilized, but fNIRS can also be performed on moving subjects, infants, and applied to studies of higher cortical functions such as language and cognition. Like fMRI, fNIRS measures local oxygenation response to neural activity, but rather than relying on the paramagnetic properties of hemoglobin, it measures the differences in light absorption characteristics of oxy- and deoxy-hemoglobin.[28] The biphasic character of the optical signal, caused by the opposite changes in oxygenation and deoxygenation, was demonstrated using both NIRS[31,32] and fMRI.[33,34] Again, the scientists attribute the first phase to an increase in deoxyhemoglobin concentration, while the second delayed phase is attributed to an increase in oxyhemoglobin.

fNIRS was also successfully employed in clinical practice in combination with electroencephalography (EEG). Thus, this method allowed the spatio-temporal reconstruction of the changes in tissue oxygenation/oxygen consumption rate and blood circulation response, and linking these changes to epileptic activity recorded by the EEG.[31] Currently, fNIRS can offer a non-invasive and safe, portable, and relatively low-cost method for monitoring brain metabolism and hemodynamics over a few hours. It provides unique insights into neuroscience research and treatment of various brain pathologies.[32]

## 16.5 Diffuse Optical Tomography (DOT)

Diffuse optical tomography (DOT) is based on the measurement of absorption of photons by the brain tissue which can be related to local oxygenation. Like fNIRS, DOT employs a number of near-infrared emitters and detectors positioned on the surface of subject's head. In contrast to fNIRS, DOT can perform 3D tomographic reconstruction and allows scientists to visualize local oxygenation in subcortical regions of the brain. As with all the other tomographic techniques, DOT requires image reconstruction to associate activity patterns with specific anatomical structures of the brain.[34,35]

DOT provides imaging of cerebral blood circulation and local oxygenation with the unique capability of long-term bedside imaging under clinical settings, especially in infants.[36,37] It allows the determination of the two critical physiological parameters: index of oxygen extraction fraction and metabolic rate of cerebral oxygen consumption.[36] Unfortunately, light is strongly scattered and absorbed in the brain tissue. Most of the data analysis algorithms in DOT are based on the techniques of inverse problems.[36,37] Various approaches to create accurate and efficient DOT data analysis algorithms were tried, however the spatial resolution of DOT still remains limited.[36,37]

## 16.6 Photoacoustic Imaging (PAI)

Photoacoustic imaging (PAI), also called optoacoustics, is another 3D optical imaging method.[38–40] It is based on the photoacoustic effect, which is the generation of sound waves following photon absorption by sample material. In a PAI imaging experiment, photons absorbed by the brain tissue result in rapid heating, which is subsequently converted into mechanical waves due to thermoelastic expansion. Functional PAI is usually applied to the monitoring of oxygenation, deoxygenation and blood flow, employing the relatively high absorption of light photons by hemoglobin.[41,42]

In PAI measurements, hemoglobin concentration and oxygenation state, *i.e.* oxy- and deoxy-hemoglobin, are the dominant absorbing compounds of the blood. The differences in peak absorption of oxy- and deoxy-hemoglobin allows the separation of these components in imaging.[42] Like in other optical imaging methods, based on hemoglobin absorption, absorption of photons in PAI is determined by the structure of the sample: intensity of ultrasonic waves is proportional to photon absorption and reflects physiological properties of the tissue.[42,43]

The imaging depth of PAI is limited by light penetration and scattering, which are quite high for brain tissue. The spatial resolution is determined by several factors, including the light intensity, wavelength, and ultrasound transducer properties.

Current temporal resolution in PAI is not great, due to the technical features of the imaging setup and mechanical laser scanning. Both oxy- and deoxy-hemoglobin are the main light absorbers in the red and near-infrared part of the spectrum, and this makes PAI applicable for functional brain mapping of local oxygenation and blood circulation.

Functional PAI usually uses more than one light source - lasers of different wavelengths. The two common wavelengths are 570 nm and 578 nm, where the absorption differences between blood and background brain tissues are high. The wavelength of 570 nm corresponds to the isobestic point, at which oxy- and deoxy-hemoglobin have the same molar extinction. Hence, the photoacoustic signal acquired at this wavelength reflects the total hemoglobin concentration. And 578 nm corresponds to the dominant absorption wavelength for oxyhemoglobin, which allows precise measurement of both oxy- and deoxy-hemoglobin.

The capabilities of PAI can be improved by applying multispectral optoacoustic tomography (MSOT), which allows high-resolution imaging deep inside brain tissue.[44–46] This employs illumination of the tissue with pulses of light at multiple wavelengths, and imaging detection of ultrasonic waves generated by the photoacoustic effect of the environment which surrounds absorbing molecules. Further off-line analysis allows the generation of spectral signatures of oxygenated and deoxygenated hemoglobin and other light-absorbing agents, estimation of their concentrations and calibration of the system. MSOT demonstrated that in an animal model of cancer, tumor cells injected into the brain can be visualized through the intact skull.[45]

Thus, PAI provides 3D functional maps of brain activity and offers several compelling features for functional brain mapping. Thus, it was successfully used in a neuroimaging setup with concurrent EEG recordings. Such a combination was used to visualize the onset, spread, and termination of local epileptic seizures elicited by intracortical administration of 4-aminopyridine (4-AP) in the acute model of cortical epilepsy.[45] 4-AP, a potassium channel inhibitor, is a common convulsant used to generate seizures in such animal models, and this imaging study was the first demonstration of non-invasive real-time visualization of synchronized epileptic seizures in the cortical and subcortical areas of the mouse brain.[45]

PAI microscopy was also used to generate 3D images of brain tissue with high spatial resolution, invasively and non-invasively, revealing blood oxygenation dynamics of multiple cerebral blood vessels. Thus, it was able to non-invasively monitor oxygenation dynamics of cortical vessels during experimental sessions of hyperoxia and hypoxia.[40,42] It also allowed high-speed

3D imaging of the animal brain, complementary to fMRI and IOS.[46] Functional imaging of blood oxygenation, blood flow, and oxygen metabolism in the brain were also recorded.[46,47]

Similar to PAI, acousto-optic imaging (AOI) is also a version of the hybrid imaging technique which is used to measure the optical contrast by means of a coherent light source and ultrasonic waves. However, the physical basis of AOI is different. The investigated object is illuminated with a coherent light leading to a speckle formation in the re-emitted light, while ultrasonic waves modulate light in the tissue being tested. To achieve high spatial resolution (which is currently >1 mm), good focusing of the ultrasound is required.[47] The use of acousto-optic scanning microscopy can significantly improve spatial resolution. This real-time, two-photon imaging method provides a useful tool for *in vivo* functional brain imaging, monitoring oxygen dynamics in the cortical tissue.[48]

High light scattering by the tissue limits the resolution of AOI. However, high-resolution acousto-optic tomography based on the optical contrast allows imaging even several millimeters deep into the tissues, with high frequency ultrasound providing enhanced resolution.[49–51] For AOI data analysis special algorithms were developed, but the application of this technology for functional brain mapping is still complicated.[52,53]

## 16.7 Phosphorescence Quenching Techniques

Since it is well known that neural activity is associated with metabolic changes, including local oxygenation and deoxygenation of brain tissue, this provides a physical basis for functional brain imaging techniques that use various oxygen-sensitive probes and markers. The main source of the energy in biological systems including mammalian brain is oxidative phosphorylation, which uses glucose, oxygen and produces $CO_2$, adenosine triphosphate (ATP) and lactate, which are consumed by many different cells.[54,55] High energy demand during neural activity causes significant changes in local tissue oxygenation and oxygen consumption rate which correlate with local neuronal activity of brain tissue.[54] Therefore, monitoring of concentration of dissolved (*i.e.* free) oxygen concentration in brain tissue and its dynamics by means of exogenous phosphorescent probes which are reversibly quenched by molecular oxygen, can be used to gain information on local neuronal activity in the brain. A number of such oxygen-sensitive phosphorescent materials have been developed in recent years for such applications, and evaluated in functional brain imaging studies performed on model animals (mice).

### 16.7.1 Soluble Phosphorescent Oxygen-sensitive Probes

Several highly efficient *intravascular* phosphorescent probes have been used in a number of *in vivo* imaging studies, including phosphorescence lifetime based imaging of live animal brain under 2-photon excitation.

Initially, Green 2W probe, which comprised a Pd-porphyrin dye bound to serum albumin and injected systemically into mouse vasculature, was used for phosphorescence lifetime based imaging of grafted tumors (EMT-6 mammary carcinoma).[55] This probe was later replaced by the more advanced Oxyphor probes based on Pd-porphyrin and Pd-benzoporphyrin dendrimers.[56] The most recent probe developed by Vinogradov's group, called PtP–C343, has a very complex structure and consists of a phosphorescent Pt(II)-*meso*-tetraarylporphyrin moiety, a shielding dendrimer based on polyarylglycine, several coumarin-343 light antennae units and peripheral oligoethylene-glycol residues linked to the dendrimer. PtP–C343 has improved solubility and biocompatibility, high brightness under two-photon excitation, extended clearance time. The phosphorescent Pt-porphyrin reporter is located in the core of this nanostructure, and is shielded from potential quenchers and interfering specie by the dendrimeric poly(arylglycine) chains, which protect the porphyrin and control the rate of oxygen quenching.[54] Such a probe is injected directly into the blood stream of anesthetized rodents at the start of the experiment and can be imaged immediately after the injection for a few hours.[54]

Modern *in vivo* two-photon fluorescence microscopes allow imaging of the cerebral cortex in anesthetized animals through glass-sealed cranial windows - at depths up to 600 microns. They can generate structural images of the cortical vasculature as well as functional images of relatively small parts of the neocortex. Phosphorescence lifetime imaging microscopy (PLIM) provides a quantitative readout of oxygen concentration. However, temporal resolution of such an imaging method is limited by the relatively slow signal acquisition rates and scanning speed. One the other hand, this method provides submicron and single cell spatial resolution. Valuable information about cerebral oxygenation was obtained by combining the 2-photon imaging and fMRI data.[55,56] Two-photon imaging of oxygen by PLIM was also correlated with BOLD-related cerebrovascular signal.[54–56]

In the above studies, the phosphorescent oxygen-sensitive probes were administered systemically (intravascularly) and at relatively high doses.[57] Another limitation of these imaging platforms, particularly the *in vivo* two-photon PLIM systems, are their slow image acquisition rates - many seconds if not minutes to obtain a decent 2D image.[54,58–61] The latter limits the applicability of these systems to simple point measurements. Even though multiple points can be probed within a tissue sample, this is not suitable for functional brain imaging where millisecond frame rates and high-resolution images are usually required.[15] Also since the intravascular probes stay predominantly in blood vessels with limited penetration in the interstitial space, the bulk of the tissue remains dark or very dim in such $O_2$ imaging.

The above limitations can be overcome by the use of cell-penetrating phosphorescent probes, which are applied locally, and fast wide-field imaging systems with intensity based CCD camera readout. Thus, in our initial study,[15] the near-infrared phosphorescent nanoparticle based probe

NanO2-IR, was applied topically on the surface brain tissue of anaesthetized mice. NanO2-IR is a direct analog of the cell-permeable nanoparticle based probe NanO2, which operates in the visible spectral range and which was originally designed for sensing and imaging of $O_2$ in *in vitro* cell and tissue models.[62,63]

The NanO2-IR probe based on Pt(II)-benzoporphyrin dye has longwave excitation and emission bands (560–630 nm and 730–800 nm, respectively), and is well suited for live optical imaging of animal tissue. After its topical application with a micropipette on the exposed surface of the brain and short-term incubation this probe effectively stains and also diffuses inside the cortical tissue (up to 50–100 microns). NanO2-IR probe has enabled us to perform real-time imaging of oxygenation within the somatosensory cortex with temporal resolution of about 5 ms per frame.[15] In this case, a standard CCD-camera based system was fully appropriate for imaging of anesthetized animals through a standard cranial window. The obtained data showed that cell-permeable phosphorescent probes can be used as simple and efficient contrast agents for staining live brain tissue and tracing changes in local oxygen concentration in somatosensory cortex, particularly in mice exposed to whisker stimulation. They can be used routinely in functional brain imaging experiments to generate functional pseudocolor maps of oxygen concentration from recorded imaging data.

Temporal profiles of oxygen maps generated with such phosphorescent probes also typically produce a negative peak that reaches minimum in about 2–3 s after stimulation. Subsequently, the optical signal in the activated cortical region returns to the baseline in 8–10 s after the stimulus onset. The second phase of the response (positive) reaches its maximum within 7–10 s or even later. Thus, brain functional mapping using a phosphorescent oxygen-sensitive probe shows that local oxygenation and deoxygenation are relatively slow metabolic processes, which are indirectly related to neural activity. Comparing these results with those produced using the voltage-sensitive dye[14,15] reveals that $O_2$ response, although following a similar pattern as the potentiometric response, proceeds on a different time scale and has a different shape. Voltage-sensitive dye imaging technique (VSDi) allows direct monitoring of the neural activity, since voltage-sensitive dye molecules bind to the neural membranes and transform changes in membrane potential into fluorescence signals. Therefore, temporal resolution of this technique is much higher and usually limited by technical specification of the imaging system. In contrast to VSDi, oxygen-sensitive probe reflects neural activity indirectly. It transforms into an optical signal not neural activity, but changes in the oxygen consumption in the different regions of the brain. These changes occur at a much slower pace. Thus, enhanced neuronal activity causes a local increase in oxygen consumption only after a few hundred milliseconds or even few seconds, and these marked changes are detected *via* the optical signal from the oxygen probe.[14,15]

### 16.7.2 Phosphorescent Microparticle-based Probes

Mitochondrial dysfunction and microvascular abnormalities in the brain are also associated with sepsis. Thus, it was shown that during systemic endotoxemia mitochondrial dysfunction in the cells of mouse cortex increases their sensitivity to hypoxemia, and that hypothermia can improve the outcome of such treatment.[64] In this study, changes in oxygenation on the surface of mouse brain tissue were monitored *in vivo* by confocal one-photon multi-parametric fluorescence microscopy. In this case, the phosphorescent oxygen-sensitive microbeads (size ~7 μm) impregnated with PtPFPP dye were spread on the dura by applying 5 μl of 5 mg $ml^{-1}$ aqueous suspension. Besides cortical oxygenation, cerebral mitochondrial redox state was also imaged *via* changes in fluorescence of endogenous flavoprotein/FAD, in response to changes in inspired oxygen. The results demonstrated the increased sensitivity of cortical mitochondria to hypoxemia accompanied by a decrease in cortical oxygenation, confirming the therapeutic effect of hypothermia which can be protective in some pathological conditions of the brain.[63,64]

Using similar experimental settings, changes in mouse brain tissue oxygenation, redox state (flavoprotein autofluorescence), mitochondrial membrane potential, (TMRM dye applied topically) and perfused vessel density (FITC-dextran, 70 kDa injected intravenously) were also monitored in acute *in vivo* experiments.[65] This study revealed that reduced cerebral blood flow impairs cortical mitochondrial function. Oxygen imaging also demonstrated the protective effect of terlipressin (terlipressin + 1LR and terlipressin + 2LR), in the recovery of cerebral blood flow, tissue oxygenation and mitochondrial function.[65] The authors also measured changes in oxygen concentration in cortical capillary, mitochondrial membrane potential and redox state. The phosphorescent oxygen-sensitive microparticles are easier to use than the cell-permeable and intravascular probes (see above), they provide high phosphorescence intensity signals which are easy to measure and multiplex. However, they allow monitoring of $O_2$ concentration only in discrete locations on the surface of brain tissue, *i.e.* where the particles stay.

### 16.7.3 Planar Phosphorescent Oxygen Sensors

Planar phosphorescent oxygen sensors were first tested in animal models with grafted solid tumors.[66] In this study, a dual sensor was prepared by sequential coating of an oxygen-impermeable polyester substrate (Mylar® film, 175 micron thick) with liquid precursor "cocktails" containing indicator dye and carrier polymer dissolved in organic solvent. The first coating layer used as a temperature sensor consisted of Ru(II)-tris-(1,10-phenanthroline)-hexafluorophosphate embedded in poly(acrylonitrile) matrix, while the second (top) oxygen-sensitive layer contained Pt(II)–octaethylporphyrin in a polystyrene matrix. Both sensor coatings were approximately 3–5 micron thick, and Ru-based sensor was excited at 470 nm and PtOEP sensor - at

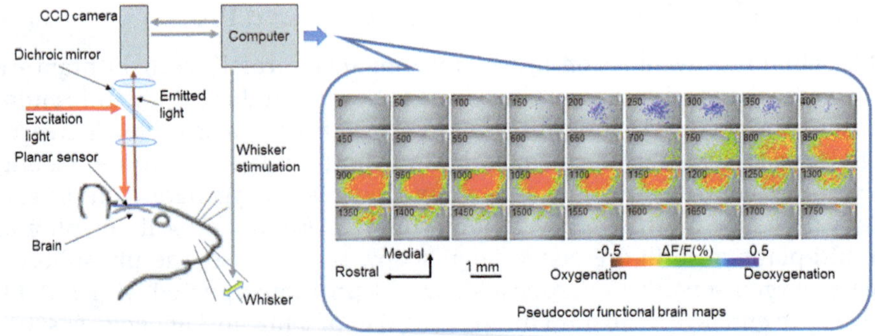


**Figure 16.1** Typical experimental set-up for *in vivo* functional brain imaging with phosphorescent oxygen sensors. Planar film with sensor coating is brought into contact with brain tissue and covered with a cranial window. Then the brain surface is illuminated with 630 nm light and emission at above 690 nm is collected with a CCD camera. Pseudocolor maps of functional activity generated from recorded images show local changes in $O_2$ in the somatosensory cortex in response to single whisker stimulation. Time after stimulus onset (in milliseconds) is indicated in top left corner. Phosphorescent signal scale is ΔF/F(%).

535 nm with emissions measured at 590 nm and 645 nm, respectively. The researchers showed that phosphorescence lifetime based imaging with such planar sensors can be effective for measuring muscle tissue oxygenation in live animals.[66] Nevertheless, this sensor material and imaging technique had limitations, particularly sensor complexity, strong absorbance in a broad spectral region which prevented the analysis of fine architecture of tested tissue and use of other probes. This has not been demonstrated in brain functional imaging.

In our recent study,[16] we applied simpler planar sensors based on a flexible oxygen-impermeable polymeric film with a phosphorescent oxygen-sensitive coating on the working side (Figure 16.1). These transparent sensors, when applied topically on mouse exposed brain surface, adhered well to the tissue and allowed dynamic, real-time phosphorescence imaging on a CCD-camera based experimental system described above, to trace oxygenation and deoxygenation patterns in different regions of the brain in response to whisker stimulation. They are more convenient and efficient for dynamic 2D imaging of the oxygen levels in cerebral cortex in animal experiments,[16] than the soluble probes or microparticle sensors described previously.[63,67–70]

## 16.8 Conclusions

In summary, brain functional imaging using phosphorescent oxygen-sensitive probes and sensors provides a useful tool for fundamental and clinical neuroscience. Further development of such minimally invasive and non-invasive *in vivo* methods is a rapidly expanding area of biomedical research.

Future trends also include coupling of this techniques with other imaging modalities and technologies, and development of multi-parametric metabolic imaging platforms. The latter are tailored to a number of useful animal models and ultimately targeted at clinical applications with human patients, with disease areas such as Parkinson's, local brain seizures, brain injury, hypoxia and stroke, sepsis, *etc.*

## References


1. C. Rink and S. Khanna, *Antioxid. Redox Signaling*, 2011, **14**, 1889.
2. P. R. Angelova, V. Kasymov, I. Christie, S. Sheikhbahaei, E. Turovsky, M. N. Korsak, J. Zwicker, A. G. Teschemacher, G. L. Ackland, G. D. Funk, S. Kasparov, A. Y. Abramov and A. V. Gourine, *J. Neurosci.*, 2015, **35**, 10460.
3. R. S. Frackowiak and K. J. Friston, *J. Anat.*, 1994, **184**, 211.
4. R. S. Frackowiak, T. Jones, G. L. Lenzi and J. D. Heather, *Acta Neurol. Scand.*, 1980, **62**, 336.
5. P. Bickler, J. Feiner, M. Rollins and L. Meng, *Anesth. Analg.*, 2017, **124**, 72.
6. R. D. Hoge and G. B. Pike, *J. Chem. Neuroanat.*, 2001, **22**, 43.
7. M. Jueptner and C. Weiller, *Neuroimage*, 1995, **2**, 148.
8. S. Bahar, M. Suh, M. Zhao and T. H. Schwartz, *Neuroreport*, 2006, **17**, 499.
9. S. Shariff, M. Suh, M. Zhao, H. Ma and T. H. Schwartz, *Epilepsy Behav.*, 2006, **8**, 363.
10. A. M. Dymond and P. H. Crandall, *Brain Res.*, 1976, **102**, 191.
11. J. Kealy, R. Bennett, B. Woods and J. P. Lowry, *Behav. Brain Res.*, 2017, **326**, 59.
12. K. Masamoto, N. Takizawa, H. Kobayashi, K. Oka and K. Tanishita, *Brain Res.*, 2003, **979**, 104.
13. K. A. Morone, J. S. Neimat, A. W. Roe and R. M. Friedman, *Neurophotonics*, 2017, **4**, 031220.
14. V. Tsytsarev, H. Arakawa, S. Borisov, E. Pumbo, R. S. Erzurumlu and D. B. Papkovsky, *J. Neurosci. Methods*, 2013, **216**, 146.
15. V. Tsytsarev, F. Akkentli, E. Pumbo, Q. Tang, Y. Chen, R. S. Erzurumlu and D. B. Papkovsky, *J. Neurosci. Methods.*, 2017, **281**, 1.
16. K. C. Brennan and A. W. Toga, *In vivo Optical Imaging of Brain Function*, ed. R. D Frostig, CRC Press/Taylor & Francis, Boca Raton (FL), 2nd edn, 2009, ch. 13.
17. R. D. Frostig and C. H. Chen-Bee, *In vivo Optical Imaging of Brain Function*, ed. R. D Frostig, CRC Press/Taylor & Francis, Boca Raton (FL), 2nd edn, 2009, ch. 9.
18. J. Ribot, S. Tanaka, K. O'Hashi and A. Ajima, *Eur. J. Neurosci.*, 2008, **27**, 2773.
19. T. Tani, J. Ribot, K. O'Hashi and S. Tanaka, *Eur. J. Neurosci.*, 2012, **35**, 44.
20. M. Gabbay, C. Brennan, E. Kaplan and L. Sirovich, *Neuroimage*, 2000, **11**, 313.
21. J. Ribot, S. Tanaka, H. Tanaka and A. Ajima, *J. Neurosci. Methods*, 2006, **153**, 8.

22. A. Sornborger, C. Sailstad, E. Kaplan and L. Sirovich, *Neuroimage*, 2003, **18**, 610.
23. T. Yokoo, B. W. Knight and L. Sirovich, *Neuroimage*, 2001, **14**, 1309.
24. S. G. Kim and S. Ogawa, *J. Cereb. Blood Flow Metab.*, 2012, **32**, 1188.
25. J. Upadhyay, G. Pendse, J. Anderson, A. J. Schwarz, R. Baumgartner, A. Coimbra, J. Bishop, J. Knudsen, E. George, I. Grachev, S. Iyengar, D. Bleakman, R. Hargreaves, D. Borsook and L. Becerra, *Neuroimage*, 2010, **49**, 2275.
26. M. A. Kamran, M. M. Mannan and M. Y. Jeong, *Front. Hum. Neurosci.*, 2016, **10**, 261.
27. V. Scarapicchia, C. Brown, C. Mayo and J. R. Gawryluk, *Front. Hum. Neurosci.*, 2017, **11**, 419.
28. A. M. Chiarelli, F. Zappasodi, F. Di Pompeo and A. Merla, *Neurophotonics*, 2017, **4**, 041411.
29. M. Ferrari and V. Quaresima, *Neuroimage*, 2012, **63**, 921.
30. F. Irani, S. M. Platek, S. Bunce, A. C. Ruocco and D. Chute, *Clin. Neuropsychol.*, 2007, **21**, 9.
31. K. Peng, D. K. Nguyen, T. Tayah, P. Vannasing, J. Tremblay, M. Sawan, M. Lassonde, F. Lesage and P. Pouliot, *Epilepsy Res.*, 2014, **108**, 491.
32. P. Y. Lin, N. Roche-Labarbe, M. Dehaes, S. Carp, A. Fenoglio, B. Barbieri, K. Hagan, P. E. Grant and M. A. Franceschini, *J. Visualized Exp.*, 2013, **14**, e4379.
33. N. Roche-Labarbe, S. A. Carp, A. Surova, M. Patel, D. A. Boas, P. E. Grant and M. A. Franceschini, *Hum. Brain Mapp.*, 2010, **31**, 341.
34. S. Nioka and Y. Chen, Optical technology developments in biomedicine: history, current and future, *Transl. Med. UniSa*, 2011, **1**, 51–150.
35. J. P. Culver, T. Durduran, D. Furuya, C. Cheung, J. H. Greenberg and A. G. Yodh, *J. Cereb. Blood Flow Metab.*, 2003, **23**, 911.
36. C. W. Lee, R. J. Cooper and T. Austin, *Pediatr. Res.*, 2017, **82**, 376.
37. Y. Hoshi and Y. Yamada, *J. Biomed. Opt.*, 2016, **21**, 091312.
38. L. Li, L. Zhu, Y. Shen and L. W. Wang, *J. Biomed. Opt.*, 2017, **22**, 76017.
39. L. D. Liao, M. L. Li, H. Y. Lai, Y. Y. Chen and N. V. Thakor, *Conf. Proc. IEEE Eng. Med. Biol. Soc.*, 2013, 1799–1802.
40. L. Wang, K. Maslov and L. W. Wang, *Proc. Natl. Acad. Sci. U. S. A.*, 2013, **110**, 5759.
41. G. F. Lungu, M. L. Li, X. Xie, L. W. Wang and G. Stoica, *Int. J. Oncol.*, 2007, **30**, 45.
42. N. C. Burton, M. Patel, S. Morscher, W. H. Driessen, J. Claussen, N. Beziere, T. Jetzfellner, A. Taruttis, D. Razansky, B. Bednar and V. Ntziachristos, *Neuroimage*, 2013, **65**, 522.
43. S. Gottschalk, T. F. Fehm, X. L. Dean-Ben, V. Tsytsarev and D. Razansky, *Neurophotonics*, 2017, **4**, 011007.
44. V. Ntziachristos and D. Razansky, *Recent Results Cancer Res.*, 2013, **187**, 133.
45. J. Yao, L. Wang, J. M. Yang, K. I. Maslov, T. T. Wong, L. Li, C. H. Huang, J. Zou and L. V. Wang, *Nat. Methods*, 2015, **12**, 407.

46. J. Yao and L. V. Wang, *Laser Photon. Rev.*, 2013, **7**, 758–778.
47. S. G. Resink, A. C. Boccara and W. Steenbergen, *J. Biomed. Opt.*, 2012, **17**, 040901.
48. Z. Zhou, D. Chen, Z. Huang, S. Wang and S. Zeng, *Adv. Exp. Med. Biol.*, 2016, **923**, 393.
49. S. Sakadzic and L. W. Wang, *Opt. Lett.*, 2004, **29**, 2770.
50. S. Sakadzic and L. V. Wang, *Phys. Rev. E: Stat., Nonlinear, Soft Matter Phys.*, 2002, **66**, 026603.
51. S. Sakadzic and L. V. Wang, *Phys. Rev. E: Stat., Nonlinear, Soft Matter Phys.*, 2006, **74**, 036618.
52. E. W. Stein, K. Maslov and L. V. Wang, *J. Biomed. Opt.*, 2009, **14**, 020502.
53. S. Sakadzic, E. Roussakis, M. A. Yaseen, E. T. Mandeville, V. J. Srinivasan, K. Arai, S. Ruvinskaya, A. Devor, E. H. Lo, S. A. Vinogradov and D. A. Boas, *Nat. Methods*, 2010, **7**, 755.
54. L. Gagnon, S. Sakadzic, F. Lesage, J. J. Musacchia, J. Lefebvre, Q. Fang, M. A. Yucel, K. C. Evans, E. T. Mandeville, J. Cohen-Adad, J. R. Polimeni, M. A. Yaseen, E. H. Lo, D. N. Greve, R. B. Buxton, A. M. Dale, A. Devor and D. A. Boas, *J. Neurosci.*, 2015, **35**, 3663.
55. S. A. Vinogradov, L. W. Lo, W. T. Jenkins, S. M. Evans, C. Koch and D. F. Wilson, *Biophys. J.*, 1996, **70**, 1609.
56. T. V. Esipova, A. Karagodov, J. Miller, D. F. Wilson, T. M. Busch and S. A. Vinogradov, *Anal. Chem.*, 2011, **83**, 8756.
57. L. Gagnon, A. F. Smith, D. A. Boas, A. Devor, T. W. Secomb and S. Sakadzic, *Front. Comput. Neurosci.*, 2016, **10**, 82.
58. O. S. Finikova, A. Y. Lebedev, A. Aprelev, T. Troxler, F. Gao, C. Garnacho, S. Muro, R. M. Hochstrasser and S. A. Vinogradov, *ChemPhysChem*, 2008, **9**, 1673.
59. J. Lecoq, A. Parpaleix, E. Roussakis, M. Ducros, Y. Goulam Houssen, S. A. Vinogradov and S. Charpak, *Nat. Med.*, 2011, **17**, 893.
60. D. G. Lyons, A. Parpaleix, M. Roche and S. Charpak, *elife*, 2016, **5**, pii: e12024.
61. A. Parpaleix, Y. Goulam Houssen and S. Charpak, *Nat. Med.*, 2013, **19**, 241.
62. K. Xu, D. A. Boas, S. Sakadzic and J. C. LaManna, *Adv. Exp. Med. Biol.*, 2017, **977**, 149.
63. A. Fercher, S. M. Borisov, A. V. Zhdanov, I. Klimant and D. B. Papkovsky, *ACS Nano*, 2011, **5**, 5499.
64. K. I. Chisholm, K. K. Ida, A. L. Davies, D. B. Papkovsky, M. Singer, A. Dyson, I. Tachtsidis, M. R. Duchen and K. J. Smith, *Adv. Exp. Med. Biol.*, 2016, **876**, 233.
65. K. I. Chisholm, K. K. Ida, A. L. Davies, I. Tachtsidis, D. B. Papkovsky, A. Dyson, M. Singer, M. R. Duchen and K. J. Smith, *J. Cereb. Blood Flow Metab.*, 2016, **36**, 1955.
66. P. Babilas, G. Liebsch, V. Schacht, I. Klimant, O. S. Wolfbeis, R. M. Szeimies and C. Abels, *Microcirculation*, 2005, **12**, 477.

67. A. Zitova, J. Hynes, J. Kollar, S. M. Borisov, I. Klimant and D. B. Papkovsky, *Anal. Biochem.*, 2010, **397**, 144.
68. C. Wotzlaw, A. Bernardini, U. Berchner-Pfannschmidt, D. Papkovsky, H. Acker and J. Fandrey, *Am. J. Physiol.: Cell Physiol.*, 2011, **301**, C266.
69. J. Samonina-Kosicka, D. H. Weitzel, C. L. Hofmann, H. Hendargo, G. Hanna, M. W. Dewhirst, G. M. Palmer and C. L. Fraser, *Macromol. Rapid Commun.*, 2015, **36**, 694.
70. K. K. Ida, K. I. Chisholm, L-M. S. Malbouisson, D. B. Papkovsky, A. Dyson, M. Singer, M. R. Duchen and K. J. Smith, *Brit. J. Anaesth.*, 2018, in press, DOI: 10.1016/j.bja.2017.11.074.

CHAPTER 17

# Applications of Phosphorescent $O_2$ Sensors in Food and Beverage Packaging Systems

MALCO C. CRUZ-ROMERO[a], CAROLINE A. KELLY[b], DMITRI B. PAPKOVSKY[b] AND JOSEPH P. KERRY*[a]

[a]Food Packaging Group, School of Food & Nutritional Sciences, University College Cork, Cork, Ireland; [b]Biophysics and Bioanalysis Group, School of Biochemistry and Cell Biology, University College Cork, Cork, Ireland
*E-mail: joe.kerry@ucc.ie

## 17.1 Introduction

Modern consumers have expectations and demands concerning retailed food products that far exceed those which were required by shoppers some 10 to 20 years ago. Issues such as product convenience, freshness, quality attributes, sensory properties, product presentation, nutritional characteristics, product containment, safety issues and shelf-life stability must be addressed in presenting food products to the consumer, as not only is each factor important in its own right, but important also due to the interplaying relationship that exists between these factors.[1]

The global food industry is very competitive, dynamic and trend-driven. While consumer preferences for food products vary by local market; global consumer trends for health, convenience and indulgence are driving growth

Detection Science Series No. 11
Quenched-phosphorescence Detection of Molecular Oxygen: Applications in Life Sciences
Edited by Dmitri B. Papkovsky and Ruslan I. Dmitriev

Published by the Royal Society of Chemistry, www.rsc.org

areas within the developed food sector. The extremely demanding and competitive food market is continually looking for novel, sophisticated, and stable products which retain to the greatest extent possible, the eating quality of the fresh product, as well as safer and better tasting food products. The role that packaging plays in addressing all of this and more besides, cannot be underestimated.[1] The fundamental aspects of all packaging materials and systems is that in an economic manner, they must contain, protect, preserve, inform (throughout the entire distribution process from point of manufacture to points of consumer usage) and provide convenience (at many different levels) while acknowledging the constraints placed upon their usage from both legal and environmental perspectives. As these fundamental principles apply to all forms of packaging materials and systems, it follows that irrespective of the specific level at which the packaging is industrially applied (primary-sales packaging, secondary-collation and handling packaging or tertiary-transport packaging), all must conform to these same principles.[2] Packaging materials used for packaging food products must conform to the fundamental principles outlined above and satisfy all that is required of the product from both technical (containment, protection and preservation) and sales (cost, convenience, sales information, labelling information, legal requirements, environmental requirements) perspectives.[3] In our own Packaging Group at University College Cork (UCC), we describe all of the packaging materials and formats utilised in primary, secondary and tertiary packaging as first-level packaging because all the conventional packaging materials used contain, protect, preserve, and inform at the most basic level. However, smart and intelligent packaging are defined as second level packaging because the formats and materials used support first-level packaging by layering the additional ability to further enhance preservation or provide additional information or convenience.[4] For the purpose of this contribution, this is how these forms of packaging will be described in the context of presenting new packaging technologies, materials and formats for food products.

The food packaging area is one in which availability and demand is continuously subject to changes due to modifications in consumer habits, distribution chain developments, availability of new technologies, new materials, increasing international distribution and increasing environmental constraints. Packaging formats for food products vary significantly and are linked with the method of production, storage, and consumer preference. Packaging of food products need to offer not only sufficient protection, but also flexibility and ease of handling in order to reduce unnecessary waste.[3] Factors to be considered when choosing a packaging system for application to food products, include; properties of the packaging material, processing conditions employed, physical, chemical, and biological characteristics of the food products intended for packaging, possible food product/package interactions and general containment issues; intended product markets and specified shelf-life. Other factors to take into consideration, include; environment to which the food product is exposed during distribution and storage;

product end-use, package waste and disposal and costs related to the package throughout the distribution chain.[5]

Smart packaging is a broad term encompassing a range of relatively new packaging concepts, most of which can be placed in one of the two principle categories: active packaging and intelligent packaging. Active packaging refers to the incorporation of certain additives into packaging materials, or incorporated within the packaging materials themselves with the aim of maintaining or extending product quality and shelf-life. Packaging may be termed active when it performs some desired role in food preservation other than providing an inert barrier to external conditions.[6] Active packaging has been defined as packaging that changes the condition of the product to extend shelf-life or to improve safety or sensory properties, while maintaining the quality of product.[4]

The trend towards food products being as natural and simple as possible, free of preservatives and unwanted additives has driven the development of alternative preservation methods such as the use of MAP.[6] The gases used in MAP for food products are principally carbon dioxide ($CO_2$), oxygen ($O_2$) and nitrogen ($N_2$). $CO_2$ is used for its bacteriostatic and fungistatic properties and it is particularly effective against moulds and Gram-negative, aerobic spoilage bacteria such as *Pseudomonas* spp., but it is less effective in controlling yeast and lactic acid bacteria[3] which cause spoilage either in the form of visible growth or by the generation of $CO_2$, which cause the package to expand and have a blown pack appearance.[6]

MAP techniques do not always facilitate complete removal of $O_2$. $O_2$ that permeates through the packaging film or is trapped within the food product or between slices of food products cannot be totally removed by MAP.[1] The presence of $O_2$ in a package can trigger or accelerate oxidative reactions which will result in deterioration of the food products. From a microbiological perspective, $O_2$ facilitates the growth of aerobic bacteria and moulds. From a physicochemical perspective oxidative reactions can result in adverse qualities being associated with the food product in question, such as; production of off-odour and off-flavours, development of undesirable colour modifications and reduction in nutritional quality. One solution to this problem is the use of $O_2$ absorbers/scavengers when packaging food products, as such technologies can remove residual $O_2$ within the pack or that entering through the packaging materials themselves and consequently, assist in maintaining the optimal quality and freshness of food products, while at the same time, increasing product shelf-life.[3]

Oxygen scavengers are the most commercially important sub-category of active packaging for food products and the most well-known take the form of small sachets containing various iron-based powders containing an assortment of catalysts.[1] These chemical systems often react with water supplied by the food to produce a reactive hydrated metallic reducing agent that scavenges $O_2$ within the food package and irreversibly converts it to a stable oxide. The iron powder is separated from the food by keeping it in a small, highly $O_2$

permeable sachet that is labelled 'Do not eat' and includes a diagram illustrating this warning. Non-metallic $O_2$ scavengers have also been developed to alleviate the potential for metallic taints being imparted to food products. Non-metallic scavengers include those that use organic reducing agents such as ascorbic acid, ascorbate salts or catechol. They also include enzymatic $O_2$ scavenger systems using either glucose oxidase or ethanol oxidase, which could be incorporated into sachets, adhesive labels or into packaging film surfaces, card and closure liner.[7]

The main advantage of using such scavengers is that they are capable of reducing $O_2$ levels to less than 0.01%, which is much lower than the typical 0.3–3.0% residual $O_2$ levels achievable by MAP.[1] $O_2$ scavengers can be used alone or in combination with MAP, but commercially, it is more common to remove most of $O_2$ content using MAP technology and then use the $O_2$ scavenger to eliminate the residual $O_2$ present in the packaged food product. Ageless® (Mitsubishi Gas Chemical Co., Japan) is the most common $O_2$ scavenging system based on iron oxidation. Other examples of $O_2$ absorbing sachets include $O_2$™ ($O_2$ Zero, UK), ATCO® (EMCO Packaging Systems, UK; Standa Industrie, France), FreshPax®, FreshMax® and Freshcard™ (Multisorb Technologies, Inc., USA), Oxyfree®, Oxylabel™ (Sorbtech International, USA) and Oxysorb (Pillsbury Co., USA).

While oxygen scavengers offer an approach to controlling $O_2$ in commercial food and beverage packages and scrubbing or scavenging systems have been available for some considerable time, uptake of such technologies by companies for implementation within commodity packs have been slow and the majority of oxygen-sensitive food and beverage products presented in retail settings today do not employ $O_2$ scavenging systems. Therefore, a different approach to oxygen presence in packs might be adopted through the use of intelligent packaging systems capable of simply detecting the presence of $O_2$ in packs.

Intelligent packaging (IP) is a packaging system that is capable of carrying out intelligent functions (such as detecting, sensing, recording, tracing, communicating, and applying scientific logic) to facilitate decision making with regard to extending shelf-life, enhancing safety, improving quality, providing information, and warning about possible problems[8] while Robertson[4] defined IP as packaging that contains an external or internal indicator to provide information about the history of the package and/or the quality of the food. Intelligent materials and articles are defined in the European regulation as materials and articles which monitor the condition of packaged food or the environment surrounding the food. Intelligent packaging (IP) systems provide the user with information on the condition of the food.[9] The information provided shall be reliable and correct. In contrast to active components, intelligent components do not have the intention to release their constituents into the food. The intelligent component may be positioned on the outer surface of the package and may be separated from the food by a functional barrier.[10]

Intelligent packaging in some way senses some properties of food it encloses or the environment in which it is maintained or stored within and is capable of informing the manufacturer, retailer and consumer of the state of these properties.[4,11] Although distinctly different from the concept of active packaging, features of intelligent packaging can be used to check the effectiveness and integrity of active packaging systems.[4] Intelligent packaging systems can be classified into three categories: (1) indicate product quality, for example, quality indicators, temperature and TTIs and gas concentration indicators; (2) provide more convenience, for example, during preparation and cooking of foods; (3) provide protection against theft, counterfeiting and tampering.

Kerry[4] stated that the development of smart packaging systems has evolved significantly over the past 30 years; however, the application of these technologies to products in the market place can still be described as being in its infancy most likely due to two primary reasons: (1) the overly cautionary and restrictive attitudes demonstrated by regulatory bodies toward these technologies and applications within this packaging area; and (2) failure of technical developers of smart packaging technologies to engage meaningfully with the retailing sector and *vice versa*, and to a lesser degree with product manufacturers. However, research in the area of smart packaging is progressing rapidly and potential applications are expected, despite the obstacles outlined above.[4] Whether first- or second-level packaging format is used for the optimal packaging of food products, all must be applied efficiently and correctly so that all packaging functions to be delivered by each level are fully achieved.[4]

Sensors are defined as devices used to detect, locate or quantify energy or matter; giving a signal for the detection or measurement of a physical or chemical property to which the devices respond.[12] A chemical sensor is a device that transforms chemical information, ranging from the concentration of a specific sample component to total composition analysis, into an analytically useful signal. The chemical information, mentioned above, may originate from a chemical reaction of the analyte or from a physical property of the system investigated.[13] A number of chemical sensors have been developed for food and food processing applications, including the optical $O_2$ sensors. $O_2$ sensors can be used to monitor the quality of packaged food products. Oxygen sensors integrated into the packaging can be used as leakage indicators or to verify the efficiency of MAP or the use of $O_2$ scavengers and ensuring the integrity of the package throughout the distribution chain. Pharmaceuticals and foods are the major consumer goods sectors where seals are essential to the physical, chemical and biological quality of the packaged product.[14] There are group of products for which seal integrity is vital to the shelf-life of the packaged product that the economic viability of these product category would not be possible as due to seal integrity failure waste would greatly increase and on-shelf availability would decrease; therefore, resulting in higher prices which consumers would not be willing to pay. For example, if we take a case such as that for the supply of sliced cured ready-to-eat meat products, In the absence of good pack-seal performance by the packaging materials chosen, the shelf-life of such products would be considerably reduced, to the point of being unavailable in the marketplace.[13] To successfully integrate gas sensors into packaging materials used for packaging food products, integration of the sensors should not affect the mechanical properties of the packaging materials used, but maintain the packaging performance during storage and monitoring. When sensors are integrated into packaging materials certain considerations require careful attention, such as: sensor integration onto or within the packaging materials, final properties of the developed packaging materials, process conditions required, impact on packaging waste management *etc.* should be taken into account.[11] It was also reported that the main barriers for the adoption of sensors by the food industry have been their non-optimal design, operational performance, limited compatibility with standard food processes, high costs and unavailability in the required volumes, and more importantly the lack of practical validation on the marketplace under real-life conditions and in on-line and at-line applications.[15]

This chapter will summarise the applications of different phosphorescent $O_2$ sensors used to monitor non-destructively $O_2$ in food and beverage packaged food products.

## 17.2 Monitoring $O_2$ Content in Modified Atmosphere Packaging of Foodstuffs

While MAP systems are relatively efficient and reliable, failures can occur due to many factors, such as; incorrect gas composition used, $O_2$ ingress due to poor barrier materials, ineffective gas flushing systems, faulty or defective heat-sealing processes or inadequate and poorly managed storage conditions and physical damage inflicted upon products during handling and transport. Conventional MAP systems provide no real-time data for evaluation of conditions within the packaging systems, and detection of MAP quality and performance is carried out using relatively expensive and destructive techniques such as headspace analysis by GC, and gas analysis using $O_2$/$CO_2$ gas analysers such as Dansensor™, Mocon Pack Check® which often need a skilled operator and consume the analyte being measured.

As well as monitoring the food product itself, it is necessary to monitor the packaging which protects the food from external environmental interference. One packaging integrity indicator termed the Ageless Eye™ (Mitsubishi Gas Chemical Company, Inc., Japan)[16] use dyes that change colour due to the presence of $O_2$. When exposed to $O_2$ (>0.5%), the Ageless Eye™ turns blue or purple, then returns to its original pink colour as the $O_2$ in the container

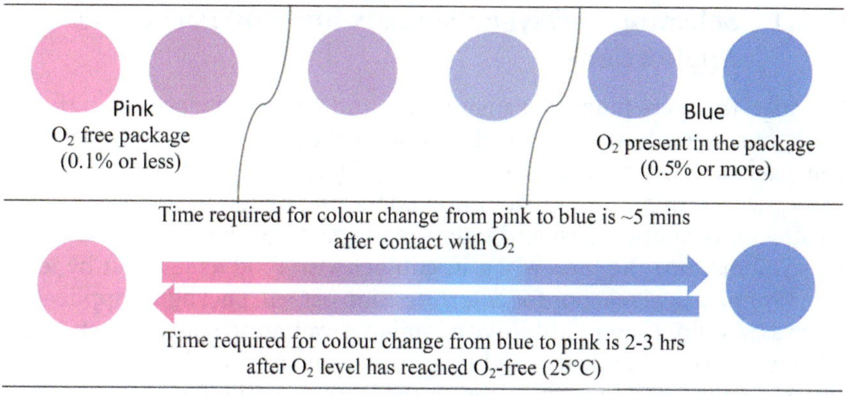


**Figure 17.1** Colour changes of Ageless Eye™ due to presence or absence of oxygen. Adapted from Mitsubishi Gas Chemical Company, Inc.[16]

is reduced (<0.1%) (Figure 17.1). However, these colour changes are affected by temperature (the lower the temperature, the slower the colour change) and the time taken to return from blue to pink takes between 2–3 h. In addition, Ageless Eye™ needs to be stored in a cool, (below 15 °C) dark place with no oxygen until ready for use.

Residual $O_2$ in a packaged food product is an important parameter for both food quality and package integrity. Even trace levels of $O_2$ can cause significant negative effects in food quality through lipid oxidation, aerobic bacteria growth, loss of taste, fast ripening and browning through enzyme catalysed reactions or oxidation of vitamin C in some beverages. Due to the importance of determining residual $O_2$ in packaged food products, the development of luminescence based solid-state sensors is one technology that has gathered momentum in recent years. These sensors utilise the principle of quenching of luminescence lifetime/intensity of dye molecules embedded in a polymeric support, by collision with $O_2$ molecules. The advantages of these sensors over conventional systems to determine residual $O_2$ in packaged food products are: quick determination, quantitative, contactless and non-destructive readouts of $O_2$ content and can be modified or adapted to fit a multitude of purposes.[17–19] Additionally, the sensors can be made inexpensive, disposable or reusable, calibration-free (or batch calibrated) and can be used on a large scale to determine rapidly, and in real time, the quality of the packaged food product and ensure the integrity of the package throughout the distribution chain. This is an improvement on electrochemical techniques such as Clark's electrode,[20] where packages must be taken at random from off the production line for determination of $O_2$ using destructive methods. Additionally, the method is sensitive to gas flow rate, stirring rate, operational temperature and affected by other gases present in the gas mix.[21]

### 17.2.1 Selection of Oxygen Sensors for Food Packaging Applications

Although there are many $O_2$ sensing dyes and materials reported in the literature, only a few have been used in food packaging applications. The selection criteria of $O_2$ sensors used for packaging applications, include: spectral properties, compatible with common packaging materials and processes, $O_2$ sensing performance, cost and safety as the package should not release their constituents into the food when in contact with food as laid out by legislation.[9] For practical reasons for determination of $O_2$ in packaging applications, the dyes should be compatible with available excitation sources such as LEDs and lasers. The chemical stability and photo-stability of the dyes are also critical factors, as in food packaging the sensors will be exposed to intense lighting and chemical conditions. When in contact with food products, the sensors should exhibit low toxicity with no leaching of the dye. Therefore the solvents used to incorporate the sensors onto packaging materials should be carefully selected. Furthermore, for large-scale food packaging applications, the sensors should be reproducible, robust and use minimal components for fabrication.[22]

For easy integration of sensors into food packaging, the sensors should preferably be planar. Integration techniques used, include: free-standing sensor inserts which can be placed in containers with the food, adhesive stickers attached to the packaging, printed sensor labels located on the inside of the packaging label and lamination where the $O_2$ sensor is laminated between the outer barrier packaging and an inner $O_2$ permeable layer. The main requirements for $O_2$ sensor integration are: easy access of the sensor to the packaging atmosphere, the availability of the interrogating instrument and protection of the sensor from the atmosphere outside of the package (*e.g.* high barrier to $O_2$ films). Additionally, integration processes will also often protect the sensor from the food product and *vice versa*. This is vital as some food components (*e.g.* meat juices) can affect sensor performance (reduced signals due to discolouration, shift of calibration due to plastification and changing of the availability of $O_2$ to the sensing dye) leading to inaccurate readings. From a food safety perspective, the use of a gas permeable film, such as LDPE, laminated on top of the sensor, between the sensor and food product, can prevent migration of the sensor components (dye and polymer matrix) into the food, and product contamination with particles that could affect the physicochemical characteristics of the food product such as texture, flavour, aroma or taste.

### 17.2.2 Characterisation of $O_2$ Sensors

Before implementation, new sensors should undergo detailed characterisation. Initial characterization involves plotting the sensors intensity/lifetime readings over a range of $O_2$ concentrations to obtain a calibration. For applications in food packaging, the sensor should show reproducible calibrations, generally over the range of 0.05–21 kPa $O_2$. Some food products such as raw meat packaged using MAP require sensors measuring up to 100 kPa $O_2$, while other products require measurement of trace $O_2$ levels (ppm range). The sensors should not have cross-sensitivity to humidity, as humidity is not controlled in many food packages.

Most solid-state sensors show a dependence on temperature and their sensitivity decreases as temperature decreases. The food industry use a variety of storage temperatures and also uses a variety of conditions during transport to various climate conditions, therefore the sensors should exhibit predictable responses over a representative range of temperatures (−20 °C to + 30 °C). Unfortunately, only a few systematic studies have been performed over such a wide temperature range.[23] The cross-sensitivity of a sensor to temperature is strong; therefore, during measurement, luminescence measurements must be corrected for temperature fluctuations in food packages.

Additionally, response time is a critical factor. The use of thin film coatings and microporous membranes with high gas permeability, allows for fast (second timescale) response times which are generally desired, especially for in-line monitoring of packaging integrity on large-scale production lines. Finally, photo-stability and long shelf-life stability of the sensors are essential if sensors are to be applied to food packaging applications.

### 17.2.3 Instruments Used to Monitor $O_2$ on Food Packaging Integrated with $O_2$ Sensors

Currently, several commercial optical $O_2$ sensors and instruments are available for industrial applications. However, only a few of these systems are suitable for packaging applications, which demand special requirements, such as operation with detachable, disposable, calibration-free sensors, a broad range of working temperatures, a portable handheld measurement device, high level of validation (*i.e.* knowledge of cross-sensitivities and interferences) and affordable costs. Comparison of the main $O_2$ sensor systems used for monitoring $O_2$ in food packaging is provided on Table 17.1.[24]

To confirm accuracy of measurement, these systems often require quick re-calibration of sensor (or whole system with instrument) before use. While the Optech™ sensors are calibration-free, they also have a validation *via* "CalCard" (or calibration vial for needle sensors). Oxydots® are $O_2$ sensors commercialised with factory calibration; however, user calibrations may be carried out if deemed necessary. PreSens sensors use a two-point calibration at 0 kPa and 21 kPa before operation. These companies offer customised sensor for specific applications and modify such sensors for trace $O_2$ analysis (PSt6, PreSens) and individual deployable sensor beads (dOxybead™, Luxcel).

For successful use of $O_2$ sensors in food packaging applications, easy integration of the sensors into food packaging is essential. Oxydot® and Presens sensors must be stuck to the packaging using silicone-based adhesive. Mocon's OpTech™ sensors provide as a self-adhesive sticker, or imPULSE™

**Table 17.1** Comparison of commercially available oxygen sensor systems commonly used for monitoring $O_2$ in food packaging applications. Adapted from Banerjee *et al.*[24]

| | Company | | |
|---|---|---|---|
| Sensor characteristics | Mocon Inc & Luxcel Ltd | Oxysense | PreSens GmbH |
| Sensor | OpTech™-$O_2$ Platinum | $O_2$xydots® | Pst3 |
| Dyes | PtTBP | $[Ru(dpp)_3]Cl_2$ | PtTFPP |
| Forms | Adhesive stickers (peelable from a card) | Sensor dots applied with silicon glue | Sensor dots applied with silicon glue |
| Measuring range | 0.001–30% permeation mode (g) 0.015–25% headspace mode (g) (0.006–10.5 mg $L^{-1}$ (l)) | 0–30% | 0–100% (g) (0–45 mg $L^{-1}$ (l)) |
| Accuracy | ± 2% or ± 150 ppm (whichever is greatest) | 5% of reading | ± 0.4% @ 20.9% $O_2$ ± 0.05% @ 0.2% $O_2$ |
| Detection limit | 0.03% (0.03 kPa) | 0.03% (gas) 15 ppm (liquid) | 0.03% (gas) 15 ppm (liquid) |
| Compensation of temperature/pressure | Automatic temperature (contactless IR sensor) and pressure compensation | Integrated temperature and pressure probes | Integrated temperature, and pressure probes |
| Operating temperature range | 5–40 °C | 0 °C–70 °C | 0–50 °C |
| Response time | < 3 s (gas) <30 s (liquid) | < 30 s | < 6 s (gas) < 40 s (liquid) |
| Price per sensor | $3 | $4.30 | $33 |

probe and a needle sensor for minimally destructive headspace analysis. Although convenient, use of the above sensors is limited by their cost, which in turn limits their inclusion for mass scale use for in-line monitoring of $O_2$ in food packaging applications. The viability of sensors to be used in food packaging are reliant on their production cost which should be less than 1cent.[22] As a result, many new studies are carried out in an effort to manufacture new optical sensors using less expensive components and that can use existing reading instrumentation or newly designed instruments.

### 17.2.4 Applications of $O_2$ Sensors for Monitoring $O_2$ Content in Packaged Food Products

Much of the proof of concept work on the use of non-destructive luminescence based $O_2$ sensors in food packaging applications was carried out in the 2000s and the concentration of $O_2$ in the packs measured using prototype

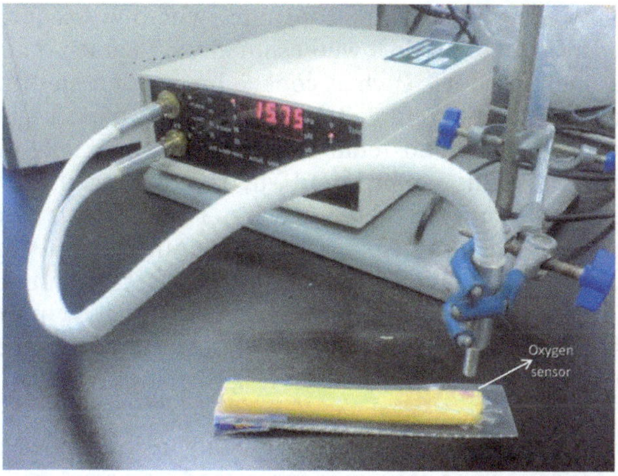


**Figure 17.2** Phase detector measuring $O_2$ on packaged sample integrated with oxygen sensor.

$O_2$ sensors and instruments (Figure 17.2);[25] however, currently $O_2$ can be measured using advanced commercially available $O_2$ sensor systems; such as Fibox (Presens), $O_2$ analyser (Oxysense) and Optech™ (Mocon).

Consumers' perceptions and evaluations of meat quality attributes such as colour and shelf-life influence purchasing decisions, and these product attributes can be affected by the packaging system used.[26] Initially, the objectives of using $O_2$ sensors integrated into packaging were to monitor packaged food products and detect faulty packages and containment failures, making food packagers aware of these issues for rectification purposes.

The muscle pigment myoglobin, responsible for fresh beef colour, can exist in one of three chemical forms, namely; deoxymyoglobin, oxymyoglobin and metmyoglobin.[27] Oxymyoglobin, the cherry-red form of the pigment is associated with freshness. Over time, oxymyoglobin is oxidised to metmyoglobin which is brown in colour and undesirable from a consumer perspective. When exposed to $O_2$ in atmospheric air, the purple muscle pigment myoglobin becomes loosely oxygenated and converts to oxymyoglobin. The use of $O_2$ in MAP is normally set as low as possible to inhibit the growth of aerobic spoilage bacteria as presence of $O_2$ may also cause oxidative rancidity (*e.g.*, fatty fish like salmon and mackerel). However, high levels of $O_2$ are used in red meat products to maintain the red colour of the meat. Elevated $O_2$ levels in the meat package (typically in the commercial range of 70–80% $O_2$) convert myoglobin into oxymyoglobin and as long as the high $O_2$ content is maintained under the created in-pack partial pressure and refrigerated temperatures, the oxymyoglobin remains relatively stable and the desirable bright red colour is maintained for a limited, but extended period.[28]

Following MAP and during subsequent storage, several quality aspects must be considered simultaneously in order to obtain an optimal product quality and shelf-life. Colour stability is also one of the important characteristics associated with the quality of MAP cured, cooked ham, since it is the primary quality cue sought by the consumer, who may use it as a freshness indicator and a signal to product wholesomeness.[29] Larsen *et al.*[30] stated that the colour of meat products such as cooked ham is of major importance for consumer acceptance. Therefore, the colour stability of packaged cured meat products during storage and retail display is commercially important.[31,32] Discolouration problems can be found when the packaged products contain small amounts of $O_2$ and are then exposed to inappropriate ultra violet (UV) light levels associated with display cabinet illumination, as well as normal retail lighting, leading to photooxidation reactions. Changing from vacuum to MAP has introduced headspace volume and residual $O_2$ in the package as new factors influence the colour of cooked meat products. The $O_2$ permeability of the packaging material also has a significant influence on the colour stability of processed meat products.[33]

While low $O_2$ MAP and vacuum packaging technologies are used for $O_2$-sensitive foods such as processed meat products, residual $O_2$ is typically present in most meat packs after sealing. Oxygen scavengers are secondary packaging devices that can help absorb this residual $O_2$, thereby minimising the negative impacts of $O_2$ on product quality.[34]

As outlined above, residual $O_2$ in the packaged food product is an important quality indicator; therefore, monitoring the presence of this analyte after MAP is very important. The $O_2$ levels within packages may indicate inadequacy of the MAP systems applied and possibly seal integrity failure. Packs on manufacturing lines, which fail to meet strict $O_2$ guidelines, should be rejected or repackaged if this is a viable alternative before they leave the factory.

Prototype $O_2$ sensors comprising of phosphorescent PtOEPK dye dissolved in polystyrene and spotted on polymeric membranes or filter paper were used to track $O_2$ levels in packaged beef and hams.[35,36] These studies demonstrated the commercial feasibility of using $O_2$ sensors as a means to detect faulty packages and containment failures. The use of disposable $O_2$ sensors integrated on individual packs were found to be very useful and informative when placed in direct contact with the food or attached to the lidding material and exposed to package headspace. Reliable optical signals, and hence quantitative information about the $O_2$ levels within the pack, were obtained.

Papkovsky *et al.*[36] monitored $O_2$ levels in MAP sliced ham products and highlighted the need for continuous monitoring of packaged food products. Although the products were packaged under an anoxic atmosphere, residual $O_2$ levels in the packs were significant (0.1–1%) and the concentration of $O_2$ gradually increased during storage. These changes in $O_2$ content during storage were explained as being due to $O_2$ exchange between $O_2$ trapped within ham slices and the package headspace, the permeability of the packaging materials and/or imperfect sealing. Screening of the packaged sliced ham products exposed multiple failures (1–5 in each batch) which could go unnoticed when using traditional headspace analysis methods, which take randomly representative packs off the line for analysis on a statistical basis. Prompt detection of such failures can save the company money, as the failed packages can be repackaged without any degradation of the food product if detected in time, but of paramount importance, pack integrity was determined and integrity has a crucial food safety link. The risk of product contamination with bacteria, or other substances, is a huge consideration in the food and beverage industries. Poor seal integrity can result in consumer complaint or illness with a consequent negative impact on brand image. Loss of business and negative publicity are strong following possibilities if package seal integrity is not managed correctly within a food manufacturing business.[14]

In another study, PtOEPK $O_2$ sensors were packaged with cheddar cheese and used as packaging integrity detection.[37] The study demonstrated again the need for individual integrity checks in packaging as 3 failures out of 67 packs tested were found.

Pt(II)-benzoporphyrin dye (PtBP) based sensors were adhered to green and brown wine bottles to assess the $O_2$ diffusion through cork-sealed colored glass bottles.[38] The author concluded that the optical sensors measured low $O_2$ concentrations and the sensor based on PtBP could be used for measurements through colored glass (*e.g.* wine bottles) due to its emission in NIR and high brightness conditions.

Fresh produce, such as vegetables and salads, are highly perishable and the respiration rates for these products are high and special care should be taken to avoid anoxia which leads to rapid deterioration of tissue.[39] Therefore, the time of processing and packaging after harvesting fresh produce should be minimal.[40] Perforated lidding is required for fresh produce in order to ensure that respiration can occur but low enough to discourage microbial growth.[41] The perforations prevent anoxia and condensation inside the packages; therefore selection of optimal packaging material, density and size of micro-perforations and atmosphere within the packs is critical for maintaining fresh produce quality over time.[39] Therefore, many of these products are packaged with little or no atmosphere modification being carried out. PtOEPK based sensors, in combination with Mocon's Optech™ instrument, was employed to provide accurate and comprehensive information on $O_2$ levels within commercially packaged bags of fresh salad leaves.[41] In addition, respiration profiles for different types of lettuce leaf were obtained non-destructively using these same sensors. Different varieties of lettuce leaves have different respiration rates; therefore, it is important to determine the respiration rate of the specific lettuce variety as the $O_2$ levels within the pack must correlate to the $O_2$ requirement of the lettuce leaves to maintain freshness and quality during storage and avoid wilting of the leaves. This is particularly important where salad leaves of different varieties are being mixed as a standard commercial practice. The study on lettuce using optical $O_2$ sensors showed that MAP can be tailored to the type of lettuce being packaged in order to extend product shelf-life.

In a different study, Borchert *et al.*[39] applied gas sensing technologies for non-destructive monitoring of headspace gases ($O_2$ and $CO_2$) during chilled storage of packaged mushrooms (*Agaricus bisporus*). To maintain mushroom quality during storage low $O_2$- (1–5%) and elevated $CO_2$-levels (<12%) are recommended. As mushrooms continue the respiration process consuming $O_2$ and releasing $CO_2$ in this study, the headspace gas composition was measured daily non-invasively using optically-based $O_2$ and $CO_2$ sensors inserted in each pack. Equilibrium levels of $O_2$ = 11.58% and $CO_2$ = 10.7% were achieved after 18.5 and 41.5 h, respectively. This study confirmed that optically-based sensor systems can be used to rapidly monitor non-destructively, headspace gases of packed mushrooms continuously, and determine the respiration rate of fresh horticultural products.

Oxygen sensors were also used in combination with active packaging in the form of ethanol emitters for monitoring and extending the shelf-life quality of MAP speciality breads. PtOEPK based sensors were used to track the effects of different ethanol-based atmosphere modifiers on MAP ciabatta bread.[42] Non-destructive optical $O_2$ sensors, which were placed within packs on-line prior to gassing and sealing, were used to monitor residual and pack ingress of $O_2$ during shelf-life evaluation of the bread packaged under MAP (10% $CO_2$ : 90% $N_2$). The $O_2$ sensors helped to identify initially containment failure of commercial packaged bread. On detailed examination of the packaging process, due to the initially high $O_2$ concentration, it was determined that the reason for the high levels of $O_2$ present in packs was due to faulty heat-sealing equipment which the commercial company was unaware of and the monitoring of $O_2$ with sensors helped to troubleshoot the packaging process. The study also reported that reduction in $O_2$ during shelf-life was shown to be related to the microbial quality of samples by propagation and respiration of moulds. The results of this study provided valuable information in terms of understanding what levels of $O_2$ are present in commercial packs of MAP bread and its impact upon quality parameters associated with bread-type products. The addition of ethanol to extend shelf-life of the bread products showed great promise in delaying the onset of inevitable mould growth. The potential for such smart packaging technologies like $O_2$ sensors and ethanol emitters in the bread industry was proposed.

The establishment and control of $O_2$ levels in packs of $O_2$-sensitive food products, such as cheese and cooked food products, is imperative in order to maintain product quality over a determined shelf-life. Sensory properties such as colour, taste and aroma have a major effect on how the food product is appraised by producers and consumers. Even a slight increase in $O_2$ can have a noticeable effect on these parameters and decrease the quality of the food. Therefore, the monitoring of $O_2$ levels and relating these levels to deterioration processes is fundamental in determining shelf-life and ensuring optimal quality of food products from factory to fork.

One such study used PtOEPK based $O_2$ sensors to correlate $O_2$ levels to lipid oxidation in vacuum- and MAP raw and cooked beef.[35] It was shown that although oxidation did not occur up to day 35 in vacuum-packed raw and cooked meats, MAP meat had an oxidative stability of 12 days, most likely due to its higher initial $O_2$ levels. The study also concluded that although $O_2$ is needed for lipid oxidation, it is not the driving factor as revealed when comparing oxidation levels in MAP cooked meat to vacuum-packed cooked meat.

Instead, it was shown that processing was the driving factor of lipid oxidation as demonstrated by higher lipid oxidation in cooked meats compared to raw meat products. This finding confirmed that cooked products are more sensitive to lipid oxidation; therefore, measures should be taken to protect these kinds of products against $O_2$.

The Optech™ $O_2$ sensors have been also used to measure the depletion of $O_2$ in raw beef and chicken fillets.[43] Both samples were packaged with a 21% headspace. The beef and chicken showed a decrease in $O_2$ levels over a 9- and 10 day storage, respectively. This confirms that aerobic bacteria, which are the most common cause of food spoilage, can modify the headspace composition, thereby depleting $O_2$ over time. However, as sensory data was not carried out in the study, there is no way to ascertain whether the food was deemed spoiled before the $O_2$ levels were depleted. As chicken is conventionally packaged in MAP with as low a level of $O_2$ as possible, it would seem likely that this product would be considered spoiled before 10 days due to other deterioration processes such as lipid oxidation.

The use of 100% $CO_2$ is favoured by many cheese packers, as the employment of this MAP approach typically encourages inhibition of microbial growth, but, more specifically, produces a cheese pack which possesses the physical appearance of a vacuum package through absorption of $CO_2$ by the cheese over time.[44] However, if the package, packaging process, storage, or distribution conditions fail to contain hard cheese products properly, then the benefits imposed by modifying the atmosphere will be ineffective, and "snug-back" or "snug-down" (a fully intentional package collapse) will not occur, thereby negatively affecting the visual appearance of the final retail pack. Additionally, when such primary packaged cheese products have been manufactured for export, the primary packaging must contend with more challenging stresses from the point of collation, handling, and distribution purposes through extended distances and environmental conditions presented as the primary packs move through the cold-chain distribution system employed in the market placement of such products. Therefore, monitoring $O_2$ levels within the package can give critical information on the status of cheese quality and shelf-life, and assist in the pinpointing of negative containment issues as they arise from the point of product manufacture to the point of retail purchase.[44]

O'Callaghan *et al.*[44] used sensors to assess the effects of package quality, packaging process efficiency, and handling and distribution on package containment of cheese packaged at industrial scale. Cheese blocks were stored on-site and experienced minimal movement (Group A) and subjected to simulated industrial distribution (Group B). Results indicated that $O_2$ content in both experimental storage groups increased over time and that the highest rate of failure occurred in Group B, with the source of failure determined to be due to the presence of leaks. The predominant reason for the formation of leaks was found to be due to the secondary packaging formats used for palletisation, distribution conditions experienced and the possible existence of calcium lactate crystals on the cheese surfaces within the primary packs. The incorporation of these non-destructive $O_2$ sensor systems to quantify $O_2$

concentrations within packaging ensured that the entire cheese supply chain was monitored continuously.

Studies have also tracked the correlation between the growth of microorganisms on food and $O_2$ levels present within food packages. PtOEPK based sensors were used to trace the $O_2$ profile of a faulty package of cheese (which is usually packaged in an anoxic environment), showing that as mould grew on the contaminated product the $O_2$ levels steadily decreased.[37] This is an important factor to consider when monitoring quality and degradative processes within food products, as initially high $O_2$ concentrations, which would usually indicate a leak, are often depleted to original levels by oxidation and microbial actions within a time frame. The reversible nature of the optical sensors demonstrates a novel and effective way of continuously monitoring fluctuations in $O_2$ levels within packs, from $O_2$ ingress within packs to $O_2$ consumption by microbial, biochemical and chemical processes.

Along with influencing some chemical reactions in solid foods, $O_2$ plays a key role in the quality of liquid products and beverages. Many chemical reactions occur during the maturation of red wine and the effect of $O_2$ (either positively or negatively) can impact the final characteristics of the wine. Therefore, $O_2$ content in wine products must be quantified and controlled.

Many wine-makers add controlled concentrations of $O_2$ to wine in order to optimise the maturation of the wine. Ugliano *et al.*[45] investigated the influence of different $O_2$ exposures on the evolution of several volatile sulphur compounds (VSCs), such as; $H_2S$, methyl mercaptan (MeSH) and ethyl mercaptan (EtSH) during Shiraz maturation. As the aim was to create $O_2$ procedures to control the evolution of VSCs in the wine, each bottle was fitted with two PreSens Pst3 sensors; one to measure headspace $O_2$ and the other to measure dissolved $O_2$ in the wine. The study findings implied $O_2$ had a direct effect on the $H_2S$ and MeSH levels, which in turn, affected the odour of the wine. By plotting the $O_2$ levels obtained by the sensors against the concentrations of VSCs, a correlation between lower $O_2$ levels and higher VSCs was found.

Similar to wine, the flavour and visual appearance of beer are dependent on various oxidation reactions taking place during processing and at packaging stages.[46] Stability of beer flavour is dependent on low $O_2$ concentrations. Although, beer should be $O_2$-free after fermentation, it is most vulnerable to oxidation at this point. Bottles are filled to the top with carbonated beer and capped with the aim to minimise $O_2$ entrapment within the bottles. When the foam dissolves in the headspace of the bottled product, in theory, $O_2$ levels residing in the headspace should be negligible. However, this process is often affected by different foam structures formed in bottle necks which trap $O_2$ to varying degrees. High $O_2$ levels lead to higher levels of carbonyls and development of undesirable flavours during storage. As beer degrades, these carbonyls increase, suggesting that beer would taste best straight after production. Oxygen sensors have been used to monitor residual $O_2$ levels in sealed bottles, oxidation levels and quality of beer over time. PtOEPK based sensors were adhered on the inside of bottles and a phase detector was used to measure the signals from the sensors. The study showed pasteurisation (a technique used to kill vegetative microorganisms) caused headspace $O_2$ to diffuse into the product, which was in turn re-released into the headspace over time. This indicated that $O_2$ was not utilised in spoilage processes; however, sensory analysis showed that beers with an initial headspace $O_2$ level of >1% were considered unacceptable by tasting panels prior to the best before dates provided and consequently, short on the shelf-life stability associated with beer. Therefore, the use of $O_2$ sensors were demonstrated to allow continuous quality monitoring with a set quality criteria such as $O_2$ levels <1% $O_2$ with readings taken in bottles prior to commencing the pasteurisation process.

Ascorbic acid (AA), otherwise known as Vitamin C, is a stabilising antioxidant which can be oxidised and broken down into undesirable by-products. When AA is oxidised over time, it converts into L-dehydroascorbic acid (DHA), which has an antioxidant activity five times lower than AA,[47] which sequentially increases browning in the juice. Additionally, as DHA increases, the biological action of vitamin C is lost. As a result, packaged orange juice products are considered to be highly perishable and $O_2$-sensitive products, resulting in $O_2$ concentrations being monitored as quality factors during production and storage. Therefore, residual $O_2$ and in turn, packaging permeability, are considered to have an effect on the sensory and nutritional quality of the juice. Van Bree *et al.*[48] and Wibowo *et al.*[49] used Oxydots® to measured headspace $O_2$ and link $O_2$ presence in headspace to the degradation of AA. Both studies found a correlation between higher $O_2$ levels and increased DHA formation. These results indicate that presence of $O_2$ should be avoided and precautions taken during the processing and packaging of orange juice to minimise AA degradation.

The effects of storage temperature and $O_2$ on lipid oxidation of rapeseed oils have been investigated using Oxydots®. Kozak *et al.*[50] monitored the rate of $O_2$ consumption during storage of rapeseed oil products, with ambient headspace at varying temperatures. The study revealed that higher temperatures increased $O_2$ consumption and led to higher formation of hydrogen peroxides (a by-product of lipid oxidation). Samples stored with nitrogen headspace showed minor hydroxide peroxide formation. Thus, the optical $O_2$ non-destructive measurement technique can be applied to any liquid oil product in order to evaluate optimal storage conditions.

An important requirement in selecting packaging systems for foods is the barrier property of the packaging material. To reduce lipid oxidation of food constituents, the packaging must provide a good barrier to $O_2$. A permeability of a packaging material provides a measure of how well a certain gas or vapour can penetrate the packaging material.[51] Therefore, the knowledge of permeability of packaging polymers to gases and vapour is crucial for the preservation of food quality. One of the methods commonly used for determining $O_2$ permeability on polymer films is the ASTM D3985 Standard Test Method for $O_2$ Gas Transmission Rate through Plastic Film and Sheeting Using a Colourimetric Sensor. In this method, a partial pressure difference is maintained by sweeping one side continuously with the test gas and maintaining an inert gas on the other side into which the test gas diffuses. The concentration of the diffusing gas can be measured by chemical analysis, gas chromatography, thermal conductivity or special electrodes.[6]

Usually, the $O_2$ permeability of films and other materials are measured on special instruments such as Mocon's Ox-Tran® Model 2/21 system or Oxysense's Oxyperm OTR system and so on. Although very accurate, these systems are time-consuming, limited to planar empty samples and have low sample throughput. The Oxyperm system has made efforts to combat this by producing an adaption for measuring bottle permeation, but is still limited by cost and the need for a trained operator.[52]

The use of disposable or reusable $O_2$ sensors and non-destructive, contact-less optical measurements can address the shortcomings of the conventional instruments and enable an extended range of gas permeation tests. These sensors can be adhered to almost any surface of an assortment of shaped packs, including; whole composite material *e.g.* trays with sealed lidding, closed bottles, packages filled with liquid or solid food. In addition, the sensors can be used to measure packaging materials being stressed on the packaging line.

In our own research group, Molinaro *et al.*[53,54] used disposable $O_2$ sensors to determine $O_2$ transmission rate (OTR) on packaging films. Briefly, $O_2$ content inside a known test cell volume chamber, flushed previously with nitrogen, was monitored non-invasively over a 24 h period with $O_2$ concentration recorded at 2 h intervals using an optical measuring system Optech™ Platinum $O_2$ sensor device and disposable $O_2$ sensor stickers obtained from Mocon. The optical $O_2$ sensor works on the principle of a phosphorescent dye that is incorporated in a polystyrene polymer membrane as adherent agent and for environmental protection. The Optech device uses LED technology to take measurement in 10 s. For $O_2$ measurement, the instrument needs to be brought in an optical contact with the sensor (5–10 mm distance) to produce an $O_2$ reading and the % of $O_2$ compensated for temperature and pressure variation. Therefore and in this case, the sensors were stuck on the top of a handmade support placed in the centre of a lower part of a Perspex™ cup at a distance of 3 mm from the testing film. Nitrogen was blown into the chamber, and the films were mounted between the upper lid and rubber ring with silicon lubricant and fixed to the lower cup by screws. The $O_2$ concentration in the cup can be measured as many times as necessary as there is no requirement for gas sampling/extraction. The cups were stored under controlled temperature (23 °C) and humidity (50%) and at a constant air velocity of 152 m min$^{-1}$ maintained over the cups to ensure uniform air movement across cells. OTR was successively calculated considering the $O_2$ concentration, the volume of the test cup chamber and the area of the film exposed for gas permeation using the following equation:

$$\mathrm{OTR} = V * \left(\frac{\mathrm{Slope}}{100}\right) * 24 * \left(\frac{10000}{A}\right) \tag{17.1}$$

where OTR is expressed as ml $m^{-2}$ per day, *V* is the volume of the receiving chamber, and *A* is the area of the film available for gas permeation.

PET carboys, which are used in home winemaking, were evaluated for $O_2$ permeability using a Mocon Optech™ instrument.[55] The $O_2$ sensor stickers were adhered to the neck of each carboy and $O_2$ content in the carboy measured regularly. The study had a dual-purpose; the assessment of the suitability of the carboy for use and demonstration of the suitability of the Mocon Optech™ system for measurement of packaging permeability.

A similar study was carried out by Diéval *et al.*[56] using Presens Fibox 3 $O_2$ meter and PSt6 sensors. Upon adherence of the sensors on glass bottles, bottles with different Normacorc closures were examined for differences in $O_2$ ingress over 250 days. For comparison, the $O_2$ concentration of the bottle necks with closures was also determined on the Mocon Ox-Tran® instrument. The Presens sensors were shown to perform well when compared the Ox-Tran® $O_2$ permeability instrument, providing a simple easy-to-use alternative to traditional $O_2$ permeability measurement devices. Wirth *et al.*[57] and Ugliano *et al.*[58] have both carried out similar studies assessing $O_2$ transmission rates of packaging materials using Presens sensors.

The permeability of food packaging material is dependent on the polymer-type, temperature, relative humidity, nature of the polymer (*e.g.* crystallinity, orientation, *etc.*), thickness of the polymer and differences in partial pressure of the gasses employed. MAP packaging materials must have good barrier properties to $O_2$ as well as good water vapour barrier, good mechanical strength and low cost. As there is no polymer which possesses all of these features, multilayers or laminates are often used. This implies that there exists a multitude of packaging materials for different conditions, which makes it very difficult for the user to choose the right packaging material.

Van Bree *et al.*[59] developed a practical software simulation tool to predict the amount of $O_2$ permeating through the packaging to the headspace during storage for a variety of multilayer packaging configurations. The software was validated using Oxydots®. Simulation processes include; the influence of different packing materials, thicknesses, temperatures and headspace on overall $O_2$ concentration within the pack. This software has the potential to eliminate some of the package material testing required for new food products.

Dissolved $O_2$ can also be measured using some optical $O_2$ sensors. In wines, the concentration of $O_2$ has a major effect on the wine's final sensory properties. If concentrations are too low, aroma defects such as reduction (dominant smell of egg or rotten vegetables[60]) can occur. In contrast, if $O_2$ levels are too high, there can be an increase in oxidation and discolouration of the wine.[61] Many of these studies utilised the PreSens sensors as they are non-invasive and can be continuously monitored.

Wine can be discoloured if exposed to excess $O_2$ *e.g.* white wine is known to turn brown and red wine an orange-red tone. However, as wines are often required before the colour and texture can be stabilised naturally by aging, controlled oxygenation of the wine is required.[62] This oxygenation process is usually carried out after fermentation. However, in recent times, in order to introduce the optimum concentrations of $O_2$ required for chemical reactions to occur, the use of rotary fermenters has been recommended. As these fermenters are closed systems, optical $O_2$ sensors from Presens were used as they can be employed to measure non-invasively $O_2$ concentration within the system. From the data obtained from the PSt3 and PSt6 sensors, correlations were made between higher $O_2$ concentrations and lower tannin concentrations; resulting in more stable wine colour and lower astringency. Both lower astringency and lower tannins increase the consumer acceptability of wines.

Ascorbic acid (AA) is often added to wine in order to reduce oxidative degradation.[63] Along with its antioxidant effects, AA has also been shown to improve wine colour and flavour. Ascorbic acid can degrade both oxidatively and non-oxidatively. If uncontrolled, the latter can lead to shortened shelf-life as it depletes sulphur dioxide which provides anti-microbial protection to the wine.[64] As data on non-oxidative degradation of AA is sparse, it can be hard to predict optimal $O_2$ levels in the wine suitable to maintain oxidative degradation (the by-products of which are scavenged by sulphur dioxide). As a solution to this, a study on non-oxidative degradation of AA was carried out by Wallington *et al.*[63] who gathered kinetic data on the process. To ensure anoxic conditions, Presens' sensors were used to measure dissolved $O_2$ (PSt3) and gaseous $O_2$ (PSt6). Although, anoxic conditions are difficult to achieve, $O_2$ concentrations were kept below the sensor's limit of detection (PSt3 $<50\ \mu g\ L^{-1}$) to ensure non-oxidative degradation of the AA would occur. Data showed that lower temperatures inhibited non-oxidative AA degradation. This implies that wines should be stored and transported at lower temperatures to limit non-oxidative degradation of AA.

Although several commercial optical $O_2$ sensor systems are currently in the market, there have been very few validation and benchmark tests performed on these systems and compared to existing instrumental methods. The Mocon Optech™ system, which has the measurement range of 0–25% $O_2$ was compared to the traditional destructive headspace analyser Dansensor™ which has a range of 0–100% $O_2$.[40] The Dansensor™ punctures the package through a septum in order to analyse the atmosphere while the Optech™ sensor can give readouts *via* a contactless reader non-destructively. The study showed good correlation between these systems for values up to 40% $O_2$; however at values higher than 40% the Optech™ started to show inaccurate readings. At the same time, Opttech™ showed a better sensitivity in the low $O_2$ range – 0.03% as opposed to 0.1% for the Dansensor™ (which can only give one decimal place readings).

Likewise, very few studies have looked in detail at sensor behaviour and calibration upon its exposure to food products in packs or sensor stability. Likewise the accuracy of $O_2$ readings in conditions of industrial operation with different packaged foods and beverages has rarely been assessed. This data is particularly important when sensors are operating in direct physical contact with the food product that may have high fat contents, strong flavours, varying water content or long-term experiments with liquid samples. Package contents can affect sensors, thereby shifting sensor calibrations and, if not compensated for, leading to inaccurate readings.[65] Examples of studies that used $O_2$ sensors in food packaging applications and their main findings are presented in Table 17.2.

**Table 17.2** Examples of studies that used $O_2$ sensors in food packaging applications and their main findings.

| Product category | Measurement | Sensor material used | Instrument used for measurement $O_2$ | Main findings | Reference |
|---|---|---|---|---|---|
| *Food products* | | | | | |
| Beef (Cooked/Raw) | Residual $O_2$ levels | PtOEPK/PS on filter paper | Phosphorescence phase detector | $O_2$ levels correlated with lipid oxidation | 35 |
| Beef/Chicken (Raw) | Residual $O_2$ levels | Optech™ stickers | Mocon Optech™ | $O_2$ consumption during storage of meat products monitored | 43 |
| Bread | Tracking $O_2$ levels/ product shelf-life | PtOEPK/PS on filter paper | Phosphorescence phase detector | Containment failure determined and impact of $O_2$ levels on quality parameters determined. | 42 |
| Cheese | Packaging integrity testing | PtOEPK/PS on filter paper | Phosphorescence phase detector | Packaging integrity failure detected, $O_2$ concentration correlated to microbiological growth | 37 |
| Cheese | Packaging integrity testing throughout the entire supply chain | Optech™ stickers | Mocon Optech™ | Containment failure detected and $O_2$ concentration determined throughout the supply chain | 44 |
| Lettuce | $O_2$ levels during storage | PtOEPK/PS on filter paper | Mocon Optech™ | Respiration rate profiles of different varieties of lettuce determined | 41 |
| Mushroom | $O_2$ levels during storage | Optech™ stickers | Mocon Optech™ | Respiration rate of mushrooms determined | 39 |
| Ham | Monitor $O_2$ levels | PtOEPK/PS on filter paper | Phosphorescence phase detector | Sealing failure and compromised packages detected | 36 |
| *Beverages* | | | | | |
| Beer | Headspace $O_2$ levels | PtOEPK/PS on filter paper | Fluorescence phase detector | Tracked $O_2$ profile in the headspace, headspace with $O_2$>1% not palatable | 46 |
| Orange juice | Monitor $O_2$ levels | $O_2$xydots® | Oxysense® 210T probe | $O_2$ correlated to AA degradation | 48 |

(*continued*)

**Table 17.2** (*continued*)

| Product category | Measurement | Sensor material used | Instrument used for measurement $O_2$ | Main findings | Reference |
|---|---|---|---|---|---|
| Rapeseed oil | Monitor $O_2$ levels | $O_2$xydots® | Oxysense® 4000B system | Storage temperature correlated to $O_2$ consumption | 50 |
| Wine | Leak detection and $O_2$ ingress | PtBP/PDMS on PE foil | Pyroscience Firesting™ | Proof of concept | 38 |
| Wine | Dissolved $O_2$ levels | Presens PSt3 sensor | Presens Fibox 3 | Decreasing $O_2$ concentration increased VSCs | 58 |
| Wine | Tracking $O_2$ levels | Presens PSt3/PSt6 sensors | Presens Fibox 3 | Increasing $O_2$ concentration decreased tannins | 62 |
| Wine | Ensuring anoxic atmosphere | Presens PSt3/PSt6 sensors | Presens Fibox 3 | Model for non-oxidative AA in wine obtained | 64 |
| *Food packaging materials* | | | | | |
| Normacorc closures | $O_2$ ingress | PSt6 | Presens Fibox 3 | Determined most efficient closure | 56 |
| Packaging materials | Validation | $O_2$xydots® | Oxysense® 210T probe | Validated packaging simulation software | 48 |
| PLA films | OTR | Optech™ stickers | Mocon Optech™ | OTR of PLA films determined | 53 |
| Gelatine films | OTR | Optech™ stickers | Mocon Optech™ | OTR of gelatine films determined | 54 |
| PET Carboys | $O_2$ permeability | Optech™- stickers | Mocon Optech™ | Carboy suitability affirmed | 50 |

## 17.3 Conclusions

The area of smart packaging continues to evolve and grow at a steady rate and the ever increasing uptake of such technologies by companies continues also, but at a much slower pace than was previously predicted. One of the most interesting areas covered by smart packaging is the measurement and control of oxygen in a whole host of diverse applications, most importantly among them, the areas of oxygen-sensitive food and beverages. Numerous commercial technologies now exist to both control oxygen within product packages and successfully monitor oxygen levels within food and beverage packs. The development of numerous, non-destructive technologies for oxygen measurement within product packages have been particularly impressive over the past 15 years. This, coupled with research demonstrating the true value of using such systems for potential commercial uptake and exploitation by the food and beverage industries has been truly enlightening. Despite all of this, much more research is required to optimise oxygen sensing systems, to make these systems more economically viable and to ensure that sensor-based systems become more discrete, more multi-functional in action, proven to be safe for food and beverage contact and facilitate ease of handling during packaging waste stream management or provide sensor systems that can be collected after use and utilised again and again for a more sustainable approach to their employment within food and beverage packages, or indeed, within other non-food applications. It will be very interesting indeed, to see how this whole area develops over the next 15 years.

## Acknowledgements

The support from the National Development Plan, through the Food Institutional Research Measure (FIRM) administered by the Department of Agriculture, Food and the Marine, Ireland is acknowledged.

## References

1. M. Cruz-Romero and J. P. Kerry, Crop-based biodegradable packaging and its environmental implications, *CAB Rev.*, 2008, **3**(082), 25.
2. M. Cruz-Romero and J. Kerry, in *Processed Meats: Improving Safety Nutrition and Quality*, ed. J. P. Kerry and J. Kerry, Woodhead Publishing Ltd, Cambridge, 2011, p. 666.
3. K. Marsh and B. Bugusu, *Food Technol.*, 2007, **61**(4), 46.
4. G. L. Robertson, in *Food Packaging: Principles and Practice*, ed. G. L. Robertson, CRC Press Taylor & Francis Group, Boca Raton, 3rd edn, 2013, p. 399.
5. J. P. Kerry, in *Innovations in Food Packaging*, ed. J. H. Han, Academic Press, London, 2nd edn, 2014, p. 549.
6. M. Cruz-Romero and J. Kerry, in *Emerging Technologies in Meat Processing*, ed. E. Cummins and J. Lyng, Wiley-Blackwell, Oxford, 2017, p. 229.

7. B. P. F. Day, in *Smart Packaging Technologies for Fast Moving Consumer Goods*, ed. J. Kerry and P. Butler, John Wiley & Sons, Ltd, Chichester, 2008, p. 233.
8. L. Yam, P. T. Takhistov and J. Miltz, *J. Food Sci.*, 2005, **70**, 1.
9. Commission Regulation (EC) No 450/2009, *Off. J. Eur. Communities*, 2009, L135/3.
10. http://ec.europa.eu/food/food/chemicalsafety/foodcontact/documents_en.htm, last accessed October 2017.
11. V. Dalmoro, J. H. Z. dos Santos, M. Pires, A. Simanke, G. B. Baldino and L. Oliveira, in *Food Packaging. Nanotechnology in the Agri-food Industry*, ed. A. M. Grumezescu, Academic Press, Oxford, 2017, vol. 7, p. 111.
12. E. Kress-Rogers, *Instrumentation and Sensors for the Food Industry*, Butterworth-Heinemann Ltd, Oxford, 1st edn, 1993.
13. A. Hulanicki, S. Geab and F. Ingman, *Pure Appl. Chem.*, 1991, **63**(9), 1247.
14. M. Dudbridge, *Handbook of Seal Integrity in the Food Industry*, John Wiley & Sons, Ltd, Chichester, 2016.
15. S. A. Hogan and J. P. Kerry, Smart packaging of meat and poultry products, in *Smart Packaging Technologies for Fast Moving Consumer Goods*, ed. J. P. Kerry and P. Butler, John Wiley & Sons Ltd, Chichester, 2008, p. 33.
16. http://www.mgc.co.jp/eng/products/abc/ageless/eye.html, last accessed October 2017.
17. X. D. Wang and O. S. Wolfbeis, *Chem. Soc. Rev.*, 2014, **43**(10), 3666.
18. D. B. Papkovsky and R. I. Dmitriev, *Chem. Soc. Rev.*, 2013, **42**(22), 8700–8732.
19. Y. Amao, *Microchim. Acta*, 2003, **143**(1), 1.
20. L. C. Clark, R. Wolf, D. Granger and Z. Taylor, *J. Appl. Physiol.*, 1953, **6**(3), 189.
21. W. Trettnak, W. Gruber, F. Reininger and I. Klimant, *Sens. Actuators, B*, 1995, **29**(1–3), 219.
22. A. Mills, *Chem. Soc. Rev.*, 2005, **34**(12), 1003.
23. D. B. Papkovsky, N. Papkovskaia, A. Smyth, J. Kerry and V. I. Ogurtsov, *Anal. Lett.*, 2000, **33**(9), 1755.
24. S. Banerjee, C. Kelly, J. P. Kerry and D. B. Papkovsky, *Trends Food Sci. Technol.*, 2016, **50**, 85.
25. D. B. Papkovsky, M. A. Smiddy, N. Y. Papkovskaia and J. P. Kerry, *J. Food Sci.*, 2002, **67**(8), 3164.
26. C. Grebitus, H. H. Jensen, J. Roosen and J. G. Sebranek, *J. Food Prot.*, 2013, **76**, 99.
27. D. H. Kropf, *Meat Focus Int.*, 1993, **1**, 269.
28. M. G. O'Sullivan, M. Cruz-Romero and J. P. Kerry, *LWT–Food Sci. Technol.*, 2011, **44**, 2193.
29. L. D. Nannerup, M. Jakobsen, F. van den Berg, J. S. Jensen, J. K. S. Møller and B. Bertelsen, *Meat Sci.*, 2004, **68**, 577.
30. H. Larsen, F. Westad, O. Sørheim and L. H. Nilsen, *J. Food Sci.*, 2006, **71**, S407.
31. J. K. S. Møller, M. Jakobsen, C. J. Weber, T. Martinussen, L. H. Skibsted and G. Bertelsen, *Meat Sci.*, 2003, **63**, 169.

32. J. Solomon, *Meat Poult.*, 2004, **9**, 38.
33. K. T. Lee, *Meat Sci.*, 2010, **86**, 138.
34. L. Vermeiren, L. Heirlings, F. Devlieghere and J. Debevere, in *Novel Food Packaging Techniques*, ed. R. Ahvenainen, Woodhead Publishing Limited, Cambridge, 2003, p. 22.
35. M. Smiddy, M. Fitzgerald, J. P. Kerry, D. B. Papkovsky, C. K. O'Sullivan and G. G. Guilbault, *Meat Sci.*, 2002, **61**(3), 285.
36. D. B. Papkovsky, M. A. Smiddy, N. Y. Papkovskaia and J. P. Kerry, *J. Food Sci.*, 2002, **67**(8), 3164.
37. F. C. O'Mahony, T. C. O'Riordan, N. Papkovskaia, J. P. Kerry and D. B. Papkovsky, *Food Control*, 2006, **17**(4), 286.
38. B. J. Müller, T. Burger, S. M. Borisov and I. Klimant, *Sens. Actuators, B*, 2015, **216**, 527.
39. N. B. Borchert, M. C. Cruz-Romero, P. V. Mahajan, M. Ren, D. B. Papkovsky and J. P. Kerry, *Food Packag. Shelf Life*, 2014, **2**, 17.
40. N. Borchert, A. Hempel, H. Walsh, J. P. Kerry and D. B. Papkovsky, *Food Control*, 2012, **28**(1), 87.
41. A. Hempel, M. Sullivan, D. Papkovsky and J. Kerry, *Foods*, 2013, **2**(2), 213.
42. A. W. Hempel, M. G. O'Sullivan, D. B. Papkovsky and J. P. Kerry, *Eur. Food Res. Technol.*, 2013, **237**(2), 117.
43. M. K. Morsy, H. H. Khalaf, A. M. Sharoba and H. H. El-Tanahi, *2nd International Conference on Biotechnology Applications in Agriculture (ICBAA)*, Benha University,Moshtohor and Hurghada, Egypt, April 2014.
44. K. A. M. O' Callaghan, D. B. Papkovsky and J. P. Kerry, *Sensors*, 2016, **16**, 916.
45. M. Ugliano, J. B. Dieval, T. E. Siebert, M. Kwiatkowski, O. Aagaard, S. Vidal and E. J. Waters, *J. Agric. Food Chem.*, 2012, **60**(35), 8561.
46. A. Hempel, M. G. O'Sullivan, D. B. Papkovsky and J. P. Kerry, *LWT–Food Sci. Technol.*, 2013, **50**(1), 226.
47. E. Nkhili and P. Brat, *Anal. Bioanal. Chem.*, 2011, **400**(5), 1451.
48. I. Van Bree, J. M. Baetens, S. Samapundo, F. Devlieghere, R. Laleman, I. Vandekinderen, B. Noseda, R. Xhaferi, B. De Baets and B. De Meulenaer, *Food Chem.*, 2012, **134**(1), 207.
49. S. Wibowo, T. Grauwet, J. S. Santiago, J. Tomic, L. Vervoort, M. Hendrickx and A. Van Loey, *Food Chem.*, 2015, **187**, 140.
50. W. Kozak and U. Samotyja, *Food Control*, 2013, **33**(1), 162.
51. R. P. Singh and D. R. Heldman, *Introduction to Food Engineering*, Academic Press, Oxford, 5th edn, 2014.
52. http://www.oxysense.com/oxygen-measurement-accessories.html, accessed on October 2017.
53. S. Molinaro, M. Cruz Romero, M. Boaro, A. Sensidoni, C. Lagazio, M. Morris and J. Kerry, *J. Food Eng.*, 2013, **117**, 113.
54. S. Molinaro, M. Cruz-Romero, A. Sensidoni, M. Morris, C. Lagazio and J. P. Kerry, *Innovative Food Sci. Emerging Technol.*, 2015, **30**, 15.
55. E. L. Gibbs, *Oxygen Permeation of BetterBottle® Carboys-direct Measurement*, 2013.

56. J. B. Diéval, S. Vidal and O. Aagaard, *Packag. Technol. Sci.*, 2011, **24**(7), 375.
57. J. Wirth, S. Caillé, J. M. Souquet, A. Samson, J. B. Dieval, S. Vidal, H. Fulcrand and V. Cheynier, *Food Chem.*, 2012, **132**(4), 1861.
58. M. Ugliano, M. Kwiatkowski, S. Vidal, D. Capone, T. Siebert, J. B. Dieval, O. Aagaard and E. J. Waters, *J. Agric. Food Chem.*, 2011, **59**(6), 2564.
59. I. Van Bree, B. De Meulenaer, S. Samapundo, A. Vermeulen, P. Ragaert, K. C. Maes, B. De Baets and F. Devlieghere, *Innovative Food Sci. Emerging Technol.*, 2010, **11**(3), 511.
60. M. Mestres, O. Busto and J. Guasch, *J. Chromatogr. A*, 2000, **881**(1–2), 569.
61. J.-M. Salmon, *LWT–Food Sci. Technol.*, 2006, **39**(9), 959.
62. J. M. McRae, M. P. Day, K. A. Bindon, S. Kassara, S. A. Schmidt, A. Schulkin, R. Kolouchova and P. A. Smith, *Tetrahedron*, 2015, **71**(20), 3131.
63. N. Wallington, A. C. Clark, P. D. Prenzler, C. Barril and G. R. Scollary, *Food Chem.*, 2013, **141**(3), 3139.
64. M. P. Bradshaw, C. Barril, A. C. Clark, P. D. Prenzler and G. R. Scollary, *Crit. Rev. Food Sci. Nutr.*, 2011, **51**(6), 479.
65. K. Eaton and P. Douglas, *Sens. Actuators, B*, 2002, **82**(1), 94.

# Subject Index